SEMICONDUCTOR PULSE CIRCUITS
WITH EXPERIMENTS

SEMICONDUCTOR PULSE CIRCUITS

WITH EXPERIMENTS

Brinton B. Mitchell

East Los Angeles College

RINEHART PRESS, San Francisco

Library of Congress Catalog Card Number: 78-100386

ISBN 0-03-083036-2

Printed in the United States of America

6   038   9876

TO LOIS

Preface

Rapid and continuing advances in the field of electronic science, as well as expansion in the electronics industry, have created a constant need for new instructional material in these areas. Nowhere has the need been greater than in the field of semiconductor pulse and switching circuits. This book fills this need for the electronic technician, the student of electronics technology in junior colleges or industrial schools, and for the laymen seeking knowledge in this area.

This book is unique because it is a comprehensive textbook on pulse and switching circuits which includes a complete laboratory experiment in each chapter. Thus the need for a separate laboratory manual has been eliminated. Each experiment has been planned to teach the reader to design, for practical use, most of the basic switching circuits. Each circuit is presented and examined in its entirety and in its simplest form. The questions and exercises, presented at the end of each chapter, have been designed to test the student's comprehension of the theory, as well as his knowledge of the practical application of this theory.

The objectives of this book are:

1. To present the background in electronic theory necessary for a thorough understanding of the operation of the most important basic pulse and switching circuits.
2. To relate this theory to the most basic practical circuit of each type.
3. To synthesize the reader's knowledge of basic circuit theory, network theorems, and semiconductor theory into an ability to design the simplest practical working circuits.
4. To enable the reader to analyze given circuits and to predict the operating parameters of each of these circuits.

The selection of the most basic and necessary pulse and switching circuits, for presentation in an introductory textbook, is certainly open to debate.

Those selected by the author are believed to represent most of "the basic circuits." The nonsaturated switch, for example, has been omitted. This omission was decided upon because transistors are manufactured today with a minimum of base charge; hence, they operate as switches in the saturated mode in all but the fastest of switching circuits. Moreover, the use of nonsaturated switches in integrated circuits is practically nonexistent.

In this book, the transistor or semiconductor device is treated as an ideal switch, whenever this is possible. For the student at the technician level, this simplification is believed to be essential so that he may concentrate on the fundamentals of the circuit.

The student, for whom this textbook is planned, should have a working knowledge of the material covered in the following areas of instruction:

1. dc circuit theory
2. ac circuit theory
3. Basic circuit theorems and attendant mathematics
4. Semiconductor theory

I wish to thank Mrs. Jean Crary, Mrs. June Marks, and Mrs. Lillian White for their invaluable secretarial assistance and patience throughout this work.

Los Angeles, California   **Brinton B. Mitchell**
January 1970

Laboratory Equipment

All of the experiments in this book may be performed with inexpensive semiconductor devices. A minimal laboratory teaching station should include the following equipment:

1. One (1) Audio signal generator, sine-square-wave frequency range, 20 Hz to 200 kHz (sine and square) output voltage, 0–10 V rms; 0–10 V p-p

2. One (1) Oscilloscope, dc time base type frequency response, dc to 450 kHz vertical sensitivity, 100 mV/cm (ac and dc)

3. One (1) Oscilloscope probe—10X attenuator, 10 MΩ

4. One (1) Vacuum-tube voltmeter, ac/dc type voltage range, 1.5 to 500 V full scale dc and ac rms resistance range, 1 Ω to 100 MΩ

5. Two (2) Semiconductor power supplies output voltage, 0–30 V output current, 0–250 mA

Note:

A representative sample of manufacturer's semiconductor specification sheets is included in Appendix B. However, the selection need not be made from the semiconductor devices presented here, as any which are appropriate for the experiments may be used. If substitutions are made, it may be necessary to obtain additional specification sheets.

Contents

Chapter 11 *SCHMITT TRIGGER*   153

Chapter 12 *MONOSTABLE MULTIVIBRATOR*   168

Chapter 13 *ASTABLE MULTIVIBRATOR*   182

SEMICONDUCTOR
PULSE CIRCUITS
WITH EXPERIMENTS

INTRODUCTION

Semiconductor pulse and switching circuits are those in which the semiconductor device acts as a switch. Many of the existing semiconductor devices are used to perform this switching function. Some of these are diodes, bipolar transistors, field effect transistors, silicon-controlled rectifiers, unijunctions, and tunnel diodes. This book explains the application of these devices.

The transistor is one of the most widely used of the active elements listed above. In the study of pulse and switching circuits, the concept of the transistor as an amplifier must temporarily be replaced by the concept of the transistor as a switch. Although the amplification parameter of the transistor is utilized in switching circuits, it plays a subordinate role in the switching circuit applications.

Electronic switching circuits are generally identified by the function performed and are part of a much larger system of circuits designed to perform some useful task. Because this is an introductory text, many of the basic switching circuits are presented as separate and complete circuits. Somewhat more complex circuits, formed by a combination of two or more of these basic circuits and designed to perform a series of functions, are discussed in the latter part of this book.

In electronic context, the term *pulse* may be defined as a voltage or current waveform which is nonsinusoidal and which usually has sharp, steep edges. This voltage and/or current waveform may or may not be periodically recurrent. Normally, the voltage and/or current waveform is produced by the switching circuit.

> *NOTE:* **The study of pulse and switching circuits involves not only the magnitude of a voltage (pulse or otherwise) but equally, or more important, it involves the shape of the voltage and/or current wave.**

Refer to Fig. 1-1(a), (b), and (c). Observe that there are three basic voltage waveforms: the step or impulse, the ramp, and the exponential. From these, almost all the other most commonly used voltage waveforms are generated.

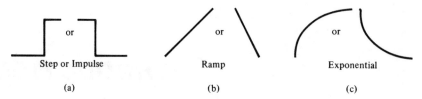

Step or Impulse	Ramp	Exponential
(a)	(b)	(c)

Figure 1-1   Basic Voltage and/or Current Waveshapes

The most often used practical voltage waveforms are the rectangular wave [shown in Fig. 1-2(a)] which is a combination of two-step waveforms and which is frequently called a *pulse;* the sawtooth wave [shown in Fig. 1-2(b)] which is a combination of two ramp waveforms or a combination of a ramp and a step waveform; the integrated wave [shown in Fig. 1-2(c)] which is a combination of two exponential waveforms; and a differentiated wave [shown in Fig. 1-2(d)] which is a combination of a step waveform and an exponential waveform.

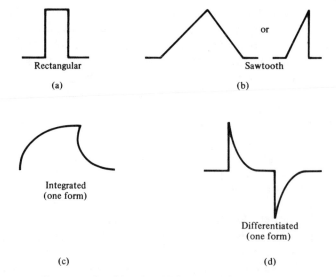

Figure 1-2   Common Combination Voltage and/or Current Waveshapes

The discussions, experiments, and queries which follow in this book are designed to answer the following questions and thereby engender the student's understanding of pulse and switching circuit theory and its application. These questions are

1. How are voltage and/or current waveshapes generated?
2. How may these voltage and/or current waveshapes be analyzed after they have been generated?
3. How may these waveshapes be used after they have been generated (how and why are they needed)?

LABORATORY REPORTS

It should be noted here that no formal data sheets are used in this book. The omission is intentional and, it is hoped, will in itself better prepare the student. The experiments herein have been written to simulate, as closely as possible, a practical working situation. An employer may not provide "data sheets" but may expect necessary data to be presented to him in neat, precise, and self-explanatory form. This formal report of the "necessary data" is called a *laboratory report* and should consist of the following:

1. A title page
   (a) The title of the experiment
   (b) The name of the person writing the report
   (c) The course title and number
   (d) The date
2. A statement of the hypothesis to be proved
3. A statement of the plan for proving this hypothesis
4. A clearly labeled and titled schematic of the circuit
5. An explanation of the theory and operation of the circuit
6. An explanation of the choice of data
7. An equipment list
   (a) The type of equipment needed
   (b) The make and model of each piece of equipment
   (c) The serial number of each piece of equipment
8. Precise data, neatly recorded (the form may vary from experiment to experiment)
9. A complete set of sample computations in which all steps are shown
10. Carefully organized graphs where they are applicable
    (a) A lettered title
    (b) Adequate margins
    (c) Easily readable and labeled scales
    (d) A schematic of the circuit
    (e) A plot of all points obtained from the data
    (f) A smooth curve (where applicable) connecting the plotted points as closely as possible
11. An analysis which compares the experimental results with the theoretical intent and an explanation of any differences between the two
12. Any additional comments, when necessary

chapter 1

THE GENERATION
AND ANALYSIS
OF SQUARE WAVES

A square waveshape of voltage, usually called a *square wave*, is a special case of the rectangular waveshape. A square wave is a recurrent rectangular waveform and is a special case because the pulse duration and the time between recurrent pulses are equal. The first nonsinusoidal waveform to be analyzed is the square wave because square-wave generators are more generally in use than are pulse generators. The explanations and diagrams that follow describe the rectangular or pulse waveform.

1.1   DEFINITION OF TERMS

The circuit shown in Fig. 1-3(a) may be used to produce the rectangular pulse waveform shown in Fig. 1-3(b). In this circuit, there are only two possible voltage levels: 10 V when switch S is in position 1, and 0 V when switch S is in position 2. The amplitude of the pulse is referred to as the *peak value*. As time increases from zero, the first edge of the pulse encountered is referred to as the *leading edge*, and the edge formed when the pulse drops is termed the *trailing edge*. The duration of the pulse is called the pulse width, symbolized t_p. The interval from the start of one pulse to the start of the next pulse is called the *pulse repetition time*, symbolized *prt*. Usually the term pulse repetition time is used only when the pulses occur at regular intervals, as shown in Fig. 1-3(b). A series of successive pulses is called a *pulse train*. In a pulse train, the number of pulses per second is called the *pulse repetition rate*, abbreviated *prr*, or *pulse repetition frequency*, abbreviated *prf*. A square wave, which is a special case of the rectangular pulse waveform, is produced when the pulse duration t_p and the pulse interval t_2 are equal.

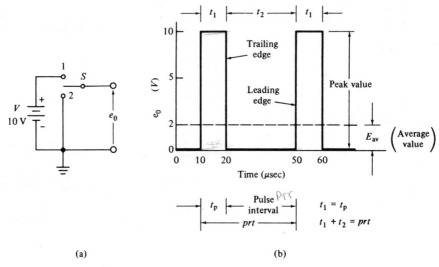

Figure 1-3   Theoretical Pulse (Rectangular Waveshape)

The pulse repetition rate (*prr*) is equal to the reciprocal of the time required for one pulse repetition time (*prt*). Hence:

$$prr = \frac{1}{prt} \quad \text{(Hz)} \tag{1}$$

The average value of any waveform is its dc component (either voltage or current). To determine the average voltage of the theoretical waveform shown in Fig. 1-3(b), divide the area (A_p) of the pulse by the pulse repetition time (*prt*). The average value of voltage is the value indicated by a dc voltmeter.

$$A_P = t_p \times E_{peak} \quad \text{(second-volt)} \quad \text{(sec--V)} \tag{2}$$
$$A_P = 10 \times 10^{-6} \times 10 = 100 \ \mu\text{sec--V}$$

$$E_{av} = \frac{A_P}{prt} \quad \text{(V)} \tag{3}$$

$$E_{av} = \frac{100 \ \mu\text{sec--V}}{40 \ \mu\text{sec}} = 2.5\text{V}$$

Should the pulse start from some value other than zero, the numerical value of A_P in Eq. 2 would be the algebraic sum of the positive area of the pulse (A^+) and the negative area of the pulse (A^-).

Figure 1-4(a) shows the waveform of Fig. 1-3(b) as it would be seen on the face of an ac oscilloscope. Figure 1-4(b) shows the same waveform as it would be viewed on the face of a dc oscilloscope. The assumption made here is that the trace line of both scopes, with no signal applied, would be in the center of the screen.

Independently of waveshape, voltage and/or current pulses may be generated from any given dc reference level. Pulse and switching circuits employ

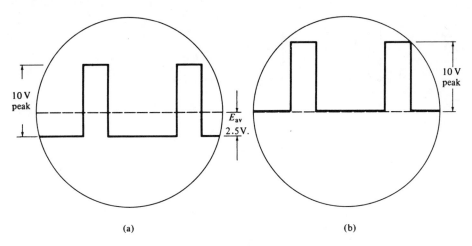

Figure 1-4 (a) ac Oscilloscope Face; (b) dc Oscilloscope Face

and generate positive and negative voltage and/or current pulses. In general, the term *positive pulse* signifies a positive voltage excursion from a zero–volt reference level. Conversely, the term *negative pulse* signifies a negative voltage excursion from a zero-volt reference level. See Fig. 1-5(a) and (b). Figure 1-5(c) through (f) illustrates a sampling of possible voltage pulses and their respective dc reference levels.

The special type of circuit designed to establish or change the dc reference level of voltage pulses is called a *clamper circuit*. This is discussed in chapter 5.

Another term used in pulse work is *duty cycle*. This is the ratio of the average value to the peak value of the voltage waveform. It is generally used in connection with transmitter power output and is normally expressed as a percentage. Compute the duty cycle of the voltage waveform shown in Fig. 1-3(b).

$$\% \text{ duty cycle} = \frac{E_{\text{av}} \times 100}{E_{\text{peak}}} \quad \text{(percent)} \qquad [4]$$

$$\% \text{ duty cycle} = \frac{2.5 \times 100}{10} = 25\%$$

$$\% \text{ duty cycle} = \frac{E_{\text{av}} \times 100}{E_{\text{peak}}} = \frac{\dfrac{A_{\text{p}} \times 100}{prt}}{E_{\text{p}}}$$

$$\% \text{ duty cycle} = \frac{t_{\text{p}} \times 100}{prt} = \frac{10 \ \mu\text{sec} \ 100}{40 \ \mu\text{sec}}$$

$$\% \text{ duty cycle} = 25\%$$

Note that the voltage waveform shown in Fig. 1-3(b) is labeled theoretical. In practice, the rectangular pulse is not absolutely rectangular but it may have the general appearance of the waveform shown in Fig. 1-6. When the

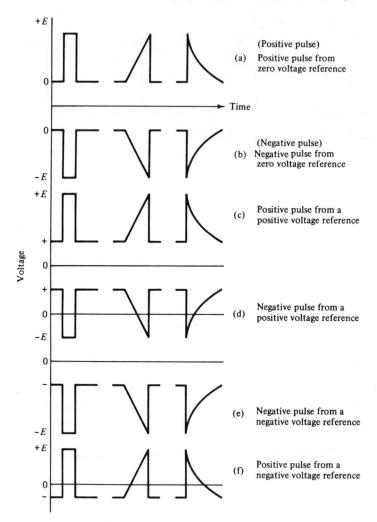

Figure 1-5   Pulse Amplitude in All Cases *E* Volts

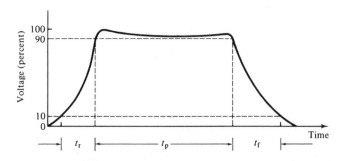

Figure 1-6   Practical Rectangular Pulse Waveform

waveform is not rectangular, one is confronted with the problem of having to define a pulse. Left to himself, each individual might define a pulse width of different value. Hence, a standard has been established, thereby enabling each of us to use the same measurement for solving for a given voltage waveform (in this case, pulse width). Note that from the vertical axis a 90 percent and a 10 percent line have been drawn. These two values are the standard for determining the characteristic of the waveform.

The pulse duration (t_p) is defined as the time that the pulse waveform is in excess of 90 percent of the maximum value of the waveform (sec).

The rise time   (t_r) is the time required for the pulse to rise from 10 percent of the maximum value of the waveform to 90 percent of the maximum value of the waveform (sec).

The fall time   (t_f) (sometimes referred to as the decay time) is the time required for the trailing edge of the waveform to decay from 90 percent of the maximum value of the waveform to 10 percent of the maximum value of the waveform (sec).

Note that some textbooks use the symbol (t_d) for pulse width. In this text the symbol (t_p) is arbitrarily used to symbolize pulse width because (t_d) is subsequently used to represent *delay time*.

For some pulse waveforms, the rise time and the fall time are so small that it is practically impossible to measure them accurately, even with a high-frequency oscilloscope. When the pulse duration is very small, the rise time and fall time are an appreciable portion of the entire time of the pulse.

1.2 METHODS OF SQUARE-WAVE GENERATION

Rectangular pulses may be generated in a number of ways. Some of these methods are as follows:

1.   A rectangular or square wave of voltage may be generated by overdriving a sine wave, class A amplifier, thereby driving the transistor into saturation on one half cycle and driving the transistor into cutoff on the other half cycle. The resultant output-voltage waveform is a close approximation of a rectangular or square wave, depending on the dc operating point of the amplifier.

2.   A square wave of voltage may be generated by using one of the many forms of the multivibrator circuit. This is a two-stage RC-coupled amplifier in which the output of the first stage is fed to the input of the second stage and the output of the second stage is then fed back as the input to the first stage. Since both stages are overdriven, the resultant output voltage is a square

wave or a rectangular waveform. The study of this type of circuit is taken up in detail in later chapters.

3.   A square wave of voltage may also be generated when a sine wave voltage source of the fundamental (first harmonic) frequency is connected in parallel to the voltage sources of the other odd harmonic frequencies of the fundamental of proper amplitude and phase. The resultant square-wave output-voltage waveform has a pulse repetition rate (*prr*) equal to the frequency of the fundamental of the sine wave.

1.3   GENERATION OF A RECTANGULAR WAVEFORM OF VOLTAGE BY THE OVERDRIVEN AMPLIFIER METHOD

A square wave of voltage may be generated from a sine wave of voltage. This is accomplished by overdriving a class A transistor amplifier with a sine wave. The amplitude of the sine wave input voltage must be large enough to ensure that the transistor is driven into saturation and into cutoff on the alternate half cycles of the input voltage. The basic class A amplifier selected for use in this chapter is the unstabilized grounded-emitter amplifier shown in Fig. 1-7(a).

Figure 1-7(b) shows how the amplifier produces a square wave by alternately driving the transistor into saturation and cutoff, thereby clipping the peaks of the sine wave. This clipping of the peaks of the sine wave produces a practical approximation of a square wave at the output of the amplifier. Because an unstabilized amplifier is used, in all probability the Q point will not be in the center of the load line; therefore, the resultant output voltage will not be a square wave but, rather, a rectangular waveform. If the Q point of the unstabilized amplifier is adjusted to the center of the load line, the output voltage of the ideal transistor is the square wave described above. This method of square-wave generation is one used in practical circuits. These circuits use a series of two or three bias-stabilized amplifiers. Therefore, the resultant output voltage is relatively immune to temperature variations.

1.4   SAMPLE OVERDRIVEN AMPLIFIER DESIGN PROBLEM

Refer to Fig. 1-7(a).

Object: To produce a square wave of voltage with a *prr* of 1000 Hz and a pulse amplitude of 20 V peak from a 1000 Hz sine wave of voltage with an amplitude of 2 V peak-to-peak.

Given: An NPN silicon transistor with the following parameters:

$$h_{fe} = 50$$
$$h_{ie} = 1000 \ \Omega$$
$$I_{co} = \text{negligible}$$

A variable dc power supply (0 to 30 V) output current, 0 to 250 mA.

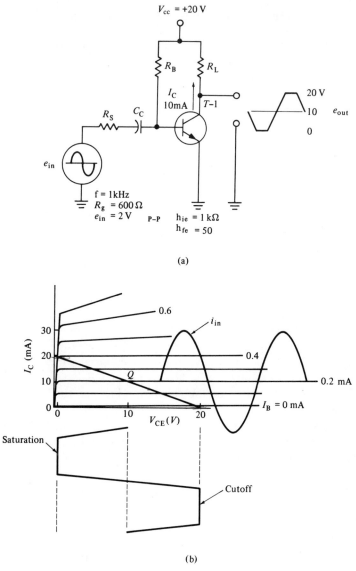

(a)

(b)

Figure 1-7 Class A Overdriven Grounded Emitter Amplifier

Solution: From the characteristic curves shown in Fig. 1-7(b), it can be seen that the choice of V_{cc} is determined by the necessary amplitude of the output-voltage pulse and hence must be +20 V. The choice of collector current is arbitrary because no external load is to be driven. A relatively low value of collector current should be selected because the input impedance of the amplifier is more linear at low collector-current values. This nonlinearity of the input impedance at higher collector-current values may produce a rectangular output-voltage waveform. Select a 10 mA collector current for the quiescent Q operating point.   With these points established, determine R_L.

$$R_L = \frac{E_{R_L}}{I_C} = \frac{V_{cc} - V_{CE}}{I_C} = \frac{20 - 10}{10 \text{ mA}} = \frac{10}{10 \text{ mA}}$$

$$R_L = 1000 \ \Omega$$

The relationship among the collector current, the base current, and the h_{fe} of the transistor is used to determine the proper value of base current necessary to produce a collector current of 10 mA.

$$I_B = \frac{I_C}{h_{fe}} = \frac{10 \text{ mA}}{50} = 0.2 \text{ mA}$$

Assume the voltage drop across the emitter-base junction to be negligible. Determine the proper value of R_B.

$$R_B = \frac{V_{cc}}{I_B} = \frac{20}{0.2 \text{ mA}} = 100 \text{ k}\Omega$$

The circuit components necessary to fulfill the ac circuit requirements may now be determined because the dc requirements, necessary to establish proper quiescent operating conditions, have been fulfilled. Following this, establish the proper value for the coupling capacitor (C_c) in the circuit. The coupling capacitor is effectively in series with the total input impedance of the amplifier. The input impedance of the amplifier must be known in order to determine the proper value for the capacitor. For practical purposes, the input impedance of the amplifier, excluding R_B, is approximately equal to h_{ie} or 1000 Ω. The total input impedance of the amplifier, including R_B, is the parallel equivalent impedance of R_B and h_{ie}. Due to the order of magnitude of the values involved, the total input impedance of the amplifier remains about 1000 Ω. If the reactance of the capacitor is one-tenth or less of the value of the input impedance of the amplifier, the ac voltage drop across the coupling capacitor may be neglected.
Hence:

$$X_c = \tfrac{1}{10} Z_{in} = \tfrac{1}{10} \times 1000 = 100 \ \Omega$$

$$C_c = \frac{1}{2\pi f X_c} = \frac{1}{6.28 \times 10^3 \times 10^2}$$

$$C_c = 1.59 \ \mu\text{F}$$

Referring to Fig. 1-7(b), the largest input base-current signal swing that a class A amplifier can accommodate is 0.4 mA peak-to-peak or 0.1414 mA rms. The transistor is driven alternately into saturation and into cutoff by arbitrarily doubling the input base current. Thus, the approximated square-wave output-voltage waveform, shown in Fig. 1-7(b), is produced. This means that the ac input base-current swing should be 0.8 mA peak-to-peak or 0.2828 mA rms.

The value of R_S necessary to satisfy these circuit requirements may now be determined. The circuit reduces to the signal source in series with the input impedance of the amplifier Z_{in} (approximately 1 kΩ).

Hence: where

$$R_T = \text{total input impedance seen by generator}$$
$$R_g = \text{impedance of generator}$$
$$R_T = R_g + R_S + Z_{in}$$
$$R_T = \frac{e_{in}}{i_b} = \frac{0.707}{0.2828 \text{ mA}} = 2.51 \text{ k}\Omega$$
$$R_S = R_T - (R_g + Z_{in})$$
$$R_S = 2.51 \text{ k}\Omega - (0.6 \text{ k}\Omega + 1 \text{ k}\Omega)$$
$$R_S = 2.51 \text{ k}\Omega - 1.6 \text{ k}\Omega$$
$$R_S = 900 \text{ }\Omega$$

Recall that the output-voltage waveform may be rectangular because of the unstabilized circuit used and because of the nonlinear spacing of the constant base currents (the nonlinear input impedance).

1.5   GENERATION OF SQUARE WAVES OF VOLTAGE BY THE ADDITION OF SINE WAVES

Recall that a square wave of voltage is produced when a sine wave voltage source of the fundamental (first harmonic) frequency is connected in parallel to the voltage sources of the other odd harmonic frequencies of the fundamental, of proper amplitude and phase. The resultant square-wave output-voltage waveform has a pulse repetition rate (*prr*) equal to the frequency of the fundamental of the sine wave. This generation may be proved graphically by the vector addition of the instantaneous values of the voltage of the fundamental (first harmonic) frequency and the instantaneous values of the voltage of all the odd harmonic frequencies, of the fundamental, of proper amplitude and phase. Refer to Fig. 1-8. This is a graphical analysis of the vector sums of the instantaneous values of voltage of a sine wave (fundamental frequency) and the third harmonic (sine wave with frequency three times the fundamental) voltage sine wave. The instantaneous voltage vectors are indicated for one instant in time with the vector sum indicated by the asterisk. If the fifth harmonic frequency sine wave voltage waveform, in phase and with an amplitude of approximately one-fifth of that of the fundamental, is added to this resultant waveform, the new resultant voltage waveshape will begin to approach that of a square wave. Theoretically, if an infinite number of odd harmonic frequency sine waves of voltage, of proper amplitude and phase, are added, the resultant voltage waveshape will be that of a perfect square wave. The vector addition of the fifth and seventh harmonic frequencies of sine waves of voltage will be assigned as problems to be proved graphically.

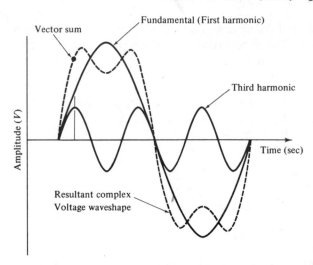

Figure 1-8   First Approximation of Square-Wave Voltage Generation from Sine Waves

If a sine wave of voltage and all the harmonics of proper amplitude and phase are added, another basic pulse voltage waveshape will result. Therefore, many of the basic voltage and/or current waveshapes may be generated by the use of the vector sums of sine waves of voltage and their harmonics.

Hence, it should be emphasized that a square wave of voltage is made up of a sine wave of voltage, the frequency of which is that of the square wave, and an infinite number of odd harmonics of sine waves of voltage. This indicates that a pulse amplifier should theoretically have a frequency response from dc to infinity in order to amplify square waves of any frequency. For example, to amplify a square wave of voltage with a pulse repetition rate of 1000 Hz, the ideal pulse amplifier should have an infinite frequency response.

THE GENERATION AND ANALYSIS OF SQUARE WAVES

OBJECT:

1. To define, determine, and measure the various parts of a square wave of voltage
2. To generate a rectangular waveform of voltage by use of an overdriven amplifier
3. To generate a square wave of voltage by use of a sine wave of voltage and its odd harmonics

MATERIALS:

1   Sine square-wave audio generator (20 Hz to 200 kHz)
1   Oscilloscope, dc time-base type; frequency response dc to 450 kHz; vertical sensitivity, 100 mV/cm
1   dc power supply (0 to 30 V; 0 to 250 mA)
1   Transistor, silicon (either NPN or PNP) with manufacturer's specification data sheet (example: 2N3903—data sheets in Appendix B)
3   Resistor substitution boxes, 15 Ω to 10 MΩ, 1 W
1   10 μF capacitor (50 V)

PROCEDURE:

1. To define, determine, and measure the various parts of a square wave of voltage:
   (a) Connect the output of a square-wave generator to the vertical input of the oscilloscope.
   (b) Adjust the frequency of the generator to obtain a resultant pulse width of 5 μsec, and adjust the peak amplitude to 5 V.
   (c) Refer to Fig. 1-5. Measure the values of t_p, t_r, t_f, and prt with the oscilloscope. Record.
   (d) On graph paper, draw the resultant scope waveform to a convenient scale. Label the measured values of t_p, t_r, t_f, and prt. Calculate the prr and the duty cycle. Record.
   (e) Repeat Steps (a) to (d) to obtain a pulse width of 0.1 msec and an amplitude of 5 V.
2. To generate a rectangular waveform of voltage by the use of an overdriven transistor amplifier:
   (a) For a given transistor, design the grounded-emitter, unstabilized, class A amplifier shown in Fig. 1-1X.

   Design this circuit to generate a rectangular-waveshape output voltage, with a 15 V peak-to-peak amplitude and with a prr of 2000 Hz. The sine wave input voltage should have a frequency of 2000 Hz with an amplitude of 10 V peak-to-peak.
   (b) Connect your designed circuit. Establish its quiescent operating point. How does the position of the Q point on the load line determine the shape of the resultant output voltage when the signal is applied? Explain.
   (c) Apply sine wave signal.

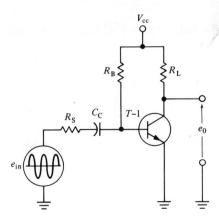

Figure 1-1X

   (d)   Observe the resultant output-voltage waveshape. Measure, graph, and label your observations.

   (e)   Repeat Step (c) of Part 1 of this experiment.

   (f)   Draw and completely label the schematic of the circuit which you evolved. Describe in detail the operation of the amplifier.

3.   To generate a square wave of voltage by the addition of sine waves of voltage:

NOTE: **The theory concept in this part of the experiment is covered by a written exercise because an actual check by performance requires special triggered sine wave generators not usually available. Such generators must permit control of the relative phase relationship between the fundamental and its harmonics.**

   (a)   Refer to Fig. 1-2X. Vectorially add the given fundamental to the given third harmonic. Use a minimum of 30 points. Draw a resultant voltage waveshape.

   (b)   To the resultant voltage waveshape drawn in Step (a), vectorially add the voltage of the fifth harmonic, and draw the new resultant voltage waveshape.

   (c)   To the resultant voltage waveshape obtained in Step (b), vectorially add the voltage of the seventh harmonic, and draw the final resultant voltage waveshape.

QUESTIONS AND EXERCISES

1.   How may the leading edge of a pulse be defined?
2.   How may the trailing edge of a pulse be defined?
3.   How is the amplitude of a pulse expressed?
4.   A rectangular pulse is made up of what basic waveshape and/or shapes?
5.   Define *rise time* and indicate its lettered symbol.
6.   Define *pulse width* and indicate its lettered symbol.
7.   Define *fall time* and indicate its lettered symbol.

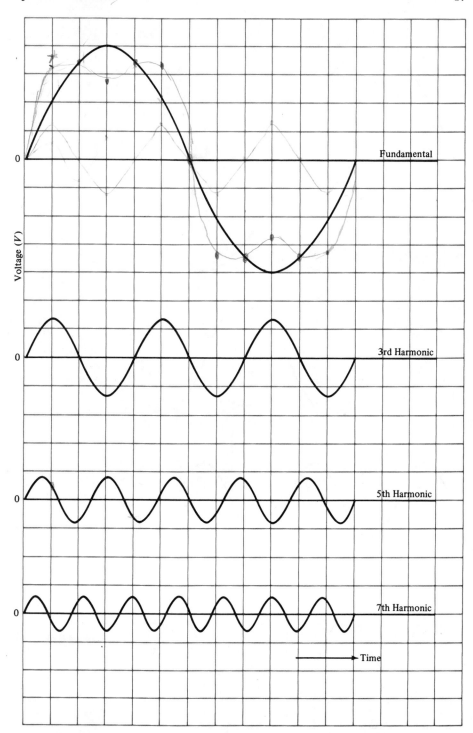

Figure 1-2X

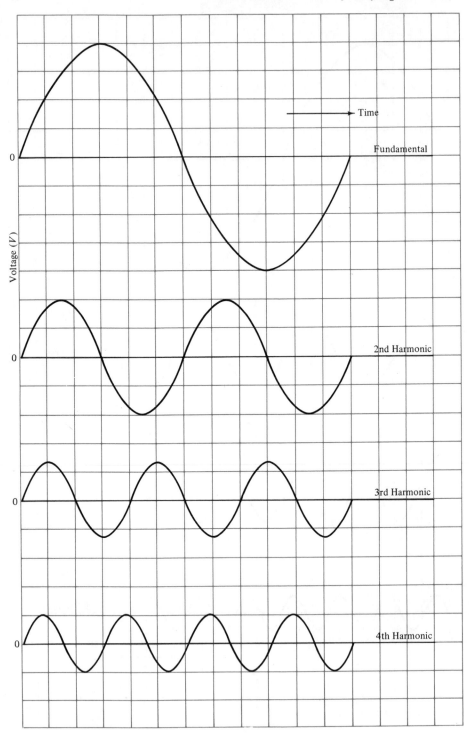

Figure 1-3X

8.   Define E_{av} and state its practical use.

9.   Which harmonics of a fundamental sine wave of voltage determine the sharpness of the corners of a resultant voltage waveshape?

10. Using the voltage waveshapes of Fig. 1-3X, determine the shape of the resultant voltage waveform by vector addition.

11. A rectangular pulse has a pulse repetition rate (*prr*) of 5000 pulses per second and a pulse width of 40 μsec.
     (a)   What is the pulse interval?
     (b)   What is the pulse repetition time?
     (c)   What is the duty cycle?
     (d)   Draw and label the waveform.

12. If the rectangular pulse of Prob. 11 is a positive pulse with an amplitude of 10 V, what would a dc voltmeter indicate when placed across this source?

13. Draw and label the voltage waveform of Prob. 12 as it would appear on the face of an ac oscilloscope. Assume that the trace line of the scope is in the center of the screen when no voltage is applied.

14. A 15 V positive rectangular pulse, from a −3 V reference, has a pulse repetition rate (*prr*) of 12,500 pulses per second and a pulse width of 10 μsec. What would a dc voltmeter indicate when placed across this source?

15. Draw and label the voltage waveform of Prob. 14 as it would appear on the face of a
     (a)   dc oscilloscope
     (b)   ac oscilloscope
     Assume that the trace line of the scope is in the center of the screen when no voltage is applied.

chapter 2

RC INTEGRATOR—LINEAR WAVESHAPING CIRCUIT

When nonsinusoidal waveforms are applied to linear networks, the resultant output waveform may be different from the input waveform. Hence, the circuit is said to shape the input-voltage waveform or to be a linear waveshaping circuit. See Fig. 2-1(a). Linear waveshaping circuits are classified according to components. The three basic classes are RC circuits, RL circuits, and RLC circuits. The RC circuit is the simplest and is classified according to the waveshape of its output voltage. The two most common of these output-voltage waveshapes are the integrated and the differentiated. The differentiated output is discussed in chapter 3; the integrated output is considered here.

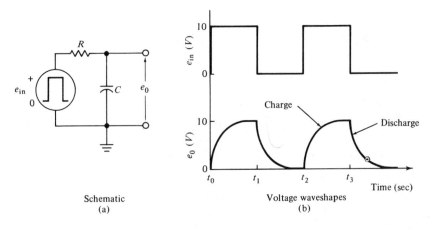

Schematic
(a)

Voltage waveshapes
(b)

Figure 2-1 RC Integrator

2.1 DEFINITION OF *RC* INTEGRATOR

An *RC* integrator circuit is a series circuit which consists of a resistor, a capacitor, and a voltage source and is one in which the output voltage is taken across the capacitor. The voltage which appears across the capacitor is referred to as the integrator output. When charging, the capacitor combines, or integrates, its original voltage with the new change in voltage. Hence, the term *integrator circuit* is generally and loosely applied to this circuit configuration. The output-voltage waveform, shown in Fig. 2-1(b), is one form of the integrator circuit produced wave. Variations of this waveform are discussed in a subsequent section of this chapter.

The integrator circuit shown in Fig. 2-1 may also be referred to as a low-pass *RC* filter. A low-pass filter passes the low frequencies and attenuates the higher frequencies, as its name implies. This characteristic of this circuit is evidenced by the exponential waveshape of the output voltage. As pointed out in chapter 1, the sharp corners of a voltage waveform are caused by the higher-frequency components of the applied step or rectangular waveform. The study of sinusoidal voltages established the fact that as the frequency increases, the reactance of the capacitor decreases, thus effectively bypassing its higher frequencies to ground. If the higher frequencies are bypassed to ground, they are not available as part of the output voltage; hence, the typical exponential-shaped output-voltage waveshape is produced.

2.2 WAVEFORM DISPLAY

Refer to Fig. 2-1(b). Notice that the input voltage and the output voltage have been plotted vertically and to the same time base. This placement is extremely important because if the input- and the output-voltage waveforms are placed horizontally adjacent, the characteristics of the output voltage as compared with those of the input voltage, at a particular instant in time, are extremely difficult to determine. The vertical presentation illustrated in Fig. 2-1(b) simplifies the comparison, and only by waveform comparison does the operation of the circuit become evident.

Therefore,   **An output voltage and/or current waveform should be drawn above or below the input (or some other reference) waveform and to the same time base.**

2.3 DESCRIPTION OF OPERATION

Let us now analyze the operation of the circuit in Fig. 2-1(a). At t_0 time an input-pulse voltage of $+10$ V is applied. From t_0 to t_1 time, the circuit resembles an *RC* series circuit to which a 10 V battery has been connected. The instant the voltage is applied, all the voltage appears across the resistor. This occurs because, in zero time, the capacitor has had no time to charge. A voltage cannot exist across the plates of a capacitor unless there are more electrons on one plate than there are on the other. Even though electrons move very rapidly through a circuit, they require a finite amount of time to do so.

Therefore, in zero time the electrons have had no time to move; hence, the capacitor is uncharged and has no voltage drop across it. Kirchhoff's second law states: the sum of the voltage drops around any closed circuit must equal the sum of the voltage sources. Hence, the +10 V source must equal the sum of the voltage drop across the capacitor and the drop across the resistor. See Fig. 2-1(a). Because the voltage drop across the capacitor is zero volts at t_0 time, the voltage drop across the resistor is forced to equal 10 V at t_0 time. See Fig. 2-1(b). As time increases from t_0 time toward t_1 time, the electrons move through the circuit and charge capacitor *C* to +10 V if the interval is sufficiently long. See e_0 voltage waveform in Fig. 2-1(b).

At t_1 time, the capacitor is charged to +10 V, and the source voltage of +10 V is removed. Therefore, the applied voltage is zero volts, and the input-voltage source ideally appears as a short circuit. Hence, the capacitor starts to discharge through the resistor *R* and completely discharges to zero volts if time permits.

The charge Eq. 5 describes the capacitor-charge curve.

$$e_C = E(1 - \epsilon^{-t/RC}) \qquad \text{Charge Eq.} \qquad [5]$$

in which

    e_C is the instantaneous value of voltage across the capacitor at a specific time t (V).

    E is the value of voltage applied to the circuit (V).

    ϵ is a constant 2.718 (base of the Napierian logarithm).

    t is the time the capacitor is allowed to charge (sec).

    R is the value of resistance in the circuit through which the electrons must travel in order to charge the capacitor (Ω).

    C is the value of capacitance in the circuit (F).

The discharge Eq. 6 describes the capacitor-discharge curve.

$$e_C = E\epsilon^{-t/RC} \qquad \text{Discharge Eq.} \qquad [6]$$

in which

    e_C is the instantaneous value of voltage across the capacitor at a specific time t (V).

    E is the value of voltage to which the capacitor has been charged and is the value of voltage from which it discharges (V).

    t is the time the capacitor is allowed to discharge (sec).

    ϵ is the constant 2.718 (base of the Napierian logarithm).

    R is the value of resistance in the circuit through which the electrons must travel in order to discharge the capacitor (Ω).

    C is the value of capacitance in the circuit (F).

The capacitor-charge Eq. 5 indicates mathematically that the capacitor can only charge to the applied voltage in (∞) infinite time. Theoretically, the capacitor may never charge to the source voltage in a finite amount of time. Practically, the capacitor is said to have charged to the applied voltage when it has charged to 99 percent of that voltage.

The capacitor-discharge Eq. 6 demonstrates the same principle. Theoretically, the capacitor may never discharge completely. Practically, the capacitor is said to have completely discharged when 99 percent of its charge has been depleted.

2.4 TIME CONSTANT

The product RC appears in both the charge and discharge equations. It is defined as a *time constant*, symbolized τ and expressed in the prime unit of time, seconds. This relationship is expressed as Eq. 7.

$$\tau = RC \qquad \text{Time Constant Eq.} \qquad [7]$$

The following formula manipulation demonstrates why the time constant (τ) is expressed in the unit of seconds.

$$\tau = RC \qquad Q = \text{quantity of charge} \qquad \text{(coulomb)}$$
$$Q = CE \qquad I = \text{current} \qquad \text{(ampere)}$$

Therefore,
$$C = \frac{Q}{E}$$

$$\tau = \frac{RQ}{E}$$

But
$$I = \frac{Q}{t}$$

Therefore,
$$Q = It$$

Hence,
$$\tau = \frac{RIt}{E}$$

Dividing numerator and denominator by I,

$$\tau = \frac{\dfrac{RtI}{I}}{\dfrac{E}{I}} = \frac{Rt}{\dfrac{E}{I}}$$

But
$$R = \frac{E}{I}$$

Hence
$$\tau = \frac{Rt}{R}$$

$$\tau = t \qquad \text{(sec)}$$

As symbolized in the charge Eq. 5, when t equals τ time, the capacitor will have charged to 63.1 percent of the applied voltage. The mathematical proof of this relationship follows:

$$e_C = E(1 - \epsilon^{-t/RC}) \qquad\qquad [5]$$

since
$$\tau = RC$$
$$e_C = E(1 - \epsilon^{-t/\tau})$$

if
$$t = \tau \qquad \text{time}$$

then
$$e_C = E - \frac{E}{\epsilon^{+1}} = E - \frac{E}{2.718}$$

$$e_C = E - 0.369E$$
$$e_C = 0.631E$$

Expressed as a percentage,

$$e_C = 63.1\% \text{ of } E \text{ in one time constant}$$

As symbolized in the discharge Eq. 6, when t equals τ time, the capacitor will have discharged down to 36.9 percent of its original voltage. The mathematical proof of this relationship follows:

$$e_C = E\epsilon^{-t/RC} \qquad\qquad [6]$$

since
$$\tau = RC$$
$$e_C = E\epsilon^{-t/RC}$$

If
$$t = \tau \qquad \text{time}$$

then
$$e_C = \frac{E}{\epsilon^{+1}} = \frac{E}{2.718}$$

$$e_C = 0.369E$$

Expressed as a percentage,

$$e_C = 36.9\% \text{ of } E \text{ in one time constant}$$

Recall that a practical full charge of a capacitor is considered to be approximately 99 percent of the applied voltage. This constitutes five (5) time constants. The mathematical proof follows:

If
$$t = 5\tau$$
then
$$e_C = E(1 - \epsilon^{-t/\tau})$$
$$e_C = E(1 - \epsilon^{-5\tau/\tau})$$

or
$$e_C = E - \frac{E}{(2.718)^{+5}} = E - \frac{E}{148}$$

$$e_C = E - 0.00677E$$
$$e_C = 0.9932E$$

or, expressed as a percentage,

$$e_C = 99.32\% \text{ of } E \qquad \text{when } t = 5\tau$$
$$e_C \approx 99\% \quad\ \text{ of } E \qquad \text{when } t = 5\tau$$

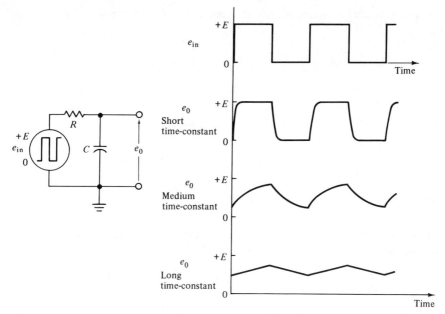

Figure 2-2   RC Integrator

Series *RC* circuits are classified according to the relationship between pulse duration and the length of the time constant. See Fig. 2-2.

A *short time constant* circuit is one in which the pulse duration is ten or more times greater than the duration of one time constant.

A *medium time constant* circuit is one in which the pulse duration may be from one-tenth to ten times that of one time constant.

A *long time constant* circuit is one in which the pulse duration may be one-tenth or less than one-tenth of the time required for one time constant.

2.5   SAMPLE PROBLEM ANALYSIS

Refer to Fig. 2-3. This schematic illustrates a medium time constant *RC* integrator circuit. From a given square-wave input voltage, predict the amplitude and waveshape of the resultant output voltage, and predict the amplitude

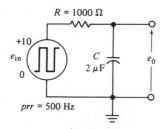

Figure 2-3   RC Integrator

and waveshape of the resultant output voltage which would be displayed by an oscilloscope.

$$\text{Input voltage} = 10 \text{ V peak } 500 \text{ Hz square wave}$$
$$R = 1000 \ \Omega$$
$$C = 2 \ \mu\text{F}$$

$$prt = \frac{1}{prr} = \frac{1}{500} = 2 \text{ msec}$$

$$t_p = \frac{prt}{2} = \frac{2 \times 10^{-3}}{2} = 1 \text{ msec}$$

$$\tau = RC = 1000 \times 2 \times 10^{-6} = 2 \text{ msec}$$

The following is the solution for the output voltage when $t = 1$ msec, as shown in Fig. 2-4.

$$e_C = E(1 - \epsilon^{-t/RC})$$

$$e_C = 10 - \frac{10}{\epsilon^{+1\times10^{-3}/2\times10^{-3}}}$$

$$e_C = 10 - \frac{10}{2.718^{+0.5}}$$

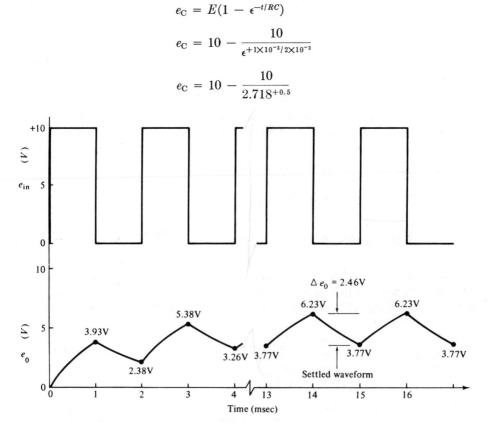

Figure 2-4

Subproblem (may be solved by slide rule):

$$2.718^{+0.5} = N$$
$$\log_{10} N = 0.5 \log_{10} 2.718$$
$$\log_{10} N = 0.5 \times 0.434$$
$$\log_{10} N = 0.2170$$

characteristic

antilog $(\log_{10} N) =$ antilog (0.2170)

mantissa

$$N = 1.65$$

$$e_C = 10 - \frac{10}{1.65}$$

$$e_C = 10 - 6.06$$
$$e_C = 3.93 \text{ V} \qquad \text{when } t = 1 \text{ msec}$$

At the end of 1 msec, the capacitor will have charged to 3.93 V, and at this time the +10 V source will have been removed. Hence, the capacitor will start to discharge from +3.93 V toward 0 V. The capacitor is allowed to discharge for 1 msec. The solution for the output voltage, when $t = 2$ msec, as shown in Fig. 2-4, follows.

$$e_C = E\epsilon^{-t/RC}$$

$$e_C = \frac{+3.93}{2.718^{+1\times10^{-3}/2\times10^{-3}}} = \frac{3.93}{2.718^{+0.5}} = \frac{3.93}{1.65}$$

$$e_C = 2.38 \text{ V} \qquad \text{when } t = 2 \text{ msec}$$

At the end of 2 msec, the input-pulse voltage of +10 V is reapplied for a duration of 1 msec. Because the capacitor will have been charged to +2.38 V, the effective change in voltage applied across the capacitor will be $(+10) - (+2.38)$ or +7.62 V. Equation 5 must be modified before it may be used to solve for the voltage across the capacitor at the end of 3 msec. This is necessary in order to compensate for the initial charge on the capacitor. Modified Eq. 5 becomes Eq. 8.

$$e_C = E - (E \pm E_o)\epsilon^{-t/RC} \qquad \text{Charge Eq. with} \qquad [8]$$
$$\text{initial charge}$$

in which

$$E_o = \text{initial charge of capacitor (V)}$$

The following is the solution for the output voltage when $t = 3$ msec, as shown in Fig. 2-4.

$$e_C = E - (E \pm E_o)\epsilon^{-t/RC}$$
$$e_C = 10 - [10 - (2.388)]\epsilon^{-1\times10^{-3}/2\times10^{-3}}$$
$$e_C = 10 - \frac{7.62}{2.718^{+0.5}} = 10 - \frac{7.62}{1.65}$$
$$e_C = 10 - 4.62$$
$$e_C = +5.38 \text{ V} \qquad \text{when } t = 3 \text{ msec}$$

For the succeeding intervals, the methods for solving for the voltage across the capacitor are the same as those established for the intervals between 1 and 2 msec and between 2 and 3 msec. The numerical value of the voltage across the capacitor for each millisecond of time is listed in Table I.

Table I

Time (msec)	e_C (V)
0	0
1	3.93
2	2.38
3	5.38
4	3.26
5	5.92
6	3.59
7	6.11
8	3.71
9	6.19
10	3.75
11	6.21
12	3.76
13	6.22
14	3.77
15	6.23
16	3.77
17	6.23

These values indicate the exponential settling of the voltage across the capacitor. The output-voltage waveform will remain constant from 14 msec in time to infinity.

If an oscilloscope is connected to the output terminals of the RC integrator circuit shown in Fig. 2-3, it will display and measure the peak-to-peak output voltage ($e_C = 2.46$ V) shown in Fig. 2-4.

The lengthy calculation for the solution of this sample problem should prepare one to understand why the simultaneous equations, which are used in an alternate method, are valid. This alternate method of solution employs a simultaneous solution of the charge and discharge equations.

2.6 TWO-EQUATION METHOD OF ANALYSIS OF SAMPLE PROBLEM

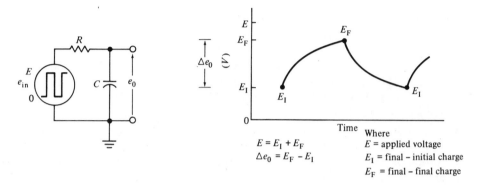

$E = E_I + E_F$

$\Delta e_0 = E_F - E_I$

Where
E = applied voltage
E_I = final – initial charge
E_F = final – final charge

Figure 2-5

The capacitor must discharge from the value of the final charge (E_F) to the final initial charge (E_I). Hence the basic discharge Eq. 6 becomes

$$e_C = E\epsilon^{-t/RC} \qquad [6]$$
$$E_I = E_F\epsilon^{-t/RC}$$

But
$$E_F = E - E_I$$

Hence
$$E_I = (E - E_I)\epsilon^{-t/RC}$$

The capacitor must start its charge from the final initial charge value (E_I) and must charge to the final charge value (E_F). Hence the basic charge Eq. 8 becomes

$$e_C = E - (E \pm E_I)\epsilon^{-t/RC} \qquad [8]$$
$$E_F = E - (E - E_I)\epsilon^{-t/RC}$$

The expedience of the two-equation method is demonstrated by the following numerical solution to the sample problem in Fig. 2-3.

$$\Delta e_o = e_C = E_F - E_I$$
$$\ast E_I = (E - E_I)\epsilon^{-t/RC}$$
$$\tau = 2 \times 10^{-3} \text{ sec} \quad R \times C$$
$$t = 1 \times 10^{-3} \text{ sec}$$
$$E_I = \frac{10 - E_I}{\epsilon^{+0.5}} = \frac{10 - E_I}{1.65}$$

$$E_{\mathrm{I}} = 6.05 - 0.605E_{\mathrm{I}}$$
$$1.605E_{\mathrm{I}} = 6.05$$
$$E_{\mathrm{I}} = 3.77 \text{ V}$$
$$\cancel{\ } E_{\mathrm{F}} = E - (E - E_{\mathrm{I}})\epsilon^{-t/RC}$$
$$E_{\mathrm{F}} = 10 - [10 - (+3.77)]\epsilon^{-0.5}$$
$$E_{\mathrm{F}} = 10 - \frac{6.23}{1.65} = 10 - 3.77$$
$$E_{\mathrm{F}} = 6.225 \text{ V}$$
$$\Delta e_{\mathrm{o}} = e_{\mathrm{C}} = E_{\mathrm{F}} - E_{\mathrm{I}}$$
$$\Delta e_{\mathrm{o}} = 6.225 - 3.77$$
$$\Delta e_{\mathrm{o}} = 2.455 \text{ V}$$

This value represents the peak-to-peak settled output voltage shown in Fig. 2-4. If an oscilloscope were connected to the output of the circuit shown in Fig. 2-3, it would display the settled waveform shown in Fig. 2-4.

In general, the term *integrated output* refers to a long time constant circuit. See Fig. 2-2. This circuit produces an approximation of a linear ramp voltage waveform. The disadvantage of this method of producing a linear ramp sweep of voltage is that the peak-to-peak output voltage is extremely small and generally requires additional amplification for practical use.

The short time constant circuit is used to develop the relationship between the rise time of a square or rectangular voltage waveform and its upper 3 dB frequency. This is accomplished by expressing rise time in terms of time constants and by using sine wave analysis to establish the upper 3 dB frequency which may then be expressed in terms of time constants. Time constants, as a common denominator, may then be used to relate rise time to the upper 3 dB frequency of the output-voltage waveform.

2.7   RELATIONSHIP OF RISE TIME TO TIME CONSTANTS

The rise time of the practical square wave of voltage as well as the output of the short time constant integrator circuit may be determined by calculating the difference between the time required for the output voltage to reach $0.9E$ and the time required for the output voltage to reach $0.1E$. Each may be determined from the basic charge Eq. 5. See Fig. 2-6.

$$t_{\mathrm{r}} = t_2 - t_1$$
$$e_{\mathrm{C}} = E(1 - \epsilon^{-t/RC}) \tag{6}$$
$$t_1 = \text{time for } e_{\mathrm{C}} = 0.1E$$
$$0.1E = E(1 - \epsilon^{-t_1/RC})$$
$$0.1 = 1 - \epsilon^{-t_1/RC}$$
$$\epsilon^{+t_1/RC} = 1.11$$

$$\frac{t_1}{RC} \log_{10} 2.718 = \log_{10} 1.11$$

$$t_1 = \frac{RC0.045}{0.434} = 0.103RC \approx 0.1RC$$

$$t_1 \approx 0.1RC$$

$$t_2 = \text{time for } e_C = 0.9E$$

$$0.9E = E(1 - \epsilon^{-t_2/RC})$$

$$0.9 = 1 - \epsilon^{-t_2/RC}$$

$$\epsilon^{+t_2/RC} = 10$$

$$t_2 = \frac{RC \log_{10} 10}{\log_{10} 2.718} = \frac{RC1.000}{0.434}$$

$$t_2 = 2.3RC$$

$$t_r = t_2 - t_1$$

$$t_r = 2.3RC - 0.1RC$$

$$t_r = 2.2RC \qquad\qquad [9]$$

Hence it has been established that the rise time is directly proportional to the time constant of the circuit.

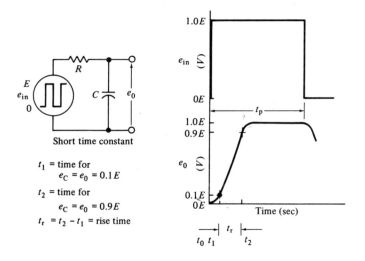

$t_1 = \text{time for}$
    $e_C = e_0 = 0.1E$

$t_2 = \text{time for}$
    $e_C = e_0 = 0.9E$

$t_r = t_2 - t_1 = \text{rise time}$

Figure 2-6 Low-Pass *RC* Circuit

2.8 ESTABLISHMENT OF UPPER 3 dB FREQUENCY BY SINE-WAVE ANALYSIS EXPRESSED IN TERMS OF TIME CONSTANTS

There is a direct connection between the rise time of a resultant output-voltage waveform and its upper frequency limit. This is due to the fact that the high-frequency harmonics of the sine wave determine the squareness of the resultant output-voltage waveform—a fact established in chapter 1. As in sine wave analysis, the 3 dB reference is employed to establish the practical limit of the upper frequency. The upper 3 dB frequency f_2 is that at which the output power has dropped to half of the mid-frequency value. Half power may also be expressed as 0.707 of the mid-frequency output voltage.

Refer to Fig. 2-6. The following derivation establishes the high frequency at which the resultant output voltage would drop to 0.707 of the mid-frequency value if a sine wave of voltage were applied to this circuit.

$$e_o = e_{in} \frac{-jX_c}{R - jX_c} \qquad \text{since } X_c = \frac{1}{\omega C} \text{ and } \omega = 2\pi f$$

$$\frac{e_o}{e_{in}} = \frac{\dfrac{1}{j\omega C}}{R + \dfrac{1}{j\omega C}} = \frac{1}{1 + j\omega RC}$$

$$f_2 = \text{upper 3 dB frequency} \qquad \text{(Hz)}$$
$$\omega = 2\pi f_2 = \omega_2$$

$$f_2 = \frac{1}{\tau_2}$$

If $\qquad \tau_2 = 2\pi RC$

then $\qquad f_2 = \dfrac{1}{2\pi RC}$

Hence $\quad \omega_2 = 2\pi f_2 = 2\pi\left(\dfrac{1}{2\pi RC}\right) = \dfrac{1}{RC}$

Since $\quad \dfrac{e_o}{e_{in}} = \dfrac{1}{1 + j\omega RC} = \dfrac{1}{1 + j\omega_2 RC}$

Substituting the value for $\omega_2 = 1/RC$,

$$\frac{e_o}{e_{in}} = \frac{1}{1 + j\left(\dfrac{1}{RC}\right)RC} = \frac{1}{1 + j1}$$

$$\frac{e_o}{e_{in}} = \frac{1}{1.414} = 0.707$$

Hence $\quad e_o = 0.707 e_{in} \qquad$ when $f_2 = \dfrac{1}{2\pi RC}$

$$f_2 = \frac{1}{2\pi RC} \text{ upper 3 dB frequency} \qquad \text{(Hz)} \qquad\qquad [10]$$

Solving Eq. 10 for RC,

$$RC = \frac{1}{2\pi f_2}$$

Substituting this value back into Eq. 9,

$$t_r = 2.2RC$$

$$t_r = 2.2\left(\frac{1}{2\pi f_2}\right) = \frac{0.35}{f_2}$$

$$t_r = \frac{0.35}{f_2} \qquad t_r = \text{rise time} \qquad \text{(sec)} \qquad \qquad [11]$$

$$f_2 = \text{upper 3 dB frequency} \qquad \text{(Hz)}$$

Equation 11 may be used to establish

(1)   The upper 3 dB frequency of a given amplifier.

This is a standard method for the establishment of the high-frequency response of a hi-fi amplifier. A square wave of voltage is applied to the input of an amplifier and the rise time of the resultant output-voltage waveform is measured. The upper 3 dB frequency of the amplifier may be determined by using the value of the measured rise time in Eq. 11.

(2)   The upper 3 dB frequency limit necessary for a proposed amplifier to reproduce a given rectangular voltage waveform.

To design an amplifier which will faithfully amplify a given rectangular voltage waveform, it is necessary to know its upper frequency limit. The measured value of the rise time of the voltage waveform to be amplified is used in Eq. 11 to establish this frequency.

LABORATORY EXPERIMENT
RC INTEGRATOR—LINEAR WAVESHAPING CIRCUIT

OBJECT:
1. To analyze the *RC* integrator circuit and to prove that the practical circuitry verifies the theory
   (a) Long time constant circuit
   (b) Medium time constant circuit
   (c) Short time constant circuit—illustrate the relationship between rise time and frequency response

MATERIALS:
1 Square-wave generator (20 Hz to 200 kHz)
1 Oscilloscope, dc time base; frequency response, dc to 450 kHz; vertical sensitivity, 100/mV/cm
1 Resistor substitution box (10 Ω to 10 MΩ, 1 W)
1 Capacitor substitution box (0.0001 to 0.22 μF, 400 V)

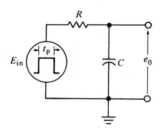

Figure 2-1X

PROCEDURE:
1. Long time constant circuit
   (a) Connect the circuit shown in Fig. 2-1X.

   $$R = 100 \text{ k}\Omega$$
   $$C = 0.01 \text{ μF}$$

   (b) Set the input square-wave voltage to 10 V peak and to a pulse duration of 100 μsec.
   (c) Determine the shape and amplitude of the output-voltage waveform by use of an oscilloscope.
   (d) Draw the input- and output-voltage waveforms on two respective graphs, on the same sheet of paper, with the same time base, and label completely.
   (e) Calculate the output voltage, using the direct two-equation method to verify the measured output voltage.
2. Medium time constant circuit
   (a) Repeat Steps (a) through (e) of Part 1. Change the circuit values to

   $$R = 100 \text{ k}\Omega$$
   $$C = 0.001 \text{ μF}$$

(b) Calculate the output voltage, using the charge and discharge equations, to verify the measured output voltage. Show all work. Include a graph which depicts the calculated points. (See sample Fig. 2-4.)

3. Short time constant circuit
   (a) Connect the circuit shown in Fig. 2-1X.

$$R = 100 \text{ k}\Omega$$
$$C = 0.0001 \text{ }\mu\text{F}$$

   (b) Set the input square-wave voltage to 10 V peak and to a pulse duration of 100 μsec.
   (c) Determine the shape and amplitude of the output-voltage waveform by use of an oscilloscope.
   (d) On graph paper, draw and label the input- and output-voltage waveforms.
   (e) Measure the rise time t_r and record it on the graph drawn in Step (d).
   (f) Calculate the upper 3 dB frequency of the square wave.

DONE

QUESTIONS AND EXERCISES

1. Define an integrator circuit.
2. What is the characteristic waveshape of the output voltage of an integrator circuit? Make a sketch of this waveshape.
3. Determine the value of R for the circuit illustrated in Fig. 2-2X, in which the voltage across the capacitor is 30 V, 3 msec after the switch is closed.

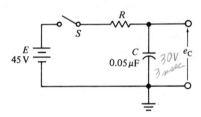

Figure 2-2X

4. Refer to Fig. 2-3X. Determine how long it will take the output voltage to reach +5 V after the switch is closed if the capacitor has an initial charge as indicated.

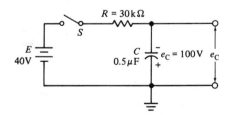

Figure 2-3X

5. In Fig. 2-4X, what will be the output voltage at the end of 12 msec if the input is a square wave with a 5 msec pulse width and an amplitude of 35 V?

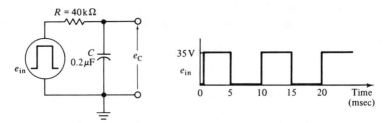

Figure 2-4X

6. Explain the relationship between rise time and the upper 3 dB frequency of a rectangular voltage waveform.
7. Explain why a capacitor may never charge completely.
8. Explain the significance of the relationship between the time constant τ and the pulse duration t_p.
9. In Fig. 2-6, if $t_r = 2$ μsec, what would be the upper 3 dB frequency limit of an amplifier which could reproduce this voltage waveform?
10. What would be the minimum rise time of a 1000 Hz square wave applied to a hi-fi amplifier if the upper 3 dB frequency of the amplifier were 20,000 Hz? (Express answer in microseconds.)
11. Define a short time constant circuit.
12. Using the circuit illustrated in Fig. 2-1X, if $R = 27$ kΩ and $C = 0.003$ μF, when a positive 5000 Hz, 15 V peak square wave is applied, what will be the amplitude of the settled output-voltage waveform?
13. Explain what is meant by a low-pass filter.
14. Explain why the rise time is directly proportional to the time constant of the circuit.
15. If the rise time of a rectangular pulse were 0.5 μsec, what would be the upper 3 dB frequency limit of an amplifier which could reproduce this voltage waveform?

chapter 3

RC DIFFERENTIATOR—LINEAR WAVESHAPING CIRCUIT

An RC differentiator circuit is a series circuit which consists of a resistor, a capacitor, and a voltage source, and it is one in which the output voltage is taken across the resistor. The voltage which appears across the resistor is referred to as the differentiator output. The voltage source is frequently a rectangular waveshape. Recall that, in contrast, an integrator circuit is one in which the output voltage is taken across the capacitor of a long time constant RC series circuit. The basic RC series circuit is, therefore, common to both. This relationship is illustrated in Fig. 3-1.

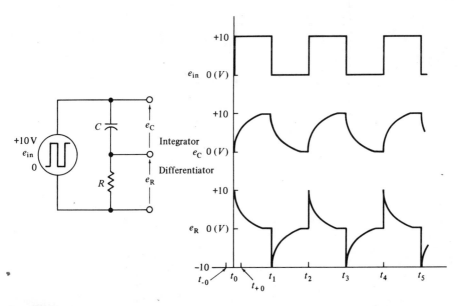

Figure 3-1

3.1   DEFINITION OF DIFFERENTIATED OUTPUT

A differentiated output is one in which the output voltage is directly proportional to the slope of the input voltage. The slope is the rate of change of input voltage with reference to time. Hence, any pulse circuit in which the output voltage is directly proportional to the rate of change of the input voltage is generally and loosely referred to as a differentiator circuit. For the output voltage to be proportional to the rate of change of the input voltage, however, the time constant of the circuit must be short ($\tau = 0.1t_\text{p}$ or less). Therefore, the short time constant circuit is the one which is technically referred to as the differentiator circuit. Practically, though, any circuit of this description, independent of time constant, is loosely referred to as a differentiator circuit. Refer to Fig. 3-2, which illustrates the typical "differentiated" output-voltage waveforms for short, medium, and long time constant circuits.

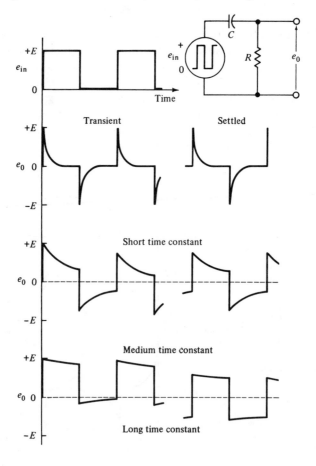

Figure 3-2

The differentiated circuit shown in Fig. 3-2 may also be referred to as a high-pass RC filter. A high-pass filter passes the high frequencies and attenuates the low frequencies, as the name implies. The characteristic of this circuit is evidenced by the tilt of the long time constant voltage waveform shown in Fig. 3-2, and it is further evidenced by the extreme case of tilt of the short time-constant voltage waveform shown in Fig. 3-2.

3.2 DESCRIPTION OF OPERATION

To understand the operation of this circuit, it is necessary to establish the relationship between time and voltage. Refer to Fig. 3-1. At t_0 time, the input voltage is 0 V, it is $+10$ V, and it is all values interjacent. The input voltage immediately preceding t_0 time is 0 V and the voltage immediately following t_0 time is $+10$ V. Therefore, it is necessary to be able to distinguish between the time immediately preceding t_0 time (designated t_{-0} time) and the time immediately following t_0 time (designated t_{+0} time).

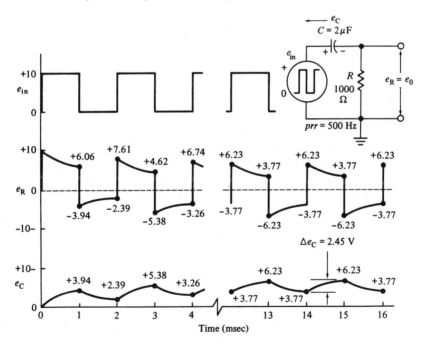

Figure 3-3

Note that the circuit in Fig. 3-3 is identical to the circuit in Fig. 2-3. In this chapter, the emphasis is upon the differentiated output of this circuit, whereas in the sample problem in chapter 2 the emphasis is upon the integrated output. Depending upon the component, across which the output is taken, this medium time constant circuit is either an integrator or a differentiator. Hence, the results of the sample problem in chapter 2 are used to explain the differentiated output of this common circuit.

Refer to Fig. 3-3. Note that at t_0 time the voltage across the capacitor is 0 V. Hence, all the voltage drop is forced to appear across resistor R. This is in accordance with Kirchhoff's voltage law, as explained in chapter 2. The input pulse of +10 V is applied for 1 msec. During this interval, the capacitor has had time to charge +3.94 V. Again, in accordance with Kirchhoff's law, the voltage drop across the resistor must equal the difference between the applied +10 V and the +3.94 V drop across the capacitor, or at t_{-1} time the voltage across the resistor is +6.06 V.

$$e_R = e_{in} - e_C$$
$$e_R = (+10) - (+3.94)$$
$$e_R = +6.06 \text{ V when } t_{-1} = 1 \text{ msec}$$

By t_{+1} time the input-voltage pulse has been removed; hence, the input voltage is 0 V, and the source appears as a short circuit. The capacitor which has charged to 3.94 V appears as a voltage source directly across the resistor. Note, in the schematic in Fig. 3-3, that the polarity of the charged capacitor is negative on the right-hand plate and positive on the left-hand plate. Therefore, the voltage drop across the resistor becomes a negative voltage with reference to ground. Hence, the output voltage at t_{+1} time is −3.94 V. The capacitor discharges from +3.94 to +2.39 V during the interval from 1 to 2 msec. During this interval, the input voltage is 0 V. Hence, the voltage across the resistor must equal the voltage across the capacitor. The polarity of the voltage drop across the resistor, however, must be negative with reference to ground. The output voltage at t_{-2} time, therefore, must be −2.39 V.

At t_{+2} time, the input-voltage pulse of +10 V is reapplied. Recall that the capacitor is charged to 2.39 V at t_{+2} time and that a capacitor resembles a voltage source at any instant in time. The voltage source and the charged capacitor (acting like a source) are in series-opposing. Hence, the voltage across the resistor equals the difference between the two, and the polarity of the larger, or e_R, is equal to +7.61 V at t_{+2} time.

During the interval between 2 and 3 msec, the capacitor charges from +2.39 V at t_{+2} time to +5.38 V at t_{-3} time. By t_{-3} time, the voltage across the resistor has decayed to an amount equal to the voltage change across the capacitor, or e_R is equal to +4.62 V at t_{-3} time. This process is repeated until a settled output voltage is obtained. An oscilloscope connected to the output of the circuit shown in Fig. 3-3 will display the settled voltage waveform.

The capacitor discharge Eq. 6 is also valid as the equation for the exponential decay of the voltage across the resistor in a series RC circuit and may be modified for this use by substituting e_R for e_C.

$$e_C = E\epsilon^{-t/RC} \qquad\qquad [6]$$
$$e_R = E\epsilon^{-t/RC}$$

where E = voltage across R at the beginning of a pulse interval (V).

In the following sample problem, Eq. 6 is used to solve for the differentiated output of the circuit shown in Fig. 3-3.

e_R at t_{-1} time

$$t_p = 1 \text{ msec}$$
$$\tau = RC = 2 \text{ msec}$$
$$e_R = +10\epsilon^{-1\times10^{-3}/2\times10^{-3}} = 10\epsilon^{-0.5}$$
$$e_R = \frac{+10}{1.65}$$
$$e_R = +6.06 \text{ V at } t_{-1} \text{ time}$$

e_R at t_{+1} time $\qquad = -e_C$
e_C at t_{+1} time $\qquad = e_C$ at t_{-1} time
e_C at t_{-1} time $\qquad = e_{in} - e_{R_{t_{-1}}}$

$$e_C = +10 - (+6.06)$$
$$e_C = +3.94 \text{ V at } t_{-1} \text{ time}$$

e_R at t_{+1} time $\qquad = -e_C$

$$e_R = -(+3.94)$$
$$e_R = -3.94 \text{ V at } t_{+1} \text{ time}$$

From t_1 to t_2 time, the output across R decays from -3.94 V toward 0 V.
Hence: e_R at t_{-2} time

$$e_R = E\epsilon^{-t/RC} = (-3.94)\epsilon^{-0.5}$$
$$e_R = -\frac{3.94}{1.65}$$
$$e_R = -2.39 \text{ V at } t_{-2} \text{ time}$$

Table II

Time $t_{(\ )}$ msec	e_R (V) $t_{-(\ )}$	$t_{+(\ )}$	e_C (V) $t_{(\ )}$
0	+10		0
1	+6.06	−3.94	+3.94
2	−2.39	+7.61	+2.38
3	+4.62	−5.38	+5.38
4	−3.26	+6.74	+3.26
5	+4.08	−5.92	+5.92
6	−3.58	+6.42	+3.59
7	+3.89	−6.11	+6.11
8	−3.77	+6.23	+3.71
9	+3.78	−6.22	+6.19
10	−3.77	+6.23	+3.75
11	+3.77	−6.23	+6.21
12	−3.77	+6.23	+3.76
13	+3.77	−6.23	+6.22
14	−3.77	+6.23	+3.77
15	+3.77	−6.23	+6.23
16	−3.77	+6.23	+3.77

At t_{+2} time the input pulse is reapplied.

$$e_{R_{t_{+2}}} = e_{in} + e_{R_{t_{-2}}}$$
$$e_{R_{t_{+2}}} = +10 + (-2.39)$$
$$e_R = +7.61 \text{ V at } t_{+2} \text{ time}$$

This process is repeated until a settled output voltage is obtained. The values of voltages are listed in Table II, as well as in Fig. 3-3.

3.3   DIRECT SOLUTION FOR SETTLED WAVEFORM

The application of four simultaneous equations to this problem provides a direct solution for the settled differentiated output. In Fig. 3-4, the settled

$$e_R = E\,\epsilon^{-\frac{t}{RC}}$$

(A) $E_2 = E_1\,\epsilon^{-\frac{t}{RC}}$ (B) $E_4 = E_3\,\epsilon^{-\frac{t}{RC}}$

(C) $E_1 - E_4 = E$ (D) $E_2 - E_3 = E$

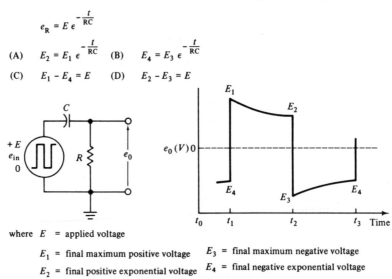

where E = applied voltage

E_1 = final maximum positive voltage E_3 = final maximum negative voltage

E_2 = final positive exponential voltage E_4 = final negative exponential voltage

Figure 3-4

output is indicated as the general case. If the value of E_1 is known, E_2 may be expressed in terms of E_1; similarly, if the value of E_3 is known, the value of E_4 may be expressed in terms of E_3. Hence, Eq. 6 may express these two relationships.

$$e_R = E\epsilon^{-t/RC} \qquad\qquad\qquad [6]$$
(A) $E_2 = E_1\epsilon^{-t/RC}$ and   (B) $E_4 = E_3\epsilon^{-t/RC}$

The total voltage at t_1 and at t_2 times, indicated in Fig. 3-4, must equal the applied voltage of the circuit E. Hence:

(C) $E_1 - E_4 = E$ and   (D) $E_2 - E_3 = E$

The numerical value for E_1, E_2, E_3, and E_4 may be obtained from the simultaneous solution of Eqs. (A), (B), (C), and (D). A mathematical expression for E_4 in terms of E_2 may be obtained by solving Eq. (C) for E_1 and substituting the resultant value for E_1 in Eq. (A). Hence:

$$E_1 = E + E_4$$
$$E_2 = (E + E_4)\epsilon^{-t/RC}$$
$$(E) \qquad E_2 = \frac{E + E_4}{\epsilon^{+t/RC}}$$

A mathematical expression for E_4 in terms of E_2 may be obtained by solving Eq. (D) for E_3 and substituting the resultant value for E_3 in Eq. (B). Hence:

$$E_3 = E_2 - E$$
$$E_4 = (E_2 - E)\epsilon^{-t/RC}$$
$$(F) \qquad E_4 = \frac{E_2 - E}{\epsilon^{+t/RC}}$$

Because Eqs. (E) and (F) contain the common unknowns E_2 and E_4, they may be solved simultaneously to establish the numerical values of E_2 and E_4. This simultaneous solution method is subsequently used to solve the same sample problem which was previously solved by the incremental method.

$$e_{\text{in}} = E = +10 \text{ V}$$
$$t = t_p = 1 \text{ msec}$$
$$\tau = RC = 2 \text{ msec}$$

(E) $\quad E_2 = \dfrac{+10 + E_4}{\epsilon^{+1\times10^{-3}/2\times10^{-3}}}$ $\qquad$ (F) $\quad E_4 = \dfrac{E_2 - (+10)}{\epsilon^{+1\times10^{-3}/2\times10^{-3}}}$

$\qquad E_2 = \dfrac{+10 + E_4}{1.65}$ $\qquad\qquad\qquad E_4 = \dfrac{E_2 - 10}{1.65}$

$\qquad E_2 = 6.06 + 0.606E_4$ $\qquad\qquad\qquad E_4 = 0.606E_2 - 6.06$

(E) $\quad E_2 - 0.606E_4 = +6.06$ $\qquad$ (F) $\quad 0.606E_2 - E_4 = +6.06$

Multiplying Eq. (E) by 0.606,

$$\begin{aligned}
(E) \qquad\qquad +0.606E_2 - 0.367E_4 &= +3.67 \\
(F) \quad (-) \qquad +0.606E_2 - 1.000E_4 &= +6.06 \\
\hline
+0.633E_4 &= -2.39 \\
E_4 &= -3.77 \text{ V}
\end{aligned}$$

Substituting the value of E_4 in Eq. (E),

$$E_2 - 0.606E_4 = 6.06$$
$$E_2 - (+0.606)(-3.77) = 6.06$$
$$E_2 = +3.78 \text{ V}$$

Substituting the values of E_2 and E_4 in Eqs. (C) and (D), respectively,

(C) $E_1 = E + E_4$
    $E_1 = (+10) + (-3.77) = 10 - 3.77$
    $E_1 = +6.23$ V

(D) $E_3 = E_2 - E$
    $E_3 = (+3.78) - (+10) = +3.78 - 10$
    $E_3 = -6.23$ V
    $E_1 = +6.23$ V
    $E_2 = +3.78$ V
    $E_3 = -6.23$ V
    $E_4 = -3.77$ V

Compare with values shown in Fig. 3-3.

The determination of the output voltage of a differentiator circuit has limited practical value. A more comprehensive knowledge of the differentiator circuit, however, may be gained by a clear understanding of the process involved in predicting the shape and magnitude of the output-voltage waveform.

In general, the term *differentiated output* refers to a short time constant circuit. See Fig. 3-2. This typical short time constant output-voltage waveform is utilized in the timing and triggering of multivibrator circuits. These applications of this circuit are discussed in subsequent chapters.

3.4 TILT TIME

Before the long time constant "differentiator" circuit may be understood, the term *tilt time*, t_t, must be defined. Refer to Fig. 3-5. In a long time constant differentiated output-voltage waveform, the exponential decay of the voltage is linear for all practical purposes. Hence, if this "linear" decay is extrapolated to the zero voltage axis, the time from the application of the pulse to that at which the linear extrapolation reaches 0 V is called tilt time, symbolized t_t, as shown in Fig. 3-5. The fractional tilt, symbolized P, may be expressed in terms of the pulse width t_p and tilt time t_t, as shown in Eq. 12.

$$P = \frac{t_p}{t_t} \qquad [12]$$

where P = fractional tilt (dimensionless)
        t_p = pulse width (sec)
        t_t = tilt time (sec)

Equation 12 is not practical for the determination of the fractional tilt P because the measurement of tilt time t_t is infeasible. The fractional tilt P may also be expressed however, in terms of easily measured voltages $(E_1 - E_2)$ and V, as shown in Eq. 13.

$$P = \frac{E_1 - E_2}{\dfrac{V}{2}} \qquad [13]$$

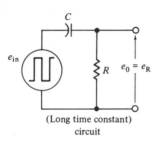

(Long time constant)
circuit

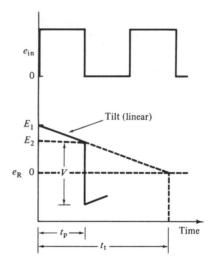

P = fractional tilt (dimensionless)
V = maximum voltage change at pulse transition time (V)
E_1 = maximum voltage excursion (V)
E_2 = decayed voltage excursion at t_p time (V)
t_p = pulse duration (sec)
t_t = tilt time (sec)

$$P = \frac{E_1 - E_2}{\frac{V}{2}} \qquad\qquad P = \frac{t_p}{t_t}$$

Figure 3-5 High-Pass RC Circuit

where P = fractional tilt (dimensionless)
               E_1 = maximum voltage excursion (V)
               E_2 = decayed voltage excursion at t_p time (V)
               V = maximum voltage change at pulse transition time (V)

Once the value of the fractional tilt P has been determined from Eq. 13, it may be inserted in Eq. 12 and tilt time t_t may be established.

3.5   RELATION OF TILT TIME TO TIME CONSTANTS

The lower 3 dB frequency of a square wave may be determined by evolving its relation to tilt time. This is accomplished by expressing tilt time and the lower 3 dB frequency in terms of time constants. Thus, the common denominator of time constants may be used to express the relation of tilt time to the lower 3 dB frequency.

Refer to the circuit in Fig. 3-5. If a sine wave of voltage were applied to this circuit, the low frequency at which the resultant output voltage would drop to 0.707 of the mid-frequency value could be obtained by the following derivation:

$$e_o = e_{in}\left(\frac{R}{R - jX_c}\right)$$

$$\frac{e_o}{e_{in}} = \frac{R}{R + \dfrac{1}{j\omega C}} = \frac{j\omega RC}{1 + j\omega RC}$$

Dividing numerator and denominator by $j\omega RC$,

$$\frac{e_o}{e_{in}} = \frac{\dfrac{j\omega RC}{j\omega RC}}{\dfrac{1 + j\omega RC}{j\omega RC}} = \frac{1}{1 + \dfrac{1}{j\omega RC}}$$

$$\omega = \omega_1 = 2\pi f_1$$

where f_1 is the lower 3 dB frequency.

If $\qquad\qquad \tau_1 = 2\pi RC$

then $\qquad\qquad f_1 = \dfrac{1}{\tau_1} = \dfrac{1}{2\pi RC}$

hence $\qquad\qquad \omega = (2\pi)\left(\dfrac{1}{2\pi RC}\right) = \dfrac{1}{RC}$

$$\frac{e_o}{e_{in}} = \frac{1}{1 + \dfrac{1}{j\left(\dfrac{1}{RC}\right)(RC)}} = \frac{1}{1 + \dfrac{1}{j1}}$$

$$\frac{e_o}{e_{in}} = \frac{1}{1 - j1} = \frac{1}{1.414} = 0.707$$

$$e_o = 0.707 e_{in}$$

This will occur when

$$f_1 = \frac{1}{2\pi RC} \qquad \text{(lower 3 dB frequency)}$$

when $\qquad\qquad \tau = RC = t_t$

$$f_1 = \frac{1}{2\pi t_t}$$

$$t_t = \frac{1}{2\pi f_1}$$

but $\qquad\qquad P = \dfrac{t_p}{t_t} = \dfrac{t_p}{\dfrac{1}{2\pi f_1}}$

for square wave
$$t_p = t_1 = t_2$$
$$T = t_1 + t_2 = 2t_p$$
$$t_p = RC$$
$$T = 2RC$$
$$f = \frac{1}{T} = \frac{1}{2RC} = \frac{1}{2t_p}$$
$$t_p = \frac{1}{2f}$$

[14]

hence
$$P = \frac{t_p}{t_t} = \frac{\dfrac{1}{2f}}{\dfrac{1}{2\pi f_1}} = \frac{\pi f_1}{f}$$

where
P = fractional tilt
f_1 = lower 3 dB frequency (Hz)
f = frequency of applied square wave (Hz)

$$P = \frac{\pi f_1}{f}$$

[15]

$$P = \frac{t_p}{t_t} = \frac{\pi f_1}{f}$$

but
$$f = \frac{1}{T} = \frac{1}{2t_p}$$

hence
$$\frac{t_p}{t_t} = \frac{\pi f_1}{\dfrac{1}{2t_p}}$$

$$\frac{1}{t_t} = 2\pi f_1$$

$$t_t = \frac{1}{2\pi f_1} = \frac{1}{6.28 f_1}$$

$$t_t = \frac{0.16}{f_1}$$

[16]

where
t_t = tilt time (sec)
f_1 = lower 3 dB frequency (Hz)

3.6 SAMPLE PROBLEM

If the input voltage to the circuit in Fig. 3-5 were a 5000 Hz square wave and the resultant output-voltage waveform measured by an oscilloscope were $V = 18$ V and $(E_1 - E_2) = 2$ V, what would be the lower 3 dB frequency of the resultant output-voltage waveform?

Solution:

$$P = \frac{E_1 - E_2}{\dfrac{V}{2}} = \frac{2}{\dfrac{18}{2}} = 0.222$$

$$T = \frac{1}{f} = \frac{1}{5 \times 10^{+3}} = 200 \; \mu\text{sec}$$

$$t_p = \frac{T}{2} = \frac{200 \times 10^{-6}}{2} = 100 \; \mu\text{sec}$$

$$t_t = \frac{t_p}{P} = \frac{100 \times 10^{-6}}{0.222}$$

$$t_t = 450 \; \mu\text{sec}$$

$$f_1 = \frac{0.16}{t_t} = \frac{0.16}{450 \times 10^{-6}}$$

$$f_1 = 356 \; \text{Hz}$$

In chapter 2, the formula used to express the relation of frequency to rise time was derived. In this chapter the formula used to express the relation of tilt time to frequency has been derived. These two formulas are extremely effective tools for the establishment of the upper and/or lower 3 dB frequency limits of a square wave or of an amplifier. Together, the technician may use them to establish the frequency response of an amplifier. The validity and effectiveness of these tools should become obvious with the careful development and derivation of these waveform and frequency relationships.

LABORATORY EXPERIMENT
RC DIFFERENTIATOR—LINEAR WAVESHAPING CIRCUIT

OBJECT:

1. To analyze the *RC* differentiator circuit and to prove that the practical circuitry verifies the theory
    (a) Short time constant circuit
    (b) Medium time constant circuit
    (c) Long time constant circuit—3 dB low-frequency response of a square wave

MATERIALS:

1 Square-wave generator (20 Hz to 200 kHz)
1 Oscilloscope, dc time base; frequency response dc to 450 kHz; vertical sensitivity, 100 mV/cm
1 Oscilloscope probe—10X attenuator, 10 MΩ
1 Resistor substitution box (10 Ω to 10 MΩ, 1 W)
1 Capacitor substitution box (0.0001 to 0.22 µF, 400 V)

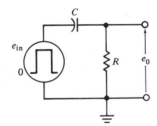

Figure 3-1X

PROCEDURE:

1. Short time constant circuit
    (a) Connect the circuit shown in Fig. 3-1X.

$$R = 100 \text{ k}\Omega$$
$$C = 0.0001 \text{ µF}$$

    (b) Set the input voltage to 20 V peak and to a pulse width of 100 µsec.
    (c) Determine the shape and amplitude of the input- and output-voltage waveform by use of an oscilloscope.
    (d) Draw the input- and output-voltage waveforms on two relative graphs, on the same sheet of graph paper, with the same time base, and label completely.
    (e) Explain why the amplitude of the output-voltage waveform is small.
2. Medium time constant circuit
    (a) Connect the circuit shown in Fig. 3-1X.

$$R = 100 \text{ k}\Omega$$
$$C = 0.001 \text{ µF}$$

   (b) Determine the shape and amplitude of the input- and output-voltage waveform by use of an oscilloscope.

   (c) Draw the input- and output-voltage waveforms on two relative graphs, on the same sheet of graph paper, with the same time base, and label completely.

   (d) Prove the results of your experiment by using the point-by-point method shown in the sample problem in Fig. 3-3.

   (e) To check the results of your point-by-point method calculations and to double-check the results of your experiment, directly determine the settled output voltages by the four-equation method.

   (f) Explain the reasons for any differences between your experimental circuit results and the results of the point-by-point and/or four-equation method.

  3. Long time constant circuit—3 dB low-frequency response of a square wave

   (a) Connect the circuit shown in Fig. 3-1X.

$$R = 100 \text{ k}\Omega$$
$$C = 0.01 \ \mu\text{F}$$

   (b) Determine the shape and amplitude of the input- and output-voltage waveform by use of an oscilloscope.

   (c) Draw the input- and output-voltage waveforms on two relative graphs, on the same sheet of graph paper, with the same time base, and label completely.

   (d) Using the four-equation method, determine the settled output voltages, and compare with the experimental results.

   (e) Determine the fractional tilt of the resultant output-voltage waveform by use of an oscilloscope.

   (f) Determine the lower 3 dB frequency of the resultant output-voltage "square wave."

QUESTIONS AND EXERCISES

1. Define a differentiator output.
2. What is the characteristic voltage waveshape of a differentiator output? Give an example.
3. Explain why the voltage amplitude of a differentiator output may be larger than the voltage amplitude of the input.
4. Refer to Fig. 3-2X. Determine how long it will take the output voltage to reach 20 V after the switch is closed.

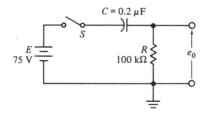

Figure 3-2X

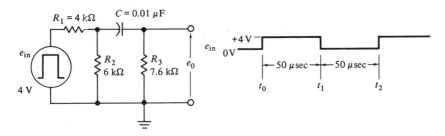

Figure 3-3X

5. Refer to Fig. 3-3X. Determine the output voltages at the various times indicated.
   (a) What is the value of e_o at t_{+0} time?
   (b) What is the value of e_o at t_{-1} time?
   (c) What is the value of e_o at t_{+1} time?
   (d) What is the value of e_o at t_{-2} time?
   (e) What is the value of e_o at t_{+2} time?
6. Define tilt.
7. Why is it necessary to establish the value of the pulse width in order to determine the lower 3 dB frequency of a square wave, even if t_p does not appear in Eq. 16?
8. Explain why sine wave analysis is a valid method for the determination of the lower 3 dB frequency of a square wave.
9. If a 1000 Hz square wave is applied to an amplifier with a fractional tilt of 0.4 at its output, what is the lower 3 dB frequency of the amplifier?
10. Refer to Eq. 13. Verify the fact that the lower 3 dB frequency of the amplifier is 500 Hz (as specified by the manufacturer), by establishing the value of $(E_1 - E_2)$ from the output-voltage waveform. The input is a 15 V peak, 10,000 Hz square wave.
11. Define a long time constant circuit.
12. Explain what is meant by a high-pass filter.
13. Refer to Fig. 3-4. When $e_{in} = +15$ V, 10 kHz square wave, $R = 68$ kΩ and $C = 0.002$ μF, what will be the amplitudes of e_o (E_1, E_2, E_3, and E_4) as seen on the face of an oscilloscope?
14. Explain how the frequency response of an audio amplifier may be determined by the use of square waves.
15. Refer to Fig. 3-4X. How long after the switch S is closed will it take the output voltage e_o to reach +20 V?

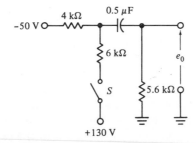

Figure 3-4X

chapter 4

CLIPPER CIRCUITS—
NONLINEAR WAVESHAPING
CIRCUIT

Nonlinear waveshaping circuits are circuits which incorporate at least one nonlinear device, a circuit component in which the volt-ampere characteristic curve is not a straight line. The most common nonlinear components are the semiconductor junction diode and the semiconductor three-element devices (example—the transistor). These components exhibit exponential volt–ampere characteristic curves.

4.1 DEFINITION OF A CLIPPER CIRCUIT

The clipper circuit is a common form of the nonlinear waveshaping circuit. Either it restricts the amplitude of a voltage and/or current waveform to some finite value or it cuts off the positive or negative peak of a voltage and/or current waveform at some finite value.

Refer to Fig. 4-1. Technically, the circuit which restricts (or limits) the amplitude of a voltage and/or current waveform is called a *limiter*, while the one which cuts off (or clips) the positive or negative peak of a voltage and/or current waveform is called a *clipper circuit*. In the latter circuit, it is only the clipped peaks which become the output voltage. Practically, however, both types of waveshaping circuits are interchangeably called limited or clipper circuits. In this text, each will be referred to as a "clipper circuit." The input-voltage waveform to a clipper circuit may have any voltage waveshape. That selected for illustration in Fig. 4-1 is a sine wave.

Clipper circuits are classified according to the active element they employ. The two most common types are those employing the semiconductor diode and those employing the transistor. The diode clipper circuits are further classified according to the placement of the diode in the circuit. This placement

52

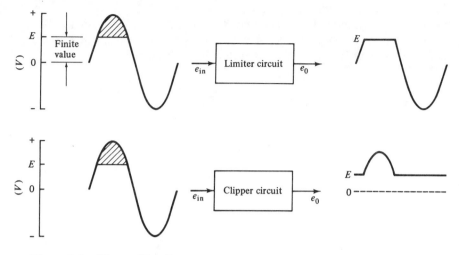

Figure 4-1 Clipper Circuit

determines whether the circuit is a series-diode, or a shunt-diode, clipper circuit.

The operation of a clipper circuit, in which a practical semiconductor junction diode is used is different from the hypothetical operation of the same circuit in which an ideal diode is assumed to be used.

4.2 IDEAL SEMICONDUCTOR DIODE

The semiconductor junction diode is the simplest circuit element which may be used as a switch. The ideal diode may be thought of as a simple ON–OFF switch. Refer to Fig. 4-2(a). The volt–ampere characteristic curve of an ideal junction diode is shown in Fig. 4-2(b). The switching action of the junction diode

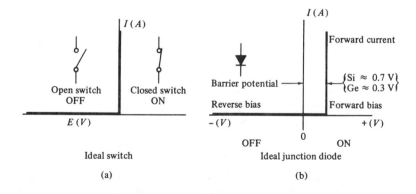

Figure 4-2 Switch Volt-Ampere Characteristic

is determined by the polarity of the voltage applied across the diode. When the latter is forward biased in excess of the barrier potential, it acts as a closed switch; when reverse biased, it acts as an open switch.

4.3   SERIES-DIODE (IDEAL) CLIPPER CIRCUIT

Refer to Fig. 4-3. The schematic is identical to that of the half-wave voltage rectifier circuit. When this circuit is used in the study of pulse circuits, it is referred to as a series-diode clipper circuit. It operates on the following

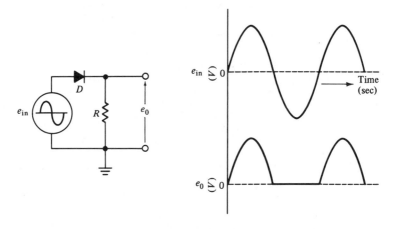

Figure 4-3   Series-Diode Clipper Circuit

principle: When there is a current flow through the resistor, there is an output voltage across resistor R. When the diode is forward biased, and hence acting as a closed switch, there is a current flow through the resistor. Therefore, the positive half cycle of the input voltage forward biases the diode; hence, the applied voltage effectively appears across the resistor R. During the negative half cycle, the diode is reverse biased and acts as an open switch; hence, no current flows through resistor R and, in turn, there is no voltage drop across it. The resultant output-voltage waveform is illustrated in Fig. 4-3.

4.4   SHUNT-DIODE (IDEAL) CLIPPER CIRCUIT

Refer to the shunt-diode clipper circuit shown in Fig. 4-4. In this circuit, there may be an output voltage only when the diode (acting as a switch) is not conducting. When the diode does conduct, it acts as a short circuit to the output; hence, the output voltage then is zero. The resultant output-voltage waveform is illustrated in Fig. 4-4.

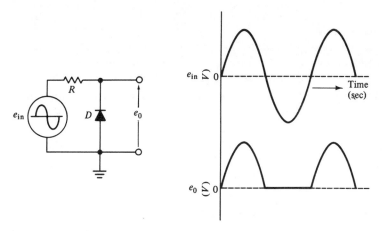

Figure 4-4 Shunt-Diode Clipper Circuit

4.5 PRACTICAL SEMICONDUCTOR JUNCTION DIODE

Refer to Fig. 4-5. Compare this volt-ampere characteristic curve for a practical junction diode with the ideal junction diode volt-ampere characteristic curve shown in Fig. 4-2. The practical diode does not conduct heavily for very small values of forward bias (for example, 0.1 V). This is due to the barrier potential formed when the diode was manufactured. If the forward bias is in excess of the barrier potential, the current will be high but not infinite. The diode, therefore, has some resistance when conducting. This resistance is

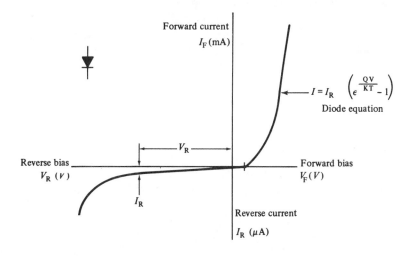

Figure 4-5 Practical Junction Diode

referred to as the dc forward resistance of the diode. It is symbolized R_f and expressed in ohms.

When the practical diode is reverse biased, a very small amount of current flows. The amplitude of the current is in the order of nanoamperes or, at most, microamperes for the normal silicon diode used in switching circuits. Therefore, the practical junction diode has a very high resistance when reverse biased. This resistance is referred to as the dc reverse resistance of the diode. It is symbolized R_r and expressed in ohms.

The volt–ampere characteristic curve of the practical junction diode is expressed by Eq. 17. When the appropriate signs are used, this equation is valid for forward, as well as reverse, current flow.

$$I = I_R(\epsilon^{qV/kT} - 1) \qquad \text{Diode Eq.} \qquad\qquad [17]$$

I = current flow through diode (A)

I_R = reverse saturation current at the reverse-bias voltage rating of the diode (A)

ϵ = a constant, 2.718 (base of the Napierian logarithm)

q = the charge of an electron = 1.602×10^{-19} coulomb (C)

k = Boltzmann's constant, 1.38×10^{-23} joules per degree centigrade

T = absolute temperature in degree Kelvin (K°)

V = applied voltage across the diode (V)

$\dfrac{kT}{q} \approx 26$ mV at 25°C (ideal)

In Eq. 17, the diode current varies with the applied voltage and the temperature. The other parameters in the equation are constants. The diode current equals the reverse saturation current within the limits of the reverse voltage rating of the diode, when reverse bias is applied. The reverse diode current varies with temperature, however.

Refer to Fig. 4-3. Recall that this basic clipper circuit is schematically simple but in practice it is fairly complex. This complexity is the result of the resistance of the diode, large when reverse biased and small when forward biased. Although it is the province of the engineer to design the circuit, the technician is expected to understand its operation and limitations and to determine a working value for the resistor R.

4.6   PRACTICAL DIODE CLIPPER CIRCUIT

In the practical diode clipper circuit shown in Fig. 4-3, the value of the resistor R must satisfy two completely divergent circuit conditions. These are

1. The amplitude of the output voltage must equal the amplitude of the input voltage when the diode is forward biased.
2. The amplitude of the output voltage must be zero when the diode is reverse biased.

Because the resistance of the diode, when the latter is forward biased, is small and the resistor R is in series with the diode, the circuit is that of a simple series voltage divider. For the output voltage to equal the input voltage, the value of R must be substantially greater than the resistance of the diode. Hence, when the diode is forward biased, the value of R should be large.

Because the resistance of the diode, when the latter is reverse biased, is large and the resistor R is in series with the diode, the circuit is that of a simple series voltage divider. For the output voltage to equal zero, the value of R must be substantially less than the resistance of the diode. Hence, when the diode is reverse biased, the value of R should be small.

Since neither of these conditions may be satisfied in the practical circuit, a compromise value for R must be used. To determine this value, solve for the geometric mean of the forward resistance of the diode (symbolized R_f) and the reverse resistance of the diode (symbolized R_r). Equation 18 expresses this relationship.

$$R = \sqrt{R_f R_r} \qquad\qquad [18]$$

R = resistance of resistor to be used in diode clipper circuit (Ω)
R_f = diode forward resistance (Ω)
R_r = diode reverse resistance (Ω)

To apply this equation to a specific circuit problem, determine the numerical values for the forward resistance (R_f) and the reverse resistance (R_r) of a specific diode. Refer to Fig. 4-6 which represents a straight-line approximation

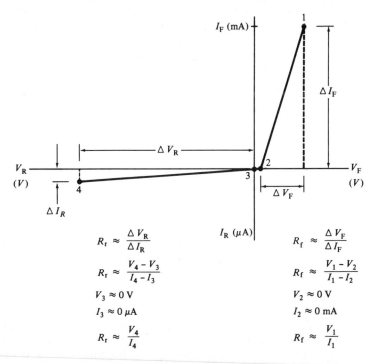

$$R_r \approx \frac{\Delta V_R}{\Delta I_R} \qquad\qquad R_f \approx \frac{\Delta V_F}{\Delta I_F}$$

$$R_r \approx \frac{V_4 - V_3}{I_4 - I_3} \qquad\qquad R_f \approx \frac{V_1 - V_2}{I_1 - I_2}$$

$$V_3 \approx 0\ \text{V} \qquad\qquad V_2 \approx 0\ \text{V}$$

$$I_3 \approx 0\ \mu\text{A} \qquad\qquad I_2 \approx 0\ \text{mA}$$

$$R_r \approx \frac{V_4}{I_4} \qquad\qquad R_f \approx \frac{V_1}{I_1}$$

Figure 4-6 Junction Diode Volt-Ampere Characteristic (Idealized)

of the volt-ampere characteristic curve of a semiconductor junction diode. The forward resistance of the diode may be approximated by the determination of the reciprocal of the slope of the forward biased portion of the chacacteristic curve of the diode. The solution for R_f, stated mathematically, is $R_f \approx \Delta V_F/\Delta I_F$. Assume point 2 in Fig. 4-6 to be at zero current and at zero voltage. The values of point 1 may be ascertained from the manufacturer's specification sheet. See Appendix B, 1N914 diode. These are $I_F = 10$ mA and $V_F = 1$ V dc. Hence:

$$R_f \approx \frac{\Delta V_F}{\Delta I_F} \approx \frac{V_1 - V_2}{I_1 - I_2} \approx \frac{1 - 0}{10 \text{ mA} - 0} \approx \frac{1}{10 \text{ mA}}$$

$$R_f \approx 100 \ \Omega$$

The reverse resistance of the diode may be approximated by determining the reciprocal of the slope of the reverse biased portion of the characteristic curve of the diode. The solution for R_r, stated mathematically, is $R_r \approx \Delta V_R/\Delta I_R$. Assume point 3 in Fig. 4-6 to be at zero current and at zero voltage. The values for point 4 may be ascertained from the specification sheet. These are $I_R = 5 \ \mu A$ and $V_R = 75$ V. Hence:

$$R_r \approx \frac{\Delta V_R}{\Delta I_R} \approx \frac{V_4 - V_3}{I_4 - I_3} \approx \frac{75 - 0}{5 \ \mu A - 0} \approx \frac{75}{5 \ \mu A}$$

$$R_r \approx 15 \ \text{M}\Omega$$

Having thus established the numerical values for R_f and R_r, for a specific diode, the value of the resistance R may be determined by the use of Eq. 18.

$$R = \sqrt{R_f R_r} = \sqrt{100 \times 15 \times 10^6} = 38.8 \ \text{k}\Omega$$
$$R = 39 \ \text{k}\Omega\text{—closest standard value}$$

The determination of R by the use of this method is approximate, but practical for the technician. The engineer more accurately determines the value of R by the use of the diode equation (Eq. 17). To be cognizant of the complexities of the circuit, the technician should be familiar with the diode equation.

4.7   BIASED SHUNT-DIODE CLIPPER

Refer to Fig. 4-7(a), a schematic of a battery-biased shunt-diode clipper. In this circuit, when the diode D is forward biased (acting as a closed switch), the output voltage equals the battery voltage ($+E$ V). When diode D is reverse biased (acting as an open switch), the output voltage equals the input voltage. If no input voltage is applied and the signal source has zero resistance, the battery voltage E reverse biases the diode. Hence, the output voltage equals the input voltage, that is, zero volts.

During the negative half cycle of the input, the signal voltage is in series-

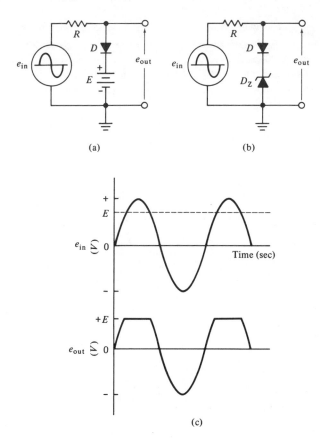

Figure 4-7   Biased Shunt Clipper

aiding with respect to the battery voltage E. The battery voltage E is of a polarity to reverse bias diode D. Hence, any increase in amplitude of the reverse-bias voltage will sustain the open switch condition of the diode. Therefore, the output voltage will equal the input voltage during the negative half cycle.

During the positive half cycle of the input, the signal voltage is in series-opposing with respect to the battery voltage E. The bias applied across the diode equals the difference between the signal voltage and the battery voltage and has the polarity of the larger voltage. Therefore, the output equals the input when the input is equal to or less than the battery voltage E. When the input is greater than the battery voltage, the diode is forward biased (acting as a closed switch); hence, the output voltage equals the battery voltage E. The resultant output voltage is shown in Fig. 4-7(c).

A zener diode may be used to replace the battery E in the shunt-diode clipper circuit in Fig. 4-7(a). The resultant zener diode circuit is shown in Fig. 4-7(b).

The volt–ampere characteristic curve of a zener diode is shown in Fig. 4-8. The zener diode operates as a conventional junction diode when the former is forward biased; when it is reverse biased, it may operate as a conventional junction diode or it may resemble a battery or a reference voltage. As the reverse bias is increased across the zener diode, the reverse current flow increases abruptly. The amplitude at which this abrupt increase occurs is the zener voltage. At this juncture, the diode operates as a closed switch. Current flow is reversed and the voltage across the diode is constant and independent

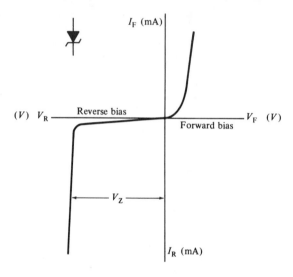

V_Z = zener voltage (V)

Figure 4-8   Zener Diode Volt-Ampere Characteristic

of current, as shown in Fig. 4-8. Hence, the current must be limited by an external resistor.

Refer to Fig. 4-7(b). Note that this zener-biased shunt-diode clipper circuit has the same output voltage as the battery-biased circuit illustrated in Fig. 4-7(a). During the negative half cycle of the input voltage, the zener diode D_Z is forward biased and operates as a closed switch. Junction diode D is reverse biased and operates as an open switch. Thus, operating in series and as switches, if either diode (switch) is open, the entire shunt branch will be open. Therefore, the output voltage equals the input voltage. During the period in which the amplitude of the positive half cycle of the input voltage is less than the zener voltage, the zener diode D_Z operates as an open switch and the junction diode operates as a closed switch. Hence, the output voltage equals the input voltage. During the period in which the amplitude of the positive half cycle of the input voltage is greater than the zener voltage, the junction diode D is forward biased. The zener diode D_Z, although reverse biased, conducts in the reverse direction and operates as a closed switch which has a constant voltage drop across it. The output voltage is constant, and the amplitude equals the zener voltage. Figure 4-7(c) illustrates this resultant output-voltage waveform. It has been assumed in this example that the battery voltage equals the zener voltage.

In general, to ensure sharp clipping action in a clipper circuit, the non-linearity of the diode must be overcome. To accomplish this, the amplitude of the input voltage must greatly exceed the barrier potential of the diode.

The technician should be aware of the disadvantages of the diode clipper circuits. When the diode in a series clipper circuit is reverse biased or OFF, the junction capacitance of the diode causes it to function as a coupling capacitor which transmits high-frequency voltage signals when the diode is OFF. The shunt-diode clipper circuit also has a disadvantage. When the diode is OFF, in a shunt-diode clipper circuit, the junction capacitance of the diode causes it to function as a low-impedance path to the high-frequency components. Hence, the corners of the output-voltage waveform are rounded.

Some functions of the clipper circuit are the removal of unwanted voltage spikes; the squaring of deteriorated square waves; and the comparison of two voltages, in which capacity the circuit is called a voltage comparer or a comparer circuit.

LABORATORY EXPERIMENT
CLIPPER CIRCUITS—NONLINEAR WAVESHAPING CIRCUITS

OBJECT:
1. To analyze series and shunt clipper circuits and to illustrate how the practical circuitry verifies the theory
2. To analyze the biased-diode clipper circuit
3. To analyze the operation of the zener diode as a reference voltage

MATERIALS:
1 Sine wave generator (20 Hz to 200 kHz)
1 Oscilloscope, dc time base; frequency response dc to 450 kHz; vertical sensitivity, 100 mV/cm
2 Silicon junction diodes with manufacturer's specification sheet (example—1N914 or equivalent)
2 Silicon zener diodes with manufacturer's specification sheet (example—1N4728: 3.3 V, 1 W or equivalent)
2 Transistor power supplies (0 to 30 V) or equivalent batteries
1 Resistor substitution box (10 Ω to 10 MΩ, 1 W)

PROCEDURE:
1. Determine the value of the resistor in the clipper circuits of Fig. 4-1X.
   (a) Determine the value of R_r and R_f from the manufacturer's specification sheet.
   (b) Substitute the numerical values obtained in Step a for R_r and R_f in Eq. 18, in order to determine the value of R.
2. Connect circuit 1 in Fig. 4-1X. For use in this circuit, select a standard color-coded resistor, the value of which is closest to that determined in Step 1.
3. Apply a sine wave input of 4 V rms at a frequency of 1000 Hz.
4. Draw a schematic of the circuit on graph paper.
5. Measure the input-voltage and output-voltage waveform displayed on the dc oscilloscope. On graph paper, draw the input-voltage waveform and, below it, to the same convenient time base, draw the output-voltage waveform as it would appear on a dc oscilloscope. Label all pertinent voltages.
6. In detail, explain how the theory justifies the function of the circuit. What factors determined the shape of the resultant output-voltage waveform?
7. For circuits 2 through 13, repeat Steps 3 through 6.

QUESTIONS AND EXERCISES

1. Draw the schematic of a series clipper circuit which will produce the same resultant output voltage as that produced by circuit 5 in Fig. 4-1X.
2. Repeat Problem 1 for circuit 6.
3. Repeat Problem 1 for circuit 7.
4. Repeat Problem 1 for circuit 8.
5. How does the output voltage of circuit 8 compare to the output voltage of circuit 9? Explain.

62

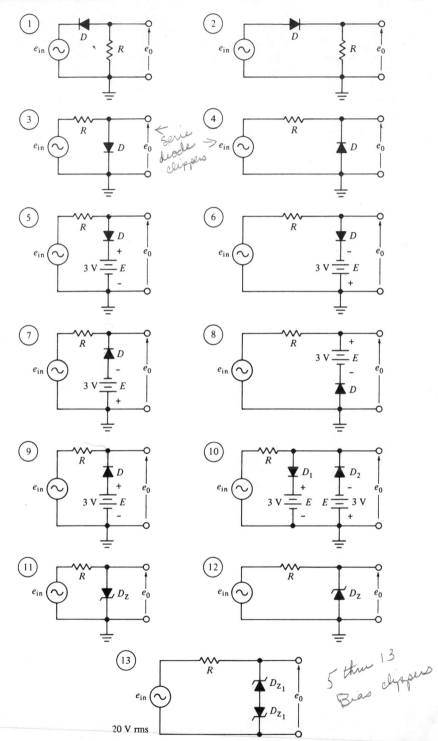

Figure 4-1X

6. Explain why the value of the forward resistance of a junction diode is not constant.
7. The solution for the proper value of the resistance to be used in a clipper circuit presents a problem. What is it, and how is it solved?
8. What is the principal disadvantage of a shunt-diode clipper circuit?
9. In what respect is a zener diode similar to a battery?
10. Under what circuit condition will a zener diode perform as a battery?
11. Determine the proper value of resistor R to be used in a clipper circuit employing a 1N916B silicon diode. (See Appendix B for parameters of 1N916B diode.)

chapter 5

CLAMPER CIRCUITS

A clamper circuit is so named because it clamps the ac voltage to a dc reference level. More specifically, it establishes a dc voltage reference level from which one extremity or peak of an ac signal swings. This circuit is also called a dc restorer or a dc inserter circuit. A clamping circuit may be thought of as an RC coupling circuit in which the time constant is switched between a high and low value. The ac signal may be a sine wave or any pulse voltage waveshape. Ideally, the circuit should change only the dc reference level and not the ac voltage waveshape.

Figure 5-1(a) illustrates a negative-voltage clamper circuit. The output voltage is negative with reference to zero, from which voltage the positive peaks originate. This output-voltage waveform is illustrated in Fig. 5-1(d).

In a clamper circuit the voltage source is in series with a charged capacitor (acting as a voltage source). The output voltage is the algebraic sum of these two voltages at any instant in time.

5.1 NEGATIVE-VOLTAGE CLAMPER CIRCUIT—IDEAL

Figure 5-1(a) is a schematic of a negative-voltage clamper circuit. The resultant output voltage is illustrated in Fig. 5-1(d). The analysis of this clamping action presupposes that no current is drawn from the output and that the diode employed is ideal.

At t_1 time, the input is +5 V with reference to ground. Diode D, acting as an ideal switch, is forward biased by this input. Hence, capacitor C charges to 5 V, the polarity of which is indicated in Fig. 5-1(b). At t_2 time the input is +3 V. The bias of the diode at t_2 time is the algebraic sum of the input voltage and the voltage across the capacitor. Hence, diode D (acting as an open switch)

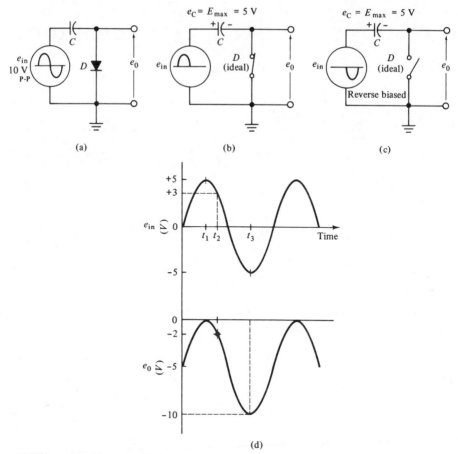

(d)

Figure 5-1   Negative-Voltage Clamper Using Ideal Diode

is reverse biased by 2 V, the output voltage of the circuit. At t_3 time the input is -5 V with reference to ground. The input voltage and the voltage across the capacitor are in series-aiding. Hence, the resultant output is -10 V.

5.2   NEGATIVE-VOLTAGE CLAMPER CIRCUIT—PRACTICAL

In the design of the practical negative-voltage clamper circuit shown in Fig. 5-2, the forward resistance of the diode R_f, the reverse resistance of the diode R_r, and the resistance of the source R_s must be taken into consideration. These parameters, in addition to the pulse repetition rate (or in the case of the sine wave, the frequency), determine the value of capacitance to be used in the circuit.

The time interval during which the diode is forward biased must be long enough to allow the capacitor to charge to the applied voltage. During this

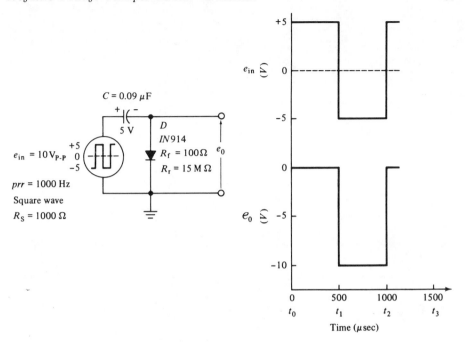

Figure 5-2   Negative-Voltage Clamper

interval, the effective circuit reduces to a short time constant differentiator circuit as shown in Fig. 5-3. The capacitor charges through the resistance of the source R_s and the forward resistance of the diode R_f. A capacitor practically charges to the applied voltage in five time constants. Hence, the interval from t_0 time to t_1 time (the duration of the input pulse t_p) must equal five time constants.

Hence: $$prt = \frac{1}{prr} = \frac{1}{1000} = 1000 \ \mu sec$$

$$t_p = \frac{prt}{2} = \frac{1000 \ \mu sec}{2} = 500 \ \mu sec$$

$$t_p = 5\tau$$

$$\tau = \frac{t_p}{5} = \frac{500 \ \mu sec}{5} = 100 \ \mu sec$$

$$\tau = (R_s + R_f)C$$

$$C = \frac{\tau}{R_s + R_f} = \frac{100 \times 10^{-6}}{1000 + 100} = \frac{100 \times 10^{-6}}{1.1 \times 10^{+3}}$$

$$C = 0.0908 \ \mu F \approx 0.09 \ \mu F$$

If C is 0.09 μF or less, it will have sufficient time to charge to the $+5$ V in t_p time.

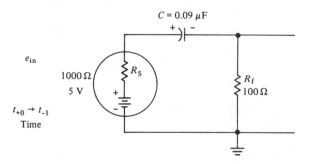

Figure 5-3   Equivalent Circuit of Negative-Voltage Clamper of Fig. 5-2 when
Diode Forward Biased During $t_{+0} \rightarrow t_{-1}$ Time

During the interval between t_1 time and t_2 time, the diode is reverse biased
and the effective circuit becomes a long time constant differentiator circuit, as
shown in Fig. 5-4. For the capacitor to operate as a voltage source, it may
not discharge any appreciable voltage in the interval during which the diode
is reverse biased. The value of the capacitor in a clamper circuit must satisfy
both forward-bias and reverse-bias circuit conditions. The value of capacitance,

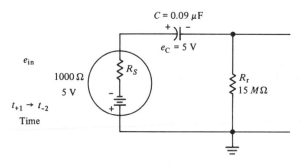

Figure 5-4   Equivalent Circuit of Negative-Voltage Clamper of Fig. 5-2 when
Diode Reverse Biased

selected to satisfy the forward-bias condition, must also satisfy the reverse-
bias condition; namely, that the capacitor will not discharge appreciably during the interval in which the diode is reverse biased. If the source voltage of
-5 V is applied for a sufficient length of time, the capacitor charges from the
5 V, of the polarity indicated in Fig. 5-4, toward 5 V of the reverse polarity.
Hence, the capacitor-charge Eq. 8 should be applied to establish the value
of the voltage across the capacitor at t_2 time.

$$e_C = E - (E \pm E_{co})\epsilon^{-t/RC} \qquad [8]$$

where $E = e_{in}$ from t_{+1} to t_{-2} time $= -5$ V
$E_{co} = $ voltage across C at t_{+1} time $= +5$ V
$t = $ time from t_{+1} to t_{-2} time $= 500$ μsec

$$R = (R_s + R_r) = 1 \times 10^{+3} + 15 \times 10^{+6} \approx 15 \text{ M}\Omega$$

$$C = 0.09 \ \mu\text{F}$$

$$e_C = -5 - [-5 - (+5)]\epsilon^{-500\times10^{-6}/15\times10^{+6}\times9\times10^{-8}}$$

$$e_C = -5 + \frac{10}{\epsilon^{+0.00037}} = -5 + \frac{10}{1.001}$$

$$e_C = -5 + 9.999 = +4.999 \text{ V}$$

$$e_C = +5 \text{ V at } t_{-2} \text{ time}$$

Therefore, the capacitor will operate effectively as the voltage source for this circuit.

The circuit illustrated in Fig. 5-2 will produce the required clamped output voltage if the amplitude of the input voltage is constant for several cycles. If the input voltage decreases in amplitude for several cycles, however, the capacitor will not have time to discharge to the lowered amplitude of the input voltage. Hence, the resultant output voltage will be clamped at a negative voltage rather than at zero. To prevent this in the practical circuit and to guarantee that the output voltage will be clamped at 0 V, regardless of the variations in the amplitude of the input voltage, the voltage across the capacitor must equal the $E_{\max}$ value of the input voltage. To accomplish this, a resistor is placed in parallel with the diode to provide a discharge path for the capacitor, as shown in Fig. 5-5.

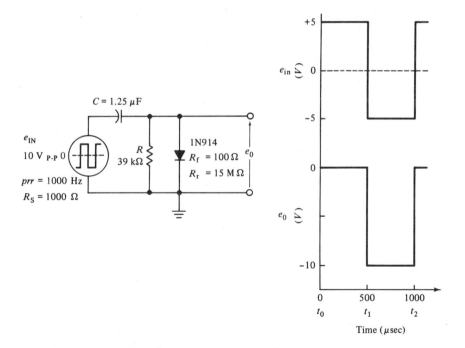

Figure 5-5   Negative-Voltage Clamper

For proper clamping to occur in the practical circuit, the reverse resistance of the diode R_r must be greater than the value of the resistor R which, in turn, must be much greater than the forward resistance of the diode R_f, or ($R_r \gg R \gg R_f$). For this condition to exist, R must be the geometric mean of R_r and R_f. The value of R may be determined by using Eq. 18

$$R = \sqrt{R_f R_r} \qquad\qquad [18]$$
$$R = \sqrt{15 \times 10^6 \times 100}$$
$$R = 39 \text{ k}\Omega$$

In the practical circuit in which the resistor is in parallel with the diode (Fig. 5-5), the discharge path and the discharge time constant differ, respectively, from the discharge path and the discharge time constant of the preceding basic circuit (Fig. 5-2). Since the value of capacitance is determined by the discharge path and time, it too is altered and a new value for C must be established. Because the capacitor discharges through the parallel resistance of R and R_r and because the value of R is much smaller than that of R_r, the equivalent parallel resistance of the combination is effectively the value of R, or 39 kΩ. Therefore, the discharge time constant of the circuit is

$$\tau = \left(R_s + \frac{R R_r}{R + R_r} \right) C$$

or

$$\tau \approx (R_s + R)C$$
$$\tau \approx (1 \times 10^{+3} + 39 \times 10^{+3})C$$
$$\tau \approx 40 \times 10^{+3}C$$

τ must be 100 times greater than the time during which the diode is reverse biased.

$$\tau = 100 t_p$$
$$\tau = 100 \times 500 \ \mu\text{sec} = 50,000 \ \mu\text{sec}$$
$$\tau = 50 \text{ msec}$$
$$C = \frac{\tau}{R} = \frac{50 \times 10^{-3}}{40 \times 10^{+3}}$$
$$C = 1.25 \ \mu\text{F}$$

Refer to Fig. 5-6 which illustrates an example of a biased clamper circuit, one in which the output voltage is clamped to some value other than zero. A fixed bias voltage must be placed in series with the parallel combination of resistor R and diode D in order to clamp a signal voltage to some value other than zero. The output of this circuit is a positively clamped voltage, clamped at a positive voltage.

The output voltage of the biased clamper circuit equals the algebraic sum of the signal voltage and the voltage across the capacitor. Therefore, the polarity and amplitude of the voltage to which the capacitor will charge must be established in order to ascertain the output voltage. See Fig. 5-6. When the input is -5 V, it is in series-aiding with respect to the bias voltage E (3 V).

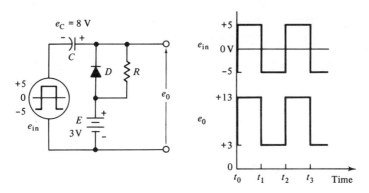

Figure 5-6

The polarity of the sum of these voltages is correct to forward bias the diode; hence, the capacitor charges to the sum of these two voltages (8 V) with the polarity indicated in Fig. 5-6.

An analysis of the circuit from the standpoint of the output voltage may be pursued because the voltage across the capacitor has been determined and thereby established as a "voltage source."

When the diode is forward biased (operating as a closed switch), the output voltage equals the bias voltage E (+3 V). When the diode is reverse biased (operating as an open switch), the output voltage equals the algebraic sum of the input voltage and the voltage across the capacitor. Thus, when a fixed voltage is established across capacitor C, the input voltage is +5 V at t_{+0} time, and this input voltage is in series-aiding with respect to the voltage across the capacitor. The polarity of this voltage reverse biases diode D. Hence, the output voltage is a +13 V at t_{+0}. At t_{+1} time the input voltage is −5 V in series-opposing the 8 V across the capacitor. The output voltage is the algebraic sum of the two, or +3 V.

The capacitor discharges slightly during the positive portion of the input voltage. Hence, the diode conducts on the negative peaks of the input-voltage waveform. This conduction recharges the capacitor and causes a slight clipping action which normally is negligible.

In the foregoing solutions, the values of the circuit components, determined by the methods specified, are only approximate but are practicable for the technician.

LABORATORY EXPERIMENT
CLAMPER CIRCUITS

OBJECT:

1. To analyze diode clamper circuits and to prove that the practical circuitry verifies the theory.
   (a) Unbiased clamper circuits
   (b) Biased clamper circuits

MATERIALS:

1 Sine square-wave generator (20 Hz to 200 kHz)

1 Oscilloscope, dc time base; frequency response dc to 450 kHz; vertical sensitivity, 100 mV/cm

1 Silicon junction diode with manufacturer's specification sheet (example—1N914 or equivalent)

1 Silicon zener diode with manufacturer's specification sheet (example—1N4728: 3.3 V, 1 W or equivalent)

1 Resistor substitution box (10 Ω to 10 MΩ, 1 W)

2 Capacitor substitution boxes (0.0001 to 0.22 μF, 400 V)

2 Batteries—1.5 V or equivalent dc power supply

PROCEDURE:

1. Determine the value of the resistor R in the clamper circuits of Fig. 5-1X.
   (a) Determine the value of R_r and R_f, from the manufacturer's specification sheet.
   (b) To determine the value of R, substitute the numerical values obtained in Step (a) for R_r and R_f in Eq. 18.
2. Determine the value of the capacitor C in the clamper circuits of Fig. 5-1X. The input voltage is a sine wave, the frequency of which is 5000 Hz.
3. Connect circuit 1, illustrated in Fig. 5-1X, using the standard values which most closely approximate the values calculated in Steps 1 and 2.
4. Apply a 10 V peak-to-peak sine wave input voltage of 5000 Hz to the circuit. Using a dc oscilloscope, observe and measure the output voltage. Record.
5. Draw a schematic of the circuit on graph paper. Draw the input-voltage waveform and below it, to the same time base, draw the output-voltage waveform observed. Label all pertinent voltages.
6. Explain in detail how the theory justifies the function of the circuit. What factors determined the shape and the reference level of the resultant output-voltage waveform?
7. For circuits 2 through 8 of Fig. 5-1X, repeat Steps 4 through 6.
8. Using a 10 V peak square wave, at a frequency of 5000 Hz, repeat Steps 2 through 7.

QUESTIONS AND EXERCISES

1. Explain the purpose of the resistor R in the clamper circuit illustrated in Fig. 5-5.

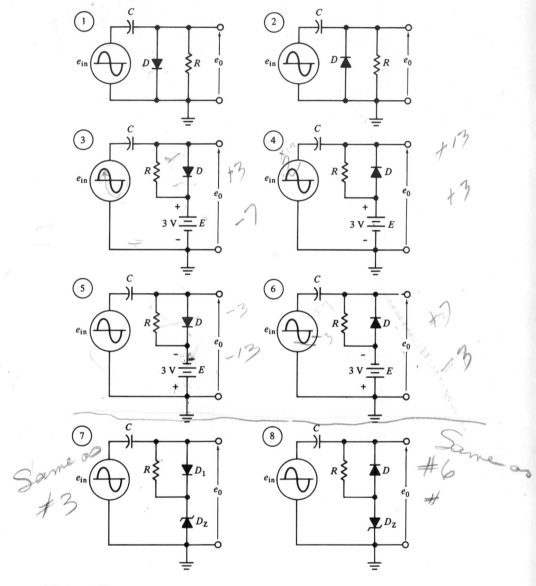

Figure 5-1X

2. Explain why the resistance of the source R_s must be small in order to ensure adequate clamping action.

3. In the circuit illustrated in Fig. 5-5, could it be possible for the capacitance to be too large? Explain.

4. In the circuit illustrated in Fig. 5-5, could it be possible for the capacitance to be too small? Explain.

5. Explain how circuit 2, illustrated in Fig. 5-1X, operates as an RC coupling circuit.

6. In a clamper circuit, is a silicon diode superior to a germanium diode? Explain.
7. Draw a schematic of a negative-voltage clamper circuit in which the output voltage may be clamped at any voltage from $+10$ to -10 V.
8. Explain why a sine wave of voltage is more difficult to clamp than a square wave of voltage.
9. In the experiment, would the value of C calculated for the 5000 Hz, sine wave input voltage be satisfactory for the 5000 Hz, square-wave input voltage? Explain.
10. In the circuit illustrated in Fig. 5-5, what effect would resistor R have on the output voltage if the value of resistance were too small? Explain.

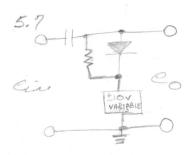

chapter 6

TRANSISTOR SWITCH

Preceding instruction in semiconductor pulse circuitry treated the transistor as it is used in a class A amplifier. The following theory treats the transistor as it is used in a switch.

6.1 IDEAL TRANSISTOR SWITCH

The transistor may be used as a switch. Ideally, if the transistor is reverse biased, it operates as an open switch as shown in Fig. 6-1(a). The open switch condition, illustrated in Fig. 6-1(a), is represented by the operating point B on the load line given in Fig. 6-1(c). This condition exists when the emitter-base junction of a transistor is reverse biased and the collector current is, therefore, reduced to zero.

When the emitter-base junction of a transistor is forward biased, the transistor will conduct. If the amplitude of the forward bias is large enough, the transistor will conduct at saturation, as shown in Fig. 6-1(b). When the transistor conducts at saturation, the collector current is limited by the load resistor R_L because the resistance of the transistor at saturation, from collector to emitter, is ideally zero. Hence, the voltage from collector to emitter is zero, independent of collector current. Therefore, the transistor resembles a closed switch from collector to emitter as shown in Fig. 6-1(b) and (c).

In a common-emitter junction transistor configuration, the base current (I_B) controls the collector current (I_C). Because $I_C = \beta I_B$, for the ideal transistor, a small base current may control a large collector current. The base current is determined by the voltage across the emitter-base junction which controls the collector current. When the input voltage is zero, or when it reverse biases the emitter-base junction, the collector current is zero. When the input voltage forward biases the emitter-base junction (0.7 V for silicon), the transistor conducts at saturation. Hence, a small input-voltage swing (0 to 0.7 V for silicon) is capable of controlling a large output collector-current swing.

Refer to Fig. 6-1(c). At point A on the load line, the collector current is

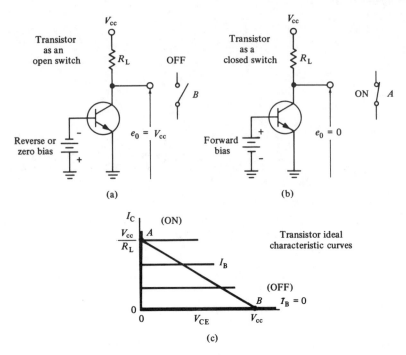

Figure 6-1   Transistor as an Ideal Switch

large, but the collector-to-emitter-voltage is zero. Hence, the transistor dissipates no power when operating as a closed switch. Conversely, at point B on the load line, the collector current is zero and the collector-to-emitter voltage is V_{cc}. Hence, the transistor dissipates no power when operating as an open switch. The ideal transistor, however, may switch a large amount of power into a load without dissipating any power. The transistor dissipates power only during the transition from the closed switch condition (ON) to the open switch condition (OFF).

6.2   PRACTICAL TRANSISTOR PARAMETERS

Figure 6-2 shows the practical volt–ampere characteristic curve of a common-emitter junction transistor. Area I represents the saturation region of operation. In this region both the emitter-base and the collector-base junctions are forward biased. Area II represents the active region of operation of the transistor. It is in this area that all the class A amplifiers operate, that the emitter-base junction is forward biased, and that the collector-base junction is reverse biased. Area III represents the cutoff region of operation. In this area the emitter-base junction and the collector-base junction are reverse biased.

The maximum power dissipation curve indicated in Fig. 6-2 represents the maximum power (symbolized P_{max}) that the transistor may dissipate. The maximum collector current permitted for the transistor is symbolized $I_{C_{max}}$. The maximum collector-emitter voltage permitted is symbolized $V_{CE_{max}}$. These values are indicated in Fig. 6-2. The three maximum values, P_{max}, $I_{C_{max}}$, and $V_{CE_{max}}$, must not be exceeded, regardless of whether the transistor is to be operated as a class A amplifier or as a switch.

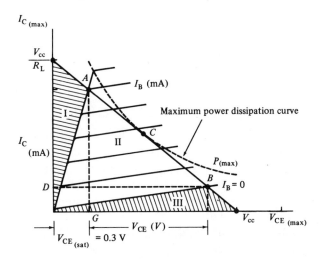

Figure 6-2 Practical Volt-Ampere Characteristic Curve

Figure 6-2 shows the practical volt–ampere characteristic curve of a common-emitter junction transistor. Point A on the load line of Fig. 6-2 represents the practical operating point for the transistor "switch" shown in Fig. 6-1(b). The collector-emitter voltage is not zero on the practical curve but 0.3 V. This voltage is referred to as the collector-emitter voltage at saturation and is symbolized $V_{CE_{sat}}$ and expressed in volts. For a silicon transistor, this value is approximately 0.3 V. The exact value for a specific transistor is a function of collector current and may vary from approximately 0.1 to 0.5 V. For a germanium transistor, a typical value for $V_{CE_{sat}}$ is 0.1 V.

Point B on the load line of Fig. 6-2 represents the practical operating point for the transistor (switch) illustrated in Fig. 6-1(a). In this circuit, the emitter-base junction is reverse biased; hence, the base current is zero. The collector current is not zero, however, but a finite value, as illustrated in Fig. 6-2 by point D. This collector current is produced by the reverse-collector-saturation current (symbolized I_{CBO}). Because the transistor is in a common-emitter configuration, I_{CBO} is amplified by $(\beta + 1)$. Hence, the value of the collector current at point D is appreciable.

Collector current which flows when the emitter is open-circuited and the collector-base junction is reverse biased is called reverse-collector-saturation current and is symbolized I_{CO} or I_{CBO}. I_{CO} is made up of two components: one, the leakage component which is the current around and across the collector-base junction surfaces (not the current which passes through the junction); the other, the saturation component, which is thermally generated. The leakage component is primarily voltage dependent and the saturation component is primarily temperature dependent. Practically, I_{CO} doubles in value for every 10°C increase in temperature for silicon and germanium.

For circuits in which the transistor is used as a switch, large-signal or dc parameters apply. Hence, the dc current gain for a transistor in a common-emitter configuration is h_{FE} rather than the small-signal ac current gain h_{fe}. The dc current gain for the transistor is

$$h_{FE} = \frac{I_C}{I_B} \qquad [19]$$

in which I_C is the specified value of collector current

I_B is the base current for the corresponding specified value of I_C

Hence, the dc current gain will vary with different values of collector current. The dc current gain is also referred to as the static value of short-circuit forward current transfer ratio. In the lists provided by manufacturers of switching transistors, a value is recorded for h_{FE} which corresponds with each value of collector current listed. The selection of a value for h_{FE}, for any particular circuit, should be made by locating the value of collector current which most closely approximates that which is used in the circuit.

6.3   IDEAL TRANSISTOR "SWITCH"—SATURATED

In the ideal saturated transistor "switch," a voltage pulse is used to control the transistor. To gain a clear understanding of this concept, observe Fig. 6-1(a) and (b), which illustrate a voltage pulse that has been simulated by two batteries in the ON and OFF circuits. Figure 6-3 illustrates the rectangular voltage pulses which are applied as control voltage to the transistor. This input-pulse voltage and, in turn, the base current (or lack of it) control the collector current of the transistor and thereby cause the latter to appear as a closed switch or as an open switch.

6.4   SAMPLE PROBLEM ANALYSIS—IDEAL TRANSISTOR SWITCH

Determine the necessary circuit parameters for the transistor switch illustrated in Fig. 6-3. The circuit is to have the following characteristics:

$$e_o = 20 \text{ V peak}$$
$$e_{in} = 5 \text{ V peak (0 to +5)}$$
$$I_C = 20 \text{ mA}$$

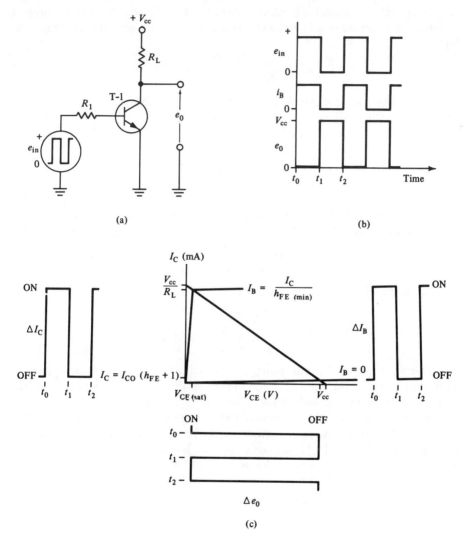

(a)

(b)

(c)

Figure 6-3 Ideal Transistor Switch

Assume an ideal silicon NPN transistor with the following specifications:

$$h_{FE} = 40$$
$$V_{BE} = 0 \text{ V} \quad \text{(ideal)}$$
$$V_{CE_{sat}} = 0 \text{ V} \quad \text{(ideal)}$$
$$I_{CBO} = 0 \quad \text{(ideal)}$$

Because the transistor operates as a switch, the value of V_{cc} determines the amplitude of output voltage. Hence, V_{cc} must be $+20$ V.

If the value of collector current is 20 mA, when the transistor is conducting at saturation (acting as a closed switch), the value of R_L may be determined as follows:

$$R_L = \frac{V_{cc}}{I_C} = \frac{20}{20 \text{ mA}} = 1 \text{ k}\Omega$$

Because the collector current is controlled by the base current, a value of base current that will assure collector-current saturation when the input pulse of $+5$ V is applied must be established. Since

$$I_C = h_{FE} I_B$$

then $$I_B = \frac{I_C}{h_{FE}} = \frac{20 \text{ mA}}{40} = 0.5 \text{ mA}$$

At t_{+0} time the input voltage is $+5$ V (forward bias). The Kirchhoff input-voltage loop equation is

$$e_{in} = E_{R_1} + V_{BE}$$
$$e_{in} = I_B R_1 + V_{BE}$$

Assuming the ideal transistor $V_{BE} = 0$ V,

$$e_{in} = I_B R_1$$
$$R_1 = \frac{e_{in}}{I_B} = \frac{5}{0.5 \text{ mA}} = 10 \text{ k}\Omega$$

Hence, when the input pulse is applied, the amount of base current necessary to make the transistor conduct at saturation will flow and cause the transistor to act as a closed switch. The transistor is said to be ON.

When the input pulse is removed at t_{+1} time, the input voltage is zero; hence no forward bias is applied. Since an ideal transistor is assumed, there is no I_{CBO} flowing. Hence, the transistor is said to be OFF. Therefore, the output voltage is equal to the source voltage V_{cc} ($+20$ V).

6.5 PRACTICAL TRANSISTOR SWITCH—SATURATED

Refer to Fig. 6-4. When an input pulse (forward bias to emitter-base junction) is applied to a practical transistor switch, the transistor cannot turn ON in zero time. Turn-ON time (symbolized t_{on}) is the time required for the collector voltage to change to 90 percent of the applied V_{cc}. The t_{on} time is made up of two parts: the delay time (symbolized t_d) and the rise time (symbolized t_r). The delay time is the time required for the output voltage to change to 10 percent of V_{cc}. The rise time is the time required for the output voltage to change from 10 percent of V_{cc} to 90 percent of V_{cc}.

Refer to Fig. 6-5. For t_{-0} time, this circuit is equivalent to part of the circuit shown in Fig. 6-4. At t_{-0} time, the input voltage reverse biases the emitter-base junction. The emitter-base junction resembles a large resistance in parallel with the junction capacitance. Hence, when the reverse-bias voltage

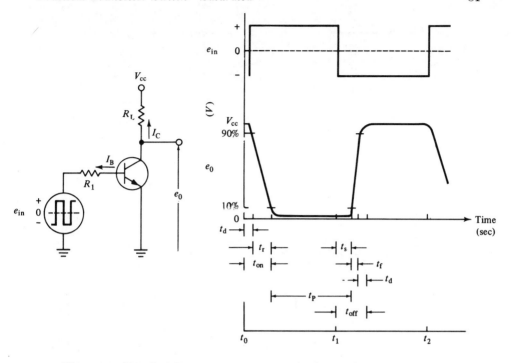

Figure 6-4 Practical Transistor Switch (Saturated)

is applied, the junction capacitance charges through R_1 to the value of the applied reverse bias, as shown in Fig. 6-5(a).

At t_{+0} time the input pulse is applied, as shown in Fig. 6-4. The equivalent circuit is shown in Fig. 6-5(b). Before the emitter-base junction may actually become forward biased, the junction capacitance must have had time to charge from $-e_{in}$ to at least zero. The time required for this to occur and the time required for the emitter current to diffuse through the base is the delay time.

The basic phenomenon which causes delay time is also responsible for the rise time, shown in Fig. 6-4(b). Figure 6-5(c) shows the equivalent circuit for t_{-0} time. In Fig. 6-4, at t_{-0} time, both the emitter-base junction and the collector-base junction are reverse biased. Hence, the emitter-base junction capacitor will be charged to $-e_{in}$ V, and the collector-base junction capacitor will be charged to the sum of e_{in} and V_{cc}. When the input pulse is applied at t_{+0} time, the collector current cannot flow until the emitter-base junction is actually forward biased (delay time). The collector current cannot reach saturation instantly because the collector-base junction capacitance must be discharged by the diffusion of carriers across the base region.

Refer to Fig. 6-4. At t_{+1} time, the input pulse is removed. The collector current, however, will remain at a value equal to or near the value flowing when the input pulse was applied at t_{-1} time. In turn, the output voltage will remain at, or near, 0 V. When the input pulse was applied at t_{-1} time, it was

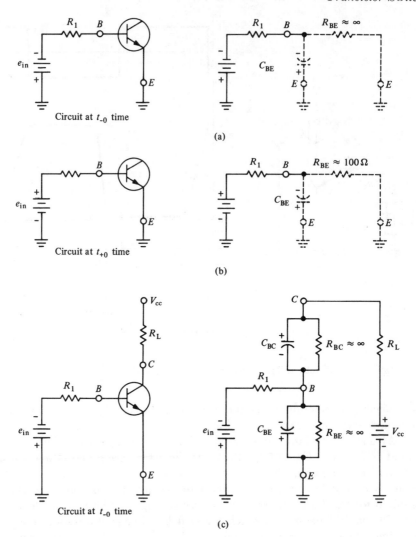

Figure 6-5   (a) Circuit at t_{-0} Time; (b) Circuit at t_{+0} Time; (c) Circuit at t_{-0}
Time

assumed that the magnitude of base current was large enough to ensure
saturation. For the transistor to be in saturation, the charge flow into the base
must be greater than the charge flow out of the base. Hence, for saturation to
exist, the collector current must be limited by the external load resistance (R_L)
and not by the base current. Therefore, when the input pulse is removed,
there is an excess of electrons stored in the base (NPN). The collector current
will continue to flow until it has removed the excess charge from the base. The
time required to do this is referred to as storage time (symbolized t_s), as shown
in Fig. 6-4.

Due to storage time, the output-pulse width is larger than the input-pulse width, as shown in Fig. 6-4. Pulse width (symbolized t_p) was defined in chapter 1.

After the input pulse has been removed, the time required for the output voltage to return to $+V_{cc}$ (NPN) is called turn-OFF time, symbolized t_{off}. Turn-off time is made up of three distinct intervals—storage time (t_s), fall time (t_f), and decay time (t_d).

Storage time is the time required to remove the *excess* base storage charge. This is accomplished by the recombination process in circuits in which the input pulse returns to 0 V. In circuits in which the input voltage reverses polarity when the input pulse is removed, however, as illustrated in Fig. 6-4, the storage time is determined by two components. In this circuit condition, the excess base storage charge is removed by the recombination process and, due to the reverse bias of the input voltage, by the reverse base-current flow. This reverse base-current flow will exist only during the interval of time needed to remove the excess base storage charge.

The fall time (t_f) is the time required for the output voltage to rise (NPN) from 10 to 90 percent of the value of V_{cc}. At the end of storage time (t_s), the excess base storage charge has been removed and the collector-base junction is reverse biased. The time required for the remaining charge (not the excess charge) to be removed from the base is defined as the fall time (t_f).

The delay time (t_d) is the time required for the collector-base junction capacitance to recharge to the voltage of the reverse bias across the collector-base junction.

The actual duration of delay time, storage time, and fall time is extremely small and may easily be of the order of nanoseconds. Figure 6-4 illustrates turn-ON time, symbolized t_{on}, and turn-OFF time, symbolized t_{off}, as listed in the specification sheets supplied by manufacturers of switching transistors. Turn-ON time is the sum of delay time and rise time. Turn-OFF time is the sum of storage time, fall time, and delay time. Hence,

$$t_{on} = t_d + t_r \qquad [20]$$
$$t_{off} = t_s + t_f + t_d \qquad [21]$$

6.6 RELATION OF COLLECTOR CURRENT AND BASE CURRENT

In transistors of a designated type, the dc current gain, symbolized h_{FE}, varies from transistor to transistor. Therefore, a circuit designed to use a designated type of transistor as a saturated switch must be constructed to function with any transistor of the given type.

Since

$$I_C = h_{FE}I_B$$

and

$$I_B = \frac{e_{in}}{R_1}$$

then

$$I_C = \frac{h_{FE}e_{in}}{R_1}$$

as shown in Fig. 6-3. Therefore, if a circuit is designed to operate with a transistor of the poorest quality (lowest value of h_{FE}) as the saturated switch, the design will ensure operation with any transistor having the same or higher value of h_{FE}. Hence, the minimum value of h_{FE}, symbolized $h_{FE_{min}}$, should be used in the design of a transistor "switch."

In practical circuitry, it is necessary that the operation of the transistor switch approach, as nearly as possible, that of the ideal transistor switch. This may be accomplished by a decrease in rise time.

Refer to Fig. 6-4. Rise time is the time required for the collector current to increase from 10 to 90 percent of the value of the saturated collector current (limited by R_L). When the transistor is OFF, the collector-base junction is reverse biased and resembles and operates as a capacitor (collector-base junction capacitance). When an input pulse is applied to the emitter-base junction, the junction operates as a forward biased diode, the resistance of which is very small (assuming that the emitter-base junction capacitance is negligible). Hence, the transistor switch, illustrated in Fig. 6-6(a), resembles the equivalent circuit illustrated in Fig. 6-6(b). For the collector current to rise from 10 to 90 percent of its value at saturation, it must charge the collector-base junction capacitance. Hence, the collector current will increase in accordance with the standard capacitor-charge Eq. 5.

$$i_C = I_C(1 - \epsilon^{-t/RC}) \qquad\qquad [5]$$

The base current is selected for the minimum value necessary to ensure collector-current saturation, as illustrated in the sample circuit problem for the ideal transistor switch. Rise time may be expressed in terms of R_L and C_{CB} (the average value of collector-base junction capacitance). Refer to Fig. 6-6(c).

$$I_C = h_{FE_{min}}I_B \qquad \text{and} \qquad I_C = \frac{V_{cc}}{R_L}$$

$$t_r = t_{on} - t_d = 90\% - 10\% = 80\%$$

$$t_r = 0.8$$

$$0.8I_C = h_{FE}I_B(1 - \epsilon^{-t_r/R_L C_{OB}}) \qquad \tau = R_L C_{CB}$$

$$\frac{0.8V_{cc}}{R_L} = \frac{V_{cc}}{R_L}(1 - \epsilon^{-t_r/\tau})$$

$$0.8 = 1 - \epsilon^{-t_r/\tau}$$

$$\epsilon^{+t_r/\tau} = 5$$

$$t_r = \tau\frac{\log_{10} 5}{\log_{10} \epsilon} = \tau\frac{0.699}{0.434}$$

$$t_r = 1.61\tau$$

Refer to Fig. 6-6(d). If the minimum value of base current (I_B) is doubled, the rise time will be reduced.

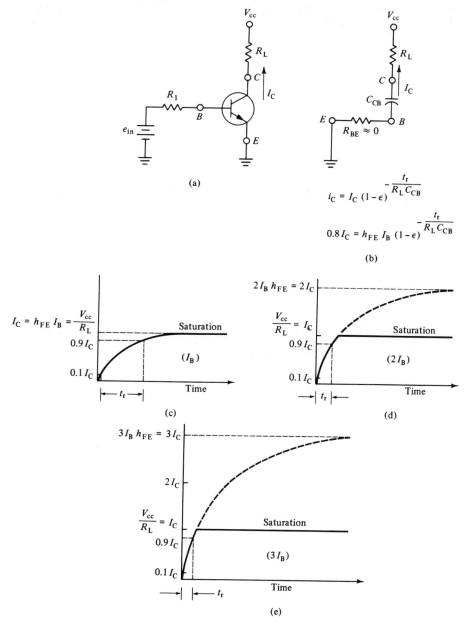

$$i_C = I_C (1 - \epsilon)^{-\frac{t_r}{R_L C_{CB}}}$$

$$0.8 I_C = h_{FE} I_B (1 - \epsilon)^{-\frac{t_r}{R_L C_{CB}}}$$

Figure 6-6 (a) Circuit of Fig. 6-4 at t_{+0} Time

$$i_C = I_C(1 - \epsilon^{-t_r/\tau})$$

$$\frac{0.8V_{cc}}{R_L} = 2h_{FE}I_B(1 - \epsilon^{-t_r/\tau})$$

$$\frac{0.8V_{cc}}{R_L} = \frac{2V_{cc}}{R_L}(1 - \epsilon^{-t_r/\tau})$$

$$\epsilon^{+t_r/\tau} = 1.67$$

$$t_r = \tau\frac{\log_{10} 1.67}{\log_{10} \epsilon} = \tau\frac{0.222}{0.434} = 0.511\tau$$

$$t_r = 0.511\tau$$

Hence, when   Base current $= 2I_{B_{min}}$, $t_r = 0.511\tau$

Base current $= I_{B_{min}}$, $t_r = 1.61\tau$

$$t_r = \frac{0.511\tau}{1.61\tau} = 0.318 \approx \frac{1}{3}$$

Hence, when base current is doubled, rise time is reduced to approximately one-third of its original value.

Refer to Fig. 6-6(e). If the minimum value of base current is tripled, the rise time will be further reduced.

$$\frac{0.8V_{cc}}{R_L} = 3h_{FE}I_B(1 - \epsilon^{-t_r/\tau})$$

$$\frac{0.8V_{cc}}{R_L} = \frac{3V_{cc}}{R_L}(1 - \epsilon^{-t_r/\tau})$$

$$t_r = 0.309\tau$$

Hence, when   Base current $= 3I_{B_{min}}$, $t_r = 0.309\tau$

Base current $= I_{B_{min}}$, $t_r = 1.61\tau$

$$t_r = \frac{0.309\tau}{1.61\tau} = 0.192 \approx \frac{1}{5}$$

Hence, when base current is tripled, rise time is reduced to approximately one-fifth of its original value.

Although the relationship established between rise time and base current is approximate, values obtained are practical. Therefore, rise time is reduced when base current is increased above the minimum value necessary for collector-current saturation. When the base current is doubled, rise time decreases to approximately one-third of its original value. When the base current is tripled, rise time decreases to approximately one-fifth of its original value.

To double or triple the value of base current, it is merely necessary to adjust the value of R_1 (Fig. 6-4) for any given input-voltage pulse. The price paid for this reduction of rise time is an increase in storage time. When the pulse is removed, the collector current cannot decrease until the excess of charge carriers, with which the base is supplied when the transistor is ON, are removed.

6.7 "SPEEDUP" CAPACITOR

In order to decrease rise time and thereby cause the transistor to turn on more rapidly, it is necessary to rapidly saturate the base with charge carriers when the input pulse is initially applied. After this initial saturation, the base current should be reduced to the minimum value necessary to ensure saturation. Refer to Fig. 6-7. The capacitor-charge current may be used as an additional source of current necessary to reduce rise time if a capacitor C is placed

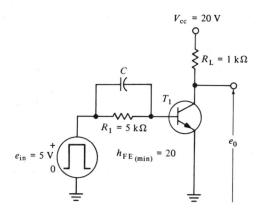

Figure 6-7

across resistor R_1 when the pulse is applied. The charging current ceases to flow when the capacitor has charged to the value of the applied input voltage. This capacitor-charging current is a component of the base current because it emanates from the base. The instant the input pulse is applied, it produces a base current which consists of two parts: (1) the capacitor-charge current and (2) the normal base current which is determined by the amplitude of the input pulse and by the value of R_1.

From the circuit shown in Fig. 6-7, each of the two components which affect the base current are separately illustrated in Fig. 6-8(a) and (b), and each should be separately analyzed. Figure 6-8(a) illustrates the capacitor-charge component of the base current and the circuit is an adaptation of the familiar differentiator, introduced in Chapter 3. Assume the resistance of the emitter-base junction to be constant. When the input pulse is applied at t_{+0} time, the entire input voltage will appear across the emitter-base junction resistance, as shown by the voltage waveform. As time increases, the capacitor will charge to the applied input voltage in accordance with the standard charge

Eq. 5. The capacitor-charge current is the base current of the transistor. As soon as the capacitor has charged to the value of the input voltage, the charging current (base current) will decay to zero, as indicated on the base-current waveform shown in Fig. 6-8(a). When the input pulse has been removed at t_{+1} time, charged capacitor C operates as a voltage source having the polarity to reverse bias the emitter-base junction. This reverse-bias action helps decrease storage time. Capacitor C provides the additional base current needed to decrease rise time when the capacitor charges. Charged, this capacitor provides the desired reverse bias necessary to reduce storage time.

Figure 6-8(b) illustrates the input circuit of the transistor switch. Assume

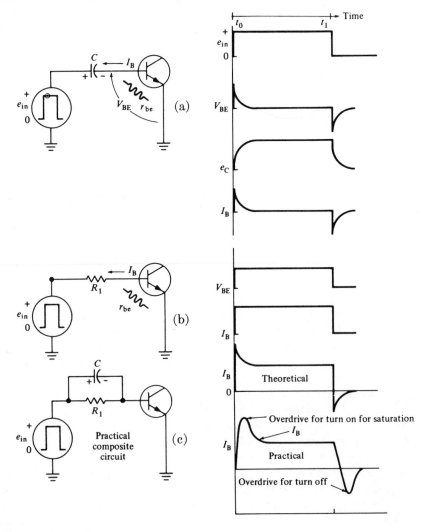

Figure 6-8

the emitter-base junction to be a fixed resistance r_{be}. This circuit is a simple resistor-type voltage divider with the base current and the emitter-base junction waveshapes as indicated.

Figure 6-8(c) illustrates the practical composite circuit and also the resultant theoretical and practical base-current waveshapes. The theoretical base-current waveshape was illustrated, assuming that the emitter-base junction capacitance was neglected. These two facts account for the difference between the practical and the theoretical current waveshapes.

The solution for the value of C (referred to as the "speedup" capacitor) is based upon the equation for the definition of current:

Refer to Fig. 6-7:

$$I = \frac{Q}{t} \quad \text{and} \quad Q = CE$$

therefore
$$I = \frac{CE}{t}$$

then
$$C = \frac{It}{E}$$

Q = quantity of charge (C)
$I = I_B$ the required overdrive base current (A)
$t = t_r$ rise time (sec)
E = amplitude of the input pulse (V)
C = capacitance (F)

6.8   SAMPLE PROBLEM—"SPEEDUP" CAPACITOR

Refer to Fig. 6-7. This circuit with no capacitor C has a rise time of 3 μsec. It is necessary to reduce rise time; therefore, a capacitor C should be added. What value of capacitance is needed to reduce the rise time to 1 μsec?

$$I_C = \frac{V_{cc}}{R_L} = \frac{20}{1 \text{ k}\Omega} = 20 \text{ mA}$$

$$I_{B_{min}} = \frac{I_C}{h_{FE_{min}}} = \frac{20 \text{ mA}}{20} = 1 \text{ mA}$$

Hence,
$$I_{B_{total}} = I_{B_{min}} + I_{B_{required \, for \, overdrive}}$$
$$2\text{mA} = 1 \text{ mA} + I_{B_{overdrive}}$$
$$I_{B_{overdrive}} = 1 \text{ mA}$$

$$C = \frac{I_B t_r}{e_{in}} = \frac{1 \times 10^{-3} \times 1 \times 10^{-6}}{5}$$

$$C = 200 \text{ pF}$$

Used 2 I_B to really overdrive (handwritten annotation)

LABORATORY EXPERIMENT
TRANSISTOR SWITCH

OBJECT:
1. To analyze the basic switching characteristics of a transistor
2. To investigate some of the methods of improving transient response

MATERIALS:

use NPN

1 Switching transistor example—(2N3638 silicon PNP)
1 Manufacturer's specification sheet for type of transistor used
1 Oscilloscope, dc time-base type; frequency response, dc to 450 kHz; vertical sensitivity, 100 mV/cm
1 Transistor power supply (0 to 30 V and 0 to 250 mA)
1 Square-wave generator (20 Hz to 200 kHz)
3 Resistor substitution boxes (10 Ω to 10 MΩ, 1 W)
1 Capacitor substitution box (0.0001 to 0.22 μF at 450 V)
1 Semiconductor diode (example—1N914 silicon junction)

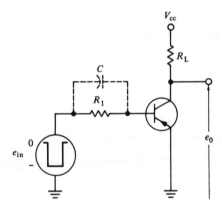

Figure 6-1X

PROCEDURE:
1. For the circuit shown in Fig. 6-1X, design a transistor switch with the following characteristics:
   Output voltage, 20 V peak
   Collector current, 20 mA
   Input voltage, 4 V peak (0 to −4 V) square wave with *prr* 10 kHz
   For design purposes, assume that the transistor is ideal.
2. Draw a schematic; label, and show all calculations necessary for the determination of the values of the circuit components.
3. Is it necessary to modify the input circuit to the transistor switch with either a clipper or a clamper circuit? Explain.
4. Connect the circuit. Use the standard color code values closest to those com-

puted. In each case, explain why a color code value larger or smaller than the computed value was selected.

5. Measure and record the input- and output-voltage waveforms. Include t_d, t_r, t_{on}, t_p, t_s, t_f, and t_{off}.

6. On graph paper, draw and label the input- and output-voltage waveforms one above the other, and to the same time base. Label waveforms with values obtained in Step 5. Include labeled schematic.

7. In order to double the value of the base current used in Step 4, compute the change in value necessary for R_1.

8. Repeat Steps 4 through 6.

9. Compare the rise time of the circuit with normal base current with that of the circuit with doubled base current. Explain the reason for the difference.

10. Change the value of R_1 to the original value computed in Step 1 (minimum value of base current necessary for collector-current saturation).

11. Determine the value of capacitance C which will reduce the value of rise time to one-third of that determined in Step 5.

12. Connect capacitor C (closest standard value) across resistor R_1, as shown in Fig. 6-1X. Explain why it was necessary to select a value of capacitance larger or smaller than that computed.

13. Repeat Steps 5 and 6.

14. Compare the three output-voltage waveforms. Explain the significance of these waveforms.

15. Increase e_{in} to 10 V peak and observe and measure storage time. Explain the significance of the results.

QUESTIONS AND EXERCISES

1. In detail, explain how the capacitor provides the theoretical base-current waveform required. Why is this possible?

2. In detail, explain why the use of the capacitor is the preferred method for improving switching time (t_{on}) as opposed to the method used in Step 9 of the experiment.

3. Explain how the capacitor may also help to reduce turn-OFF time. Why is this possible?

4. A transistor switch consists of the following:

$$C = 150 \text{ pF}, \qquad I_C = 30 \text{ mA}, \qquad h_{FEmin} = 25, \qquad V_{BEsat} = 0.5 \text{ V}$$

Assuming that the circuit is the same as the one used in the experiment and that a positive input pulse of 10 V peak is applied

(a) What type of transistor should be used?

(b) What would be the duration of rise time if a speedup capacitor were used?

5. Explain the problem created by the use of this speedup capacitor.

6. Is a clamper circuit required between the square-wave generator and the input of the switch circuit? Explain.

7. Refer to Fig. 6-1X. Explain why the lowest value of h_{FE} should be used to determine the proper value of R_1.

8. Explain why a transistor, used in a saturated switch circuit, dissipates practically no power when the transistor is conducting at saturation, or ON. *p75+76*

9. Explain why a transistor, used in a saturated switch circuit, dissipates practically no power when the transistor is not conducting, or OFF. *76*

10. When a practical transistor switch is in the OFF condition, how may the flow of the collector current be reduced from $I_C = (\beta + 1)I_{CBO}$ to $I_C = I_{CBO}$?

11. Refer to the manufacturer's specification sheet for 2N3646 transistor (see Appendix B). Should the minimum value or the maximum value of $V_{CE_{sat}}$ be used in the design of a transistor switch? Explain.

12. Refer to the manufacturer's specification sheet for 2N3646 transistor (see Appendix B). Should the minimum value or the maximum value of $V_{BE_{sat}}$ be used in the design of a transistor switch? Explain.

13. Refer to the manufacturer's specification sheet for 2N3646 transistor (see Appendix B). Should the minimum value or the maximum value of h_{FE} be used in the design of a transistor switch? Explain.

14. Refer to Fig. 6-1X. To select the color code value of resistance R_1, which most nearly approximates the calculated value, should one select a value higher or lower than the calculated value? Explain.

15. Refer to Fig. 6-1X. How would the operation of the transistor switch circuit be affected if the value of the speedup capacitor were too large?

16. Define storage time.

17. What causes storage time? Explain.

18. What is the difference between turn-ON time and rise time? Explain.

19. What are the three components of time which make up turn-OFF time? Explain.

20. Refer to Fig. 6-1X. How does excessive leakage current affect the output voltage of the transistor switch? Explain.

chapter 7

INVERTER CIRCUIT

*The inverter circuit is the practical form of the transistor
switch, and it is the basic circuit employed in all forms of
multivibrator circuits.*

7.1   DESCRIPTION OF OPERATION

In the experiment in chapter 6, rise time was reduced by the use of base-
current overdrive but at the expense of an increase in storage time. This
increase in storage time was the result of an excess in base storage charge which,
in turn, resulted from the base-current overdrive. This excess in base storage
charge holds the transistor ON after the input pulse has been removed. The
output voltage may not rise to $+V_{cc}$ (NPN) until the charge carriers in the
base have been removed. The circuit illustrated in Fig. 7-1 is designed to
reduced storage time.

To decrease the time necessary for the removal of the excess base storage
charge (storage time), it is necessary to increase the reverse bias across the
collector-base junction at the time the input pulse is removed. This is accomplished by the addition of resistor R_2 and voltage source V_{bb} to the basic
transistor switch circuit, illustrated in Fig. 6-4.

Figure 7-2 is a schematic of the switch circuit shown in Fig. 7-1, at t_{+1}
time. At t_{+1} time the input voltage is zero; hence, R_1 and R_2 form a simple
series voltage divider. The voltage drop across R_1 is of the polarity to reverse
bias the emitter-base junction, and it operates as a voltage source in series-
aiding with respect to V_{cc}, thereby increasing the reverse bias of the collector-
base junction. The reverse-bias voltage causes a negative base-current flow, as
indicated in Figs. 7-1 and 7-2. This negative base current will continue to flow
until the charge carriers in the base (minority carriers–electrons–NPN) have
been removed, at which time the negative base-current flow reduces to a
nominal value of leakage current I_{CBO}.

A specific value for R_1 and for R_2, respectively, must be established for
any circuit design in which these components are used. These values are

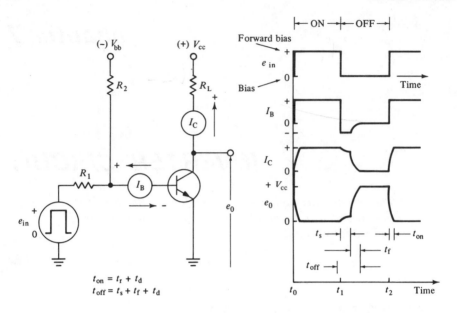

Figure 7-1

$$t_{on} = t_r + t_d$$
$$t_{off} = t_s + t_f + t_d$$

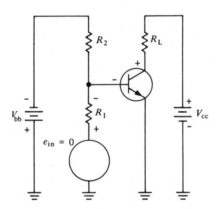

Figure 7-2 Circuit of Fig. 7-1 at t_{+1} Time

established mathematically by the simultaneous solution of two node equations derived from Kirchhoff's law, in which R_1 and R_2 are the respective unknowns. Each must satisfy two distinct circuit conditions: that the transistor is conducting at saturation when the input pulse is applied and that the transistor is cut off, or OFF, when the input pulse is removed. The unique values thus derived for R_1 and R_2 will fulfill both the ON and the OFF circuit requirements for a given amplitude of input voltage.

The schematic from which the node equation for the ON input circuit has been written is illustrated in Fig. 7-3(a); that from which the OFF input circuit has been written is illustrated in Fig. 7-3(c).

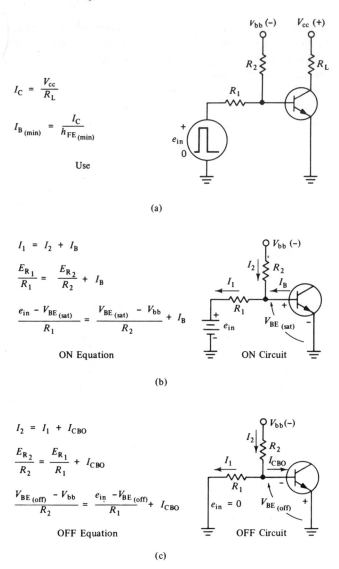

$$I_C = \frac{V_{cc}}{R_L}$$

$$I_{B\,(min)} = \frac{I_C}{h_{FE\,(min)}}$$

Use

(a)

$$I_1 = I_2 + I_B$$

$$\frac{E_{R_1}}{R_1} = \frac{E_{R_2}}{R_2} + I_B$$

$$\frac{e_{in} - V_{BE\,(sat)}}{R_1} = \frac{V_{BE\,(sat)} - V_{bb}}{R_2} + I_B$$

ON Equation   ON Circuit

(b)

$$I_2 = I_1 + I_{CBO}$$

$$\frac{E_{R_2}}{R_2} = \frac{E_{R_1}}{R_1} + I_{CBO}$$

$$\frac{V_{BE\,(off)} - V_{bb}}{R_2} = \frac{e_{in} - V_{BE\,(off)}}{R_1} + I_{CBO}$$

OFF Equation   OFF Circuit

(c)

Figure 7-3   (a) Use; (b) ON Equation; (c) OFF Equation

7.2   TRANSISTOR JUNCTION VOLTAGES

The electronic engineer is responsible for most circuit design in industry and may enlist the computer to determine the optimum value of components. Unlike the engineer, the technician must depend upon experience and a complete understanding of circuit operation in order to determine the value of components for his role in building, testing, and analyzing circuits. The ability to design a circuit which will operate reliably, independent of variations in the active component, is a necessity. For the technician, practical and simplified methods are expedient. Therefore, in an endeavor to simplify circuit design procedures, certain precepts are noted here.

In the transistor circuit, the order of magnitude of voltages used is relatively small; in fact, in some circuits the junction voltages of the transistor may not be neglected. Hence, a set of typical junction voltage values has been established. If the manufacturer's specification sheet is not available, these values may be used practically in most cases.

SILICON JUNCTION TRANSISTOR

$$V_{CE_{sat}} = 0.3 \text{ V}$$
$$V_{BE_{sat}} = 0.7 \text{ V}$$
$$V_{BE_{off}} = 0.0 \text{ V}$$

The value of I_{CBO}, of low- and medium-power silicon switching transistors, is small enough to be frequently considered negligible, even at elevated temperatures.

Because the input pulse to an inverter circuit is received from a preceding transistor stage, the practical input voltage is $V_{CE_{sat}} = 0.3$ rather than the ideal zero, as illustrated in Fig. 7-4. The output voltage of transistor T_1, as shown in Fig. 7-4(a), represents the practical pulse input voltage when the input voltage is ideally zero. Figure 7-4(b) shows transistor T_1 as the input voltage to transistor T_2. $V_{CE_{sat}}$ of transistor T_1 is of a polarity to forward bias

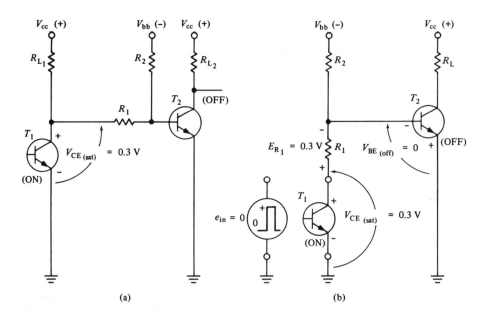

(a)   (b)

Figure 7-4

transistor T_2. Hence, for the emitter-base bias of transistor T_2 to equal the required 0 V, a voltage drop across R_1 of 0.3 V, and of the polarity indicated in Fig. 7-4(b), must be provided by the voltage source V_{bb} and the voltage divider R_1 and R_2.

7.3 SAMPLE DESIGN PROBLEM

Design a transistor inverter circuit, as shown in Fig. 7-5, with the following characteristics:

$$e_{in} = 10 \text{ V peak} \qquad (0 \text{ to } +10 \text{ V})$$
$$e_o = 10 \text{ V peak}$$
$$I_C = 10 \text{ mA}$$

Given:

Silicon transistor, $h_{FE_{min}} = 20$
2 10 V dc sources

Assume:

$V_{BE_{off}} = -0.5 \text{ V}$, as 0 V is the minimum value required
I_{CBO} to be negligible

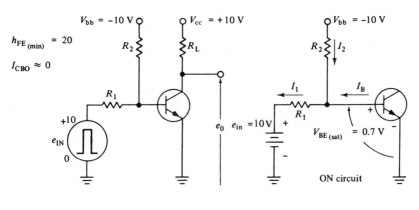

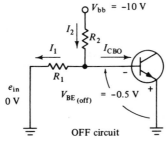

Figure 7-5

Solution:

1. Refer to node equations shown in Fig. 7-3.
2. From the circuit shown in Fig. 7-3, write the ON circuit input-node equation.
3. From the circuit shown in Fig. 7-3, write the OFF circuit input-node equation.
4. Solve resultant simultaneous equations for R_1 and R_2.

$$R_\text{L} = \frac{V_\text{cc} - V_\text{CE}_{\text{sat}}}{I_\text{C}} = \frac{10 - 0.3}{10 \text{ mA}} = \frac{9.7}{10 \text{ mA}} = 0.97 \text{ k}\Omega$$

Standard color code values 0.91 kΩ and 1 kΩ

Use $R_\text{L} = 1 \text{ k}\Omega$

$$I_\text{B} = \frac{I_\text{C}}{h_{\text{FE}_{\text{min}}}} = \frac{10 \text{ mA}}{20} = 0.5 \text{ mA}$$

ON Equation:
$$I_1 = I_2 + I_\text{B}$$

$$\frac{E_{R_1}}{R_1} = \frac{E_{R_2}}{R_2} + I_\text{B}$$

$$\frac{e_\text{in} - V_{\text{BE}_{\text{sat}}}}{R_1} = \frac{V_{\text{BE}_{\text{sat}}} - V_\text{bb}}{R_2} + I_\text{B}$$

$$\frac{(+10) - (+0.7)}{R_1} = \frac{(+0.7) - (-10)}{R_2} + 0.5 \text{ mA}$$

$$\frac{+10 - 0.7}{R_1} = \frac{+0.7 + 10}{R_2} + 0.5 \text{ mA}$$

$$\boxed{\frac{9.3}{R_1} = \frac{10.7}{R_2} + 0.5 \text{ mA}}$$ ON Equation

OFF Equation:
$$I_2 = I_1 + I_\text{CBO}$$

$$\frac{E_{R_2}}{R_2} = \frac{E_{R_1}}{R_1} + I_\text{CBO}$$

$$\frac{V_{\text{BE}_{\text{off}}} - V_\text{bb}}{R_2} = \frac{e_\text{in} - V_{\text{BE}_{\text{off}}}}{R_1} + I_\text{CBO}$$

$$\frac{(-0.5) - (-10)}{R_2} = \frac{0 - (-0.5)}{R_1} + I_\text{CBO}$$

$$\frac{-0.5 + 10}{R_2} = \frac{0 + 0.5}{R_1} + 0$$

$$\frac{9.5}{R_2} = \frac{+0.5}{R_1}$$

$$\boxed{R_2 = 19R_1} \qquad \text{OFF Equation}$$

$$\frac{9.3}{R_1} = \frac{10.7}{R_2} + 0.5 \text{ mA} \qquad \text{ON Equation}$$

$$\frac{9.3}{R_1} = \frac{10.7}{19R_1} + 0.5$$

$$\frac{9.3 - 0.563}{R_1} = 0.5$$

$$R_1 = 17.5 \text{ k}\Omega$$
$$R_2 = 19R_1 = 19(17.5 \text{ k}\Omega)$$
$$R_2 = 332 \text{ k}\Omega$$

$R_1 =$ $17.5 \text{ k}\Omega$ (computed value)

Standard color code values, 15 and 18 kΩ. Since R_1 is selected primarily for the ON circuit condition, the lower color code value should be selected to ensure base current enough for saturation.

Hence: $R_1 = 15 \text{ k}\Omega$
$R_2 =$ $332 \text{ k}\Omega$ (computed value)

Standard color code values, 330 and 390 kΩ. Because R_1 and R_2 form a voltage divider for $-V_{bb}$, and because a maximum amount of voltage is necessary to ensure cutoff, the lower of the two color code values listed for R_2 should be selected.

Hence: $R_2 = 330 \text{ k}\Omega$

7.4   SAMPLE DESIGN PROBLEM

Redesign the transistor inverter circuit illustrated in Fig. 7-5. Assume that the input pulse is produced from a preceding stage and will, therefore, swing from +0.3 to 10 V (+0.3 V $= V_{CE_{sat}}$ of preceding stage)

$$e_{in} = +0.3 \text{ to } +10 \text{ V}$$

The ON circuit input equation will remain the same:

$$\frac{9.3}{R_1} = \frac{10.7}{R_2} + 0.5 \text{ mA}$$

OFF input equation:

$$I_2 = I_1 + I_{CBO}$$

$$\frac{V_{BE_{off}} - V_{bb}}{R_2} = \frac{e_{in} - V_{BE_{off}}}{R_1} + I_{CBO}$$

$$\frac{(-0.5) - (-10)}{R_2} = \frac{(+0.3) - (-0.5)}{R_1} + I_{CBO}$$

$$\frac{-0.5 + 10}{R_2} = \frac{+0.3 + 0.5}{R_1} + 0$$

$$\frac{9.5}{R_2} = \frac{+0.8}{R_1}$$

$$\boxed{R_2 = 11.9R_1}\qquad\text{OFF Equation}$$

$$\frac{9.3}{R_1} = \frac{10.7}{11.9R_1} + 0.5$$

$$\frac{9.3}{R_1} = \frac{0.9}{R_1} + 0.5$$

$$R_1 = 16.8 \text{ k}\Omega$$

$$R_2 = 11.9R_1 = 11.9(16.8 \text{ k}\Omega)$$

$$R_2 = 200 \text{ k}\Omega$$

Standard color code values, 15 and 18 kΩ

Use $\qquad\qquad\qquad R_1 = 15 \text{ k}\Omega$

Standard color code values, 180 and 220 kΩ

Use $\qquad\qquad\qquad R_2 = 180 \text{ k}\Omega$

7.5   SAMPLE CIRCUIT ANALYSIS PROBLEM

Prove that the circuit illustrated in Fig. 7-6 will, or will not, function properly as a saturated inverter circuit.
Given:

$$\text{Silicon transistor (NPN)}, h_{FE_{min}} = 20$$
$$I_{CBO} = 0$$

Solution:

1. Prove that the base current, which actually flows in the circuit when the input pulse is applied, is equal to or greater than the minimum value of base current necessary for saturation.
2. Prove that $V_{BE_{off}}$ is actually zero, or negative (NPN), in the circuit when the input pulse is absent.

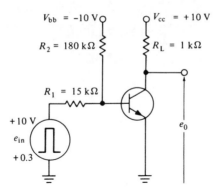

Figure 7-6

Refer to circuit and equations illustrated in Fig. 7-3.

ON circuit:

$$I_1 = I_2 + I_B$$

$$\frac{e_{in} - V_{BE_{sat}}}{R_1} = \frac{V_{BE_{sat}} - V_{bb}}{R_2} + I_B$$

$$\frac{(+10) - (+0.7)}{15\ k} = \frac{(+0.7) - (-10)}{180\ k} + I_B$$

$$\frac{9.3}{15\ k} = \frac{10.7}{180\ k} + I_B$$

$$0.62\ \text{mA} = 0.06\ \text{mA} + I_B$$

$$I_B = 0.56\ \text{mA}$$

This is greater than

$$I_{B_{min}} = \frac{I_C}{h_{FE_{min}}} = \frac{10\ \text{mA}}{20} = 0.5\ \text{mA}$$

Therefore, the circuit values fulfill the necessary ON circuit conditions.

OFF circuit:

$$I_2 = I_1 + I_{CBO}$$

$$\frac{V_{BE_{off}} - V_{bb}}{R_2} = \frac{e_{in} - V_{BE_{off}}}{R_1} + I_{CBO}$$

$$\frac{V_{BE_{off}} - (-10)}{180\ k} = \frac{(+0.3) - V_{BE_{off}}}{15\ k} + 0$$

$$15\ kV_{BE_{off}} + 150\ k = 54\ k - 180\ kV_{BE_{off}}$$

$$V_{BE_{off}} = \frac{-96\ k}{195\ k}$$

$$V_{BE_{off}} = -0.49\ \text{V}$$

which is greater than 0 V, the minimum value required for cutoff. Therefore, the circuit values fulfill the necessary OFF circuit conditions. Hence, the circuit will operate properly.

LABORATORY EXPERIMENT
INVERTER CIRCUIT

OBJECT:
1. To design an inverter circuit for a given set of circuit conditions
2. To analyze the operation of a transistor inverter circuit

MATERIALS:
1 Switching transistor (example—2N3646 silicon NPN)
1 Manufacturer's specification sheet for type of transistor used (see Appendix B for 2N3646)
1 Oscilloscope, time-base type; frequency response dc to 450 kHz; vertical sensitivity, 100 mV/cm
2 Transistor power supplies (0 to 30 V and 0 to 250 mA)
1 Square-wave generator (20 Hz to 200 kHz)
3 Resistor substitution boxes (10 Ω to 10 MΩ, 1 W)
1 Capacitor substitution box (0.0001 to 0.22 μF at 450 V)
1 Semiconductor junction diode (example—1N914 silicon)
1 Vacuum-tube voltmeter, VTVM; dc–ac type

PROCEDURE:
1. Design and analyze a transistor inverter circuit (saturated switch) with the following specifications:

$$e_{in} = 0 \text{ to } +8 \text{ V (square-wave pulse)}$$
$$prr = 5000 \text{ pulses/sec}$$
$$e_o = 15 \text{ V peak}$$
$$I_C = 15 \text{ mA}$$

Circuit to operate to temperature of 55°C

AVAILABLE:
1 NPN silicon transistor
1 Power supply $V_{cc} = 15$ V
1 Power supply $V_{bb} = 10$ V

2. Design the inverter circuit illustrated in Fig. 7-1X for the circuit conditions specified. Include all calculations. Explain why any assumptions made are valid.
3. Draw a schematic of the circuit designed in Step 2, and label with the practical color code values.
4. From the schematic drawn in Step 3, analyze the circuit designed in Step 2 to prove that it will operate satisfactorily for the required specifications when standard-value resistors are used. Include all calculations.
5. Is a clamper circuit necessary between the input-pulse source and the inverter circuit? Explain.
6. Connect the circuit designed. Check the operation of the circuit. Record all necessary data.

102

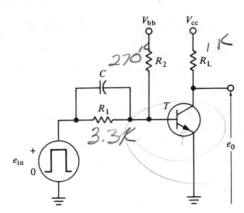

Figure 7-1X

7. Compare measured values with design values, and explain any discrepancies.
8. Compare the output-voltage pulse obtained in this experiment with the output pulse obtained in the preceding transistor switch experiment, and explain the similarities and differences.

QUESTIONS AND EXERCISES

1. What function does the source voltage V_{bb} serve in the inverter circuit illustrated in Fig. 7-1X? Explain.
2. What factors determine the amplitude of the output-voltage waveform of an inverter circuit?
3. What effect will the following conditions have upon the operation of the circuit designed in this experiment? Explain.
   (a) Input pulse less than 8 V peak
   (b) Input pulse more than 8 V peak
4. What function is served by the addition of resistor R_2 to the inverter circuit shown in Fig. 7-1X?
5. Prove mathematically that the leakage current I_{CBO} is negligible in the experimental circuit.
6. In the design of a transistor inverter circuit, a voltage other than zero is assumed for V_{BEoff}. Explain why. What polarity should it be for a PNP transistor? What polarity should this voltage be for an NPN transistor?
7. In the circuit designed for this experiment, what is the minimum value of reverse bias necessary to reduce the value of collector current to the value of I_{CBO} during the interval in which the transistor is OFF?
8. In detail, explain the operation of a transistor inverter circuit. P 73
9. Explain why the typical values for the junction voltages are sufficient for practical design purposes.
10. With the exception of the maximum ratings of a transistor, upon which transistor parameter is the design and analysis of a transistor switching circuit most dependent? Explain.
11. Refer to Fig. 7-1. Design this circuit to incorporate the following parameters:

$$V_{cc} = +20 \text{ V}, \qquad V_{bb} = -20 \text{ V}, \qquad I_C = 20 \text{ mA}$$

use an ideal silicon transistor in which $h_{FEmin} = 20$. Assume

$$V_{BE_{off}} = -0.5 \text{ V}, \qquad V_{BE_{sat}} = 0 \text{ V}, \qquad V_{CE_{sat}} = 0 \text{ V}$$

$I_{CBO} = 0$ mA, e_{in} = positive pulse which varies from 0 to $+20$ V peak

12. Refer to Fig. 7-1. Redesign the circuit designed in Prob. 11, assuming standard junction voltages.

13. Refer to Fig. 7-1. Redesign the circuit designed in Prob. 11 to incorporate the following junction voltages:

$$V_{BE_{sat}} = 0.9 \text{ V}, \qquad V_{CE_{sat}} = 0.2 \text{ V}, \qquad I_{CBO} = 10 \text{ } \mu\text{A at } 25°\text{C}$$

The positive input pulse varies from $+0.2$ to $+20$V. The circuit must operate subject to temperature conditions up to 65°C.

14. Compare the results of the inverter circuits designed according to the instructions in the following: Problems: 11, in which an ideal transistor was used; 12, in which standard junction voltages were used; and 13, in which specific junction voltages were used. From this comparison, determine which parameters most necessitate a change in the values of resistors R_1 and R_2 when a practical transistor, rather than an ideal transistor, is used.

15. Refer to Fig. 7-1. Prove that this circuit will or will not work properly as an inverter circuit when e_{in} = positive pulse from $+0.3$ to $+5$ V peak and h_{FEmin} = 25. The source voltage $V_{cc} = +15$ V and the source $V_{bb} = -10$ V.

$$R_1 = 3.9 \text{ k}\Omega, \qquad R_2 = 27 \text{ k}\Omega, \qquad R_L = 1 \text{ k}\Omega$$

Assume that standard junction voltages are used. If the circuit will not operate properly, which component or components must be changed in order to provide proper operation? To what value, or values, must the component, or components, be changed?

16. Refer to Fig. 7-1. Prove that this circuit will or will not operate properly as an inverter circuit when e_{in} = positive pulse from $+0.3$ to $+12$ V and h_{FEmin} = 10. The source voltage $V_{cc} = +12$ V and the source $V_{bb} = -12$ V.

$$R_1 = 27 \text{ k}\Omega, \qquad R_2 = 1 \text{ M}\Omega, \qquad R_L = 2.7 \text{ k}\Omega$$

Assume standard junction voltages. If the circuit will not operate properly, which component or components must be changed in order to provide proper operation? To what value, or values, must the component, or components, be changed?

chapter 8

BISTABLE MULTIVIBRATOR

The three basic classes of multivibrator circuits are the monostable, the bistable, and the astable. Each of these circuits may be further classified as saturated or nonsaturated. Also, each consists of two connected, semiconductor switch circuits. The output signal of the first switch circuit is the input signal to the second and the output signal of the second is the input signal to the first. Another division is made between multivibrator circuits which employ different types of coupling from one switch circuit to the other. The two most common types are collector coupling and emitter coupling.

The monostable and the astable multivibrator circuits are discussed in following experiments; the bistable multivibrator is defined in this one.

8.1   DEFINITION OF A BISTABLE MULTIVIBRATOR

A bistable multivibrator is an electronic circuit which has two stable states. It is commonly called a *flip–flop* circuit. The circuit may employ one or more active elements, usually two, in its simplest form. The active elements may be tunnel diodes, four-layer diodes, field effect transistors, or junction transistors. To analyze the saturated bistable multivibrator circuit, in which the junction transistor is the active element, is the purpose of this experiment.

As the name bistable multivibrator suggests, this circuit will remain in one of its two stable states until acted upon by an external force (voltage), at which time it will switch to the other stable state in which it will remain until acted upon by another external force (voltage). To understand this switching action, it is necessary to trace the operation of a specific circuit. Therefore, please refer to Fig. 8-1 for this purpose.

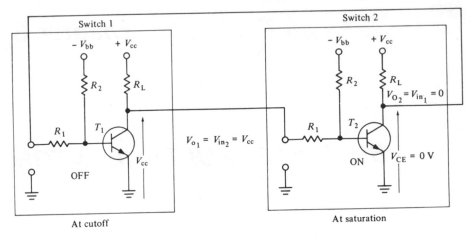

Figure 8-1   Bistable Multivibrator

8.2   DESCRIPTION OF OPERATION

Switches 1 and 2 are identical inverter circuits. Assume that switch 1 is OFF and that switch 2 is ON. Transistor T_1 is at cutoff; hence, the collector-to-emitter voltage of T_1 is $+V_{cc}$. This $+V_{cc}$ becomes the input voltage to switch 2, and the former is of the polarity to forward bias transistor T_2. Assuming that the two switch circuits have been designed properly, the forward bias applied to transistor T_2 will be sufficient to cause transistor T_2 to conduct at saturation. Hence, the collector-emitter voltage of T_2 is fed back as the input voltage to switch 1. If no input voltage is applied to switch 1, the negative voltage on the base of T_1, with reference to ground, reverse biases transistor T_2, holding it at cutoff or OFF. This negative voltage at the base of transistor T_1, with reference to ground, is obtained from the voltage drop across R_1. The voltage drop across R_1 is supplied from the voltage source V_{bb} and, in turn, from the resistive voltage divider, which consists of resistors R_1 and R_2.

Switch 1 will remain in this OFF condition and switch 2 will remain in this ON condition indefinitely. Before these stable states will reverse, an external voltage must be applied, in order to turn the OFF switch ON or to turn the ON switch OFF. A short circuit, momentarily applied between the base and emitter of the ON switch 2, will remove the forward bias. When the forward bias is removed, transistor T_2 will start to cut off. As this occurs, the collector-emitter voltage will start to rise from $V_{CE_{sat}}$ (ideally zero) toward V_{cc}. This positive-rising voltage, from the collector to emitter of T_2, is the input voltage to transistor T_1 (forward bias). Hence, transistor T_1 will start to conduct and, in turn, the output voltage will drop from $+V_{cc}$ toward $V_{CE_{sat}}$ (ideally zero). Zero input voltage to switch 2 will cause transistor T_2 to be cut off, or to be OFF. The two switches will then remain in this stable state; namely, switch 1, ON and switch 2, OFF. Because the circuit will remain in either of the stable states indefinitely or until acted upon by an external force (voltage), it is called a bistable multivibrator.

8.3 VOLTAGE GAIN DURING SWITCHING INTERVAL

The bistable multivibrator circuit employs positive feedback during the interval required for the switching action to take place. Positive feedback is sometimes referred to as regeneration. Regenerative action is the most important characteristic of a multivibrator circuit.

Positive feedback action will not occur until initiated by an externally applied trigger voltage. The momentary removal of forward bias, caused by shorting the emitter-base junction of the ON transistor, constitutes a form of triggering which starts the regenerative cycle. When the forward bias to the ON transistor T_2 is removed, T_2 starts to come out of saturation and enters the active (class A) operating region of the transistor V_{CE}—I_C characteristic curves. Refer to Fig. 6-2. For the transistor to go from this ON condition to the OFF condition, the operating point must traverse the active region of the V_{CE}—I_C characteristic curve in going from $V_{CE_{sat}}$, when ON, to V_{cc}, when OFF. During this transition interval, the collector-to-emitter voltage of transistor T_2 starts to rise from $V_{CE_{sat}}$ toward V_{cc}. The transistor switch is an inverter circuit, one in which the input voltage (emitter-base voltage) is 180° out of phase. Therefore, the switch may be considered a class A amplifier during the transition period.

When the forward bias applied to the ON transistor T_2 is decreased (in this case actually removed), the signal on the base of transistor T_2 is negative-going and is amplified and inverted. This causes the output voltage of transistor T_2 to be a positive-going signal and an amplified one. The amplified positive-going signal from the output of switch 2 becomes the input signal to switch 1. The amplified positive-going signal on the base of transistor T_1 is amplified and inverted by transistor T_1. Hence, the output signal from switch 1 is a negative-going amplified signal. This negative-going amplified signal from the output of switch 1 is the input signal to switch 2. When the forward bias applied to transistor T_2 was initially decreased (negative-going signal), a large negative-going signal was produced which increased and aided the initial change. This positive feedback or regeneration is large enough to drive the transistor to the extreme, in this case, cutoff for T_2 and saturation for T_1.

8.4 NEED FOR TRIGGERING

The basic circuit, as shown in Fig. 8-2, can serve no useful purpose without the addition of a triggering circuit to provide the voltage necessary to reverse the stable states. When the collector-coupled bistable multivibrator, shown in Fig. 8-2, is connected, one transistor will conduct at saturation and the other will remain at cutoff, due to a slight difference in value between comparable components. Therefore, unless triggering circuitry is added, the circuit will remain in this state indefinitely. Triggering for the bistable multivibrator is discussed in a following chapter; discussion in this chapter is confined to the design and analysis of the basic circuit.

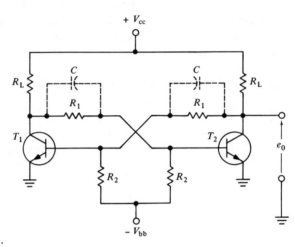

Figure 8-2   Bistable Multivibrator

8.5   SAMPLE DESIGN PROBLEM

From the circuit illustrated in Fig. 8-2, design a saturated collector-coupled bistable multivibrator with the following characteristics:

$$e_o = 10 \text{ V peak}$$
$$I_C = 30 \text{ mA}$$
$$\text{Circuit to work up to } 65°C$$

Given

$$\text{2 silicon NPN transistors,} \quad h_{FE_{min}} = 30$$
$$I_{CBO} = 0.5 \ \mu A \text{ at } 25°C$$

2 10 V dc sources

Assume:

$$\text{Typical values for junction voltages}$$
$$V_{BE_{off}} = -0.5 \text{ V}$$

Determine:

$$R_1, R_2, R_L$$

Solution:

1. Determine the value of R_L from the ON transistor as shown in Fig. 8-3.
2. From the complete schematic shown in Fig. 8-2, draw a schematic of the ON circuit (assume T_2 is ON and T_1 is OFF), as shown in Fig. 8-4.

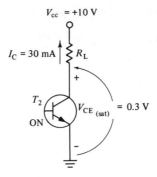

Figure 8-3

3. From the ON circuit shown in Fig. 8-4, write the ON circuit node equation.
4. From the complete schematic shown in Fig. 8-2, draw a schematic of the OFF circuit (assume T_2 is ON and T_1 is OFF), as shown in Fig. 8-5.
5. From the OFF circuit shown in Fig. 8-5, write the OFF circuit node equation.

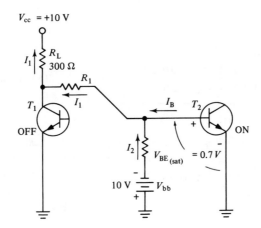

Figure 8-4 ON Circuit

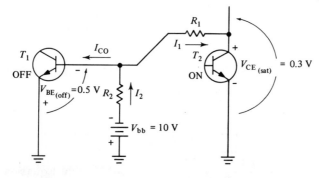

Figure 8-5 OFF Circuit

6. Solve the ON and OFF circuit equations simultaneously for R_1 and R_2.

$$R_L = \frac{V_{cc} - V_{CE_{sat}}}{I_C} = \frac{(+10) - (+0.3)}{30 \text{ mA}}$$

$$R_L = \frac{9.7}{30 \text{ mA}} = 324 \text{ }\Omega$$

Standard color code values—300 and 330 Ω.
Select $R_L = 300 \text{ }\Omega$ to ensure the 30 mA I_C required.
ON circuit equation:

$$I_1 = I_2 + I_B \qquad\qquad I_B = \frac{I_C}{h_{FE_{min}}} = \frac{30 \text{ mA}}{30}$$

$$I_B = 1 \text{ mA}$$

$$\frac{E_{R_L} + E_{R_1}}{R_L + R_1} = \frac{E_{R_2}}{R_2} + I_B$$

$$\frac{V_{cc} - V_{BE_{sat}}}{R_L + R_1} = \frac{V_{BE_{sat}} - V_{bb}}{R_2} + I_B$$

$$\frac{(+10) - (+0.7)}{0.3 \text{ k} + R_1} = \frac{(+0.7) - (-10)}{R_2} + 1 \text{ mA}$$

$$\boxed{\frac{9.3}{0.3 \text{ k} + R_1} = \frac{10.7}{R_2} + 1 \text{ mA}} \qquad \text{ON Equation}$$

OFF circuit equation:

$$I_2 = I_1 + I_{CBO} \qquad\qquad I_{CBO} = 0.5 \text{ }\mu\text{A @ } 25°\text{C}$$

$$\frac{E_{R_2}}{R_2} = \frac{E_{R_1}}{R_1} + I_{CBO}$$

1.0	35
2.0	45
4.0	55

$$I_{CBO} = 8.0 \text{ }\mu\text{A @ } 65°\text{C}$$

$$\frac{V_{BE_{off}} - V_{bb}}{R_2} = \frac{V_{CE_{sat}} - V_{BE_{off}}}{R_1} + I_{CBO}$$

$$\frac{(-0.5) - (-10)}{R_2} = \frac{(+0.3) - (-0.5)}{R_1} + 0.008 \text{ mA}$$

$$\uparrow$$
$$\text{small neglect}$$

$$\frac{9.5}{R_2} = \frac{0.8}{R_1}$$

$$\boxed{R_2 = 11.9 R_1} \qquad \text{OFF Equation}$$

$$\frac{9.3}{0.3 \text{ k} + R_1} = \frac{10.7}{R_2} + 1 \text{ mA} \qquad \text{ON Equation}$$

$$\frac{9.3}{0.3 \text{ k} + R_1} = \frac{10.7}{11.9R_1} + 1 \text{ mA}$$

$$\frac{9.3}{0.3 \text{ k} + R_1} = \frac{0.9}{R_1} + 1 \text{ mA}$$

$$1 \text{ mA } R_1^2 - 8.1R_1 + 0.27 \text{ k} = 0$$

$$X = \frac{-b \pm \sqrt{b^2 - 4ac}}{2a} \qquad\qquad \text{Quadratic Eq.} \qquad [22]$$

$$R_1 = \frac{8.1 \pm \sqrt{(8.1)^2 - (4)(1 \text{ mA})(0.27 \text{ k})}}{(2)(1 \text{ mA})}$$

$$R_1 = \frac{8.1 \pm \sqrt{64.4}}{0.002} = \frac{8.1 \pm 8.1}{0.002}$$

$$R_1 = 8.1 \text{ k}\Omega$$

Standard color code values—6.8 and 8.2 kΩ
Select $R_1 = 6.8$ kΩ to ensure base current enough to meet the ON circuit
        requirement.

$$R_2 = 11.9R_1 = (11.9)(8.1 \text{ k}\Omega)$$
$$R_2 = 96.4 \text{ k}\Omega$$

Standard color code values—82 and 100 kΩ
Select $R_2 = 82$ kΩ in order to fulfill the OFF circuit requirement by ensuring the maximum reverse-bias voltage across resistor R_1.

NOTE: **Refer to Fig. 8-4. In many applications of this circuit, the voltage drop across R_L, which is produced by I_1, may be neglected because I_1 is small and therefore $V_{CE_{off}} \approx V_{cc}$.**

8.6   SAMPLE CIRCUIT ANALYSIS PROBLEM

Prove that the circuit shown in Fig. 8-6 will, or will not, function properly as a saturated collector-coupled bistable multivibrator.

Solution:

1. Prove that the ON transistor will conduct at saturation.
    (a) From the given circuit shown in Fig. 8-6, draw a schematic of the ON circuit (assume T_2 is ON and T_1 is OFF), as shown in Fig. 8-7.
    (b) From the ON circuit, shown in Fig. 8-7, write the ON circuit node equation.
    (c) Solve the ON circuit node equation for base current. This value must be equal to, or greater than, that necessary to sustain the actual collector-current flow.
    (d) From the ON circuit, shown in Fig. 8-7, determine the actual collector-current flow. Compute the value of base current necessary to produce this collector current. This value of base current must be less

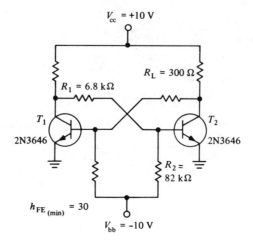

Figure 8-6

than that computed in Step 3 in order for the ON transistor to conduct at saturation.

$$I_1 = I_2 + I_B$$

$$\frac{V_{cc} - V_{BE_{sat}}}{R_1 + R_L} = \frac{V_{BE_{sat}} - V_{bb}}{R_2} + I_B$$

$$\frac{(+10) - (+0.7)}{6.8\text{ k} + 0.3\text{ k}} = \frac{(+0.7) - (-10)}{82\text{ k}} + I_B$$

$$\frac{9.3}{7.1\text{ k}} = \frac{10.7}{82\text{ k}} + I_B$$

$$I_B = 1.18\text{ mA}$$

$$(I_1 + I_C) = \frac{V_{cc} - V_{CE_{sat}}}{R_L} = \frac{(+10) - (+0.3)}{0.3\text{ k}} = 32.3\text{ mA}$$

$$I_1 = I_2 \qquad \text{assuming } I_{CBO} \approx 0$$

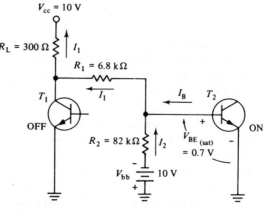

Figure 8-7   ON Circuit

$$I_1 = \frac{V_{bb} - V_{CE_{sat}}}{R_1 + R_2} = \frac{(-10) - (+0.3)}{6.8\ k + 82\ k} = \frac{10.3}{88.8\ k}$$

$$I_1 = 0.117\ mA$$

$$I_C = (I_1 + I_C) - I_1 = 32.3\ mA - 0.117\ mA$$

$$I_C = 32.2\ mA$$

$$I_B = \frac{I_C}{h_{FE_{min}}} = \frac{32.2\ mA}{30} = 1.07\ mA$$

Because $I_B = 1.18$ mA (actual base-current flow), which is greater than the 1.07 mA required, the ON transistor conducts at saturation.

2.   Prove that the OFF transistor will be cut off.

   (a)   From the given circuit, shown in Fig. 8-6, draw a schematic of the OFF circuit (assume T_2 is ON and T_1 is OFF), as shown in Fig. 8-8.

   (b)   From the OFF circuit shown in Fig. 8-8, write the OFF circuit node equation.

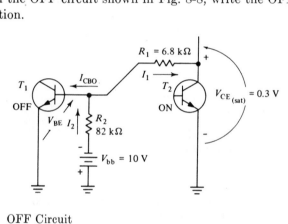

Figure 8-8   OFF Circuit

   (c)   Solve the OFF circuit node equation for $V_{BE_{off}}$. If the value of emitter-base voltage is zero, or negative (reverse bias—NPN), the OFF transistor will be cut off and hence fulfill the necessary OFF circuit requirement.

$$I_2 = I_1 + I_{CBO}$$

$$\frac{V_{BE_{off}} - V_{bb}}{R_2} = \frac{V_{CE_{sat}} - V_{BE_{off}}}{R_1} + I_{CBO}$$

$$\frac{V_{BE_{off}} - (-10)}{82\ k} = \frac{(+0.3) - V_{BE_{off}}}{6.8\ k} + \underset{\underset{\text{negligible}}{\uparrow}}{0.008}$$

$$6.8\ kV_{BE_{off}} + 68\ k = 24.6\ k - 82\ kV_{BE_{off}}$$

$$V_{BE_{off}} = -0.488\ V$$

Thus the circuit fulfills the OFF circuit requirement. Hence, the circuit will operate properly as a saturated, collector-coupled bistable multivibrator.

LABORATORY EXPERIMENT

BISTABLE MULTIVIBRATOR
(SATURATED, COLLECTOR-COUPLED)

OBJECT:

1. To design a saturated, collector-coupled, bistable multivibrator
2. To analyze the operation of a bistable multivibrator
3. To examine the requirements for the stable states of a bistable multivibrator

Touch on transistor or ground to switch to it.

FLIP-FLOP

MATERIALS: 2N2711

2 Switching transistors (example—2N3646 silicon NPN)
1 Manufacturer's specification sheet for type of transistor used
1 Oscilloscope, dc time-base type; frequency response dc to 450 kHz; vertical sensitivity, 100 mV/cm
2 Transistor power supplies (0 to 30 V and 0 to 250 mA)
6 Resistor substitution boxes (10 Ω to 10 MΩ, 1 W)
2 Capacitor substitution boxes (0.0001 to 0.22 μF at 450 V)
1 Vacuum-tube voltmeter, VTVM; dc–ac type

PROCEDURE:

1. Design a saturated, collector-coupled, bistable multivibrator circuit with the following specifications:

$$V_{cc} = 15 \text{ V}$$
$$V_{bb} = 10 \text{ V}$$
$$I_C = 15 \text{ mA for the ON transistor}$$

2. Design the saturated bistable multivibrator, shown in Fig. 8-1X, for the circuit conditions specified. Include all calculations. Explain why any assumptions which you have made are valid.

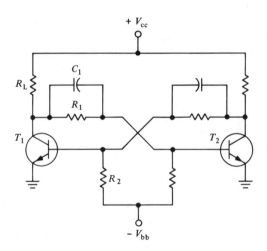

Figure 8-1X

114

3. Draw a schematic of the circuit designed in Step 2, and label it with the practical standard color code values.

4. From the schematic drawn in Step 3, analyze the circuit designed in Step 2 to prove that it will operate satisfactorily according to the required specifications when standard-value resistors are used. Include all calculations.

5. Connect the circuit designed. Check the operation of the circuit. Record all necessary data.

6. The stable states may be reversed by momentarily removing the forward bias from the ON transistor. This may be accomplished by momentarily shorting the emitter-base junction of the ON transistor. Check the operation of the circuit in this reversed stable state. Record all necessary data.

7. Compare measured values with design values, and explain any discrepancies.

QUESTIONS AND EXERCISES

1. Explain the operation of the bistable multivibrator.

2. Explain how and why the momentary removal of the forward bias from the ON transistor will completely reverse the stable states of the transistors.

3. Why are the coupling resistors R_1 bypassed by capacitors C_1? Refer to Fig. 8-1X.

4. How is the stable state of a bistable multivibrator determined?

5. Why are bistable multivibrators more sensitive to external pulses which cut off a conducting transistor than to pulses which turn an OFF transistor ON?

6. If the supply voltage V_{bb}, illustrated in Fig. 8-6, becomes an open circuit, will the circuit continue to act as a bistable multivibrator? Show the necessary calculations to substantiate your answer.

7. What determines the amplitude of the resultant output pulse which is produced when a multivibrator is triggered?

8. In a bistable multivibrator circuit, which components determine the pulse rise time?

9. Analyze the flip–flop circuit shown in Fig. 8-2X when

$$R_L = 10 \text{ k}\Omega \qquad R_1 = 100 \text{ k}\Omega$$
$$R_2 = 33 \text{ k}\Omega \qquad V_{cc} = -15\text{V}$$
$$V_{bb} = +2 \text{ V} \qquad T_1 = T_2$$
$$h_{FEmin} = 20 \qquad V_{BEoff} = 0.1 \text{ V}$$
$$I_{CBO} \approx 0$$

Assume typical silicon junction voltages. Prove that this flip–flop circuit will, or will not, operate satisfactorily.

10. By use of the circuit shown in Fig. 8-2X, design a flip–flop circuit in which the collector current for the ON transistor is 30 mA.

$$V_{cc} = -30 \text{ V} \qquad V_{bb} = +10 \text{ V}$$

Assume:
$$V_{BEoff} = 1 \text{ V}$$
$$V_{BEsat} = 0 \text{ V}$$
$$V_{CEsat} = 0 \text{ V}$$
$$I_{CBO} = 0$$
$$h_{FEmin} = 30$$

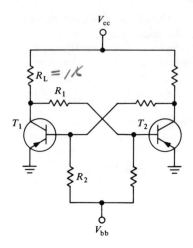

Figure 8-2X

11. By use of the circuit shown in Fig. 8-2X, design a flip–flop circuit in which the collector current for the ON transistor is 30 mA.

$$V_{cc} = -30 \text{ V} \qquad V_{bb} = +10 \text{ V} \qquad V_{BE_{off}} = +1 \text{ V}$$
$$I_{CBO} = 0 \qquad h_{FE_{min}} = 30$$

Silicon PNP transistors
Assume the following standard junction voltages:

$$V_{BE_{sat}} = -0.7 \text{ V} \qquad V_{CE_{sat}} = -0.3 \text{ V}$$

12. By use of the circuit shown in Fig. 8-2X, design a flip–flop circuit in which the collector current for the ON transistor is 30 mA. The circuit must be operable under temperature conditions up to 75°C.

$$V_{cc} = -30 \text{ V} \qquad V_{bb} = +10 \text{ V} \qquad V_{BE_{off}} = +1 \text{ V}$$
$$h_{FE_{min}} = 30 \qquad I_{CBO} = 5 \text{ } \mu\text{A} \qquad \text{at } 25°C$$
$$V_{BE_{off}} = +1 \text{ V} \qquad V_{BE_{sat}} = -0.9 \text{ V} \qquad V_{CE_{sat}} = -0.15 \text{ V}$$

Silicon PNP transistors

13. Analyze the flip–flop circuit shown in Fig. 8-2 when

$$R_L = 1.5 \text{ k}\Omega \qquad R_1 = 27 \text{ k}\Omega \qquad R_2 = 3.3 \text{ M}\Omega$$
$$V_{cc} = +15 \text{ V} \qquad V_{bb} = -20 \text{ V} \qquad h_{FE_{min}} = 20$$
$$I_{CBO} = 0$$

Silicon NPN transistors
Assume standard junction voltages.
Prove that this flip–flop circuit will, or will not, operate satisfactorily.

14. In the circuit analyzed in Prob. 13, what component, or components, must be changed in order for the circuit to operate properly? To what value, or values, must the component, or components, be changed?

15. By use of the circuit shown in Fig. 8-2, design a flip–flop circuit in which the collector current for the ON transistor is 5 mA.

Given: Silicon NPN transistors, $h_{FE_{min}} = 50$

$$V_{cc} = +12 \text{ V} \qquad V_{bb} = -12 \text{ V}$$

Assume: $V_{BE_{off}} = -0.5 \text{ V} \qquad I_{CBO} = 0$
Standard junction voltages

EMITTER-COUPLED BISTABLE MULTIVIBRATOR

The emitter-coupled bistable multivibrator circuit is similar to the collector-coupled circuit, defined in chapter 8. The differences in the two, and the unique features of the emitter-coupled circuit, will be defined in this chapter.

9.1 BASIC CIRCUIT

Refer to the circuit illustrated in Fig. 9-1. Because the configuration for a saturated circuit may be the same as that for a nonsaturated circuit, a schematic will not disclose which one is illustrated. This distinction must be made

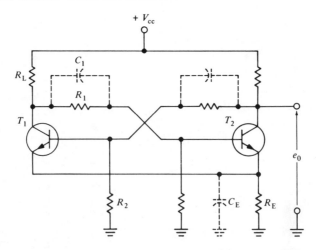

Figure 9-1 Emitter-Coupled Bistable Multivibrator

by a circuit analysis in which specific circuit values are used. Only the saturated circuit will be examined in this discussion, however, because most of the multivibrator circuits used in industry are saturated rather than nonsaturated. This is due to the availability of transistors with small storage time constants. The nonsaturated circuit is necessary only for circuits which require extremely high switching speeds.

The basic requirements for the emitter-coupled bistable multivibrator are the same as those for the collector-coupled bistable multivibrator. Recall chapter 8. When one transistor is ON, at saturation, the other transistor is OFF, at cutoff. This condition will exist indefinitely until an external voltage is applied to alter this stable state and, thereby, initiate switching action. During the switching interval, both transistors conduct in the active region of the characteristic transistor V_{CE}—I_C curves. Rapid switching action occurs because the resultant two-stage loop gain produces positive feedback.

The basic circuit of an emitter-coupled bistable multivibrator, shown in Fig. 9-1, is illustrated in Fig. 9-2, in a configuration which portrays the voltage divider networks.

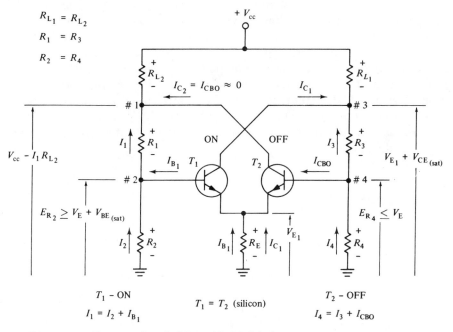

Figure 9-2 Emitter-Coupled Bistable Multivibrator

9.2 DESCRIPTION OF OPERATION

The dual relationship between the relative division of voltage within each of the voltage dividers and the voltage across the emitter resistor R_E determines the conduction state of the two transistors in this circuit.

To follow the succeeding description of switching action, assume that transistor T_1 is ON, at saturation, and that transistor T_2 is OFF, at cutoff. The emitter current of T_1 flows through the common-emitter resistor R_E and produces the voltage drop V_{E_1}. This voltage drop across R_E is of a polarity to reverse bias both transistors. Current I_2, which flows from ground through R_2 toward V_{cc}, produces a voltage drop across R_2, of the polarity indicated in Fig. 9-2. The voltage drop across R_2 is of the polarity to forward bias transistor T_1. For transistor T_1 to conduct at saturation, the voltage drop across R_2 must be greater than the sum of the voltage across R_E and $V_{BE_{sat}}$. While this condition exists, transistor T_1 conducts at saturation and thereby causes the voltage across R_E to be of the polarity to reverse bias both transistors. This voltage drop provides the necessary reverse bias to hold transistor T_2 at cutoff or OFF. Transistor T_2 is OFF, however, only if the voltage drop across R_4 is less than the voltage drop across R_E. Zero volts is the minimum value of reverse bias necessary to hold a silicon transistor at cutoff.

Refer to Fig. 9-2. If point 2 is momentarily grounded, the forward bias is removed from transistor T_1. Transistor T_1 then starts to turn OFF, and the collector voltage of T_1 rises toward V_{cc}. Hence, point 4 becomes proportionately more positive, due to the voltage divider action. At point 4, this positive voltage, with reference to ground, forward biases transistor T_2 which starts to conduct, and the latter thereby causes the voltage at point 1 to decrease, even though the momentary short, from point 2 to ground, has been removed. In turn, point 2 is proportionately less positive. Hence, the forward-bias voltage decreases and thereby decreases the conduction of transistor T_1. This action continues until transistor T_1 is at cutoff and until transistor T_2 is ON at saturation. The circuit remains in this stable state until acted upon by an external voltage.

9.3   ADVANTAGES

The advantages of the emitter-coupled bistable multivibrator over the collector-coupled bistable multivibrator are

1. The emitter-coupled circuit requires only one voltage source.
2. The emitter-coupled circuit will operate at higher temperatures because the degenerative voltage drop across the common-emitter resistor provides the necessary stability.
3. The need for steering diodes may be eliminated by applying a symmetrical triggering pulse across the common-emitter resistor. (This is discussed in detail in chapter 10.)
4. Emitter-coupled circuits may be stabilized by employing an emitter capacitor C_E, as illustrated in Fig. 9-1. This may be accomplished because the voltage drop across the emitter resistor provides the necessary reverse bias for the circuit. To stabilize this voltage during

the switching interval between stable states, a capacitor C_E is placed across the emitter resistor, and triggering pulses are applied to the bases or to collectors of the two transistors. The $R_E C_E$ time constant is five times the switching time.

9.4   DISADVANTAGES

Compared to the collector-coupled circuit, the disadvantages of the emitter-coupled bistable multivibrator are

1.   The output-voltage swing is less in the emitter-coupled circuit, due to the voltage drop across the common-emitter resistor (assuming that V_{cc} has the same value in both circuits).
2.   More power is required to operate the emitter-coupled circuit.
3.   A larger amplitude of triggering pulse is needed to operate the emitter-coupled circuit.

The Schmitt trigger circuit, a special case of the emitter-coupled bistable multivibrator, is discussed in chapter 11.

9.5   SAMPLE DESIGN PROBLEM

From the circuit shown in Fig. 9-1, design a saturated emitter-coupled bistable multivibrator with the following characteristics:

$$e_O = 10 \text{ V peak}$$
$$I_C = 20 \text{ mA}$$

Given:

2 silicon NPN transistors, $h_{FE_{min}} = 15$
$$I_{CBO} \approx 0$$

1 15 V dc source

Assume:

Typical values for junction voltages
$V_{BE_{off}} = -0.5 \text{ V}$

Determine:

$R_1, R_2, R_L,$ and R_E

Solution:

1.   Determine the value of $(R_L + R_E)$ from the ON transistor circuit shown in Fig. 9-3.
2.   Determine the value of R_E from the ON transistor circuit shown in Fig. 9-3.
3.   Determine the value of R_L from the ON transistor circuit shown in Fig. 9-3.
4.   From the complete schematic, shown in Fig. 9-1, draw a schematic of the ON circuit as shown in Fig. 9-4. Assume that T_2 is ON and that T_1 is OFF.

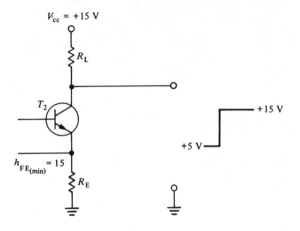

Figure 9-3

5. From the ON circuit, shown in Fig. 9-4, write the ON circuit node equation.
6. From the complete schematic, shown in Fig. 9-1, draw a schematic of the OFF circuit as shown in Fig. 9-5. Assume that T_2 is OFF and that T_1 is ON.
7. From the OFF circuit, shown in Fig. 9-5, write the OFF circuit node equation.
8. Simultaneously solve the ON and the OFF circuit equations for R_1 and R_2.

$$(R_L + R_E) = \frac{V_{cc}}{I_C} = \frac{15}{20 \text{ mA}} = 750 \text{ }\Omega$$

$$V_E = V_{cc} - [E_{R_L} + V_{CE_{sat}}]$$

$$V_E = 15 - (10 + 0.3)$$

$$V_E = 4.7 \text{ V}$$

$$I_B = \frac{I_C}{h_{FE_{min}}}$$

$$I_E = I_B + I_C$$

$$I_E = \frac{I_C}{h_{FE}} + I_C = \frac{20 \text{ mA}}{15} + 20 \text{ mA}$$

$$I_E = 1.33 \text{ mA} + 20 \text{ mA}$$

$$I_E = 21.3 \text{ mA}$$

$$R_E = \frac{V_E}{I_E} = \frac{4.7}{21.3 \text{ mA}}$$

$$R_E = 220 \text{ }\Omega\text{—standard color code value}$$

$$R_L = (R_E + R_L) - R_E$$

$$R_L = 750 - 220$$

$$R_L = 530 \text{ }\Omega$$

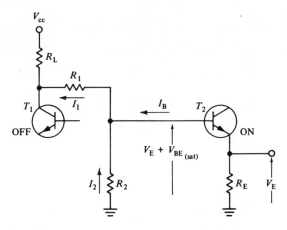

Figure 9-4 ON Circuit

Standard color code values—510 and 560 Ω
Select $R_L = 510$ Ω to ensure the 20 mA I_C required.
ON circuit equation:

$$I_1 = I_2 + I_B$$

$$\frac{E_{R_L} + E_{R_1}}{R_L + R_1} = \frac{E_{R_2}}{R_2} + I_B$$

$$\frac{V_{cc} - [V_E + V_{BE_{sat}}]}{R_L + R_1} = \frac{[V_E + V_{BE_{sat}}] - 0}{R_2} + I_B$$

$$I_B = \frac{I_C}{h_{FE_{min}}} = \frac{20 \text{ mA}}{15} = 1.33 \text{ mA}$$

$$\frac{(+15) - (4.7 + 0.7)}{0.53 \text{ k} + R_1} = \frac{(4.7 + 0.7) - 0}{R_2} + 1.33 \text{ mA}$$

$$\boxed{\frac{9.6}{0.53 \text{ k} + R_1} = \frac{5.4}{R_2} + 1.33 \text{ mA}} \qquad \text{ON Equation}$$

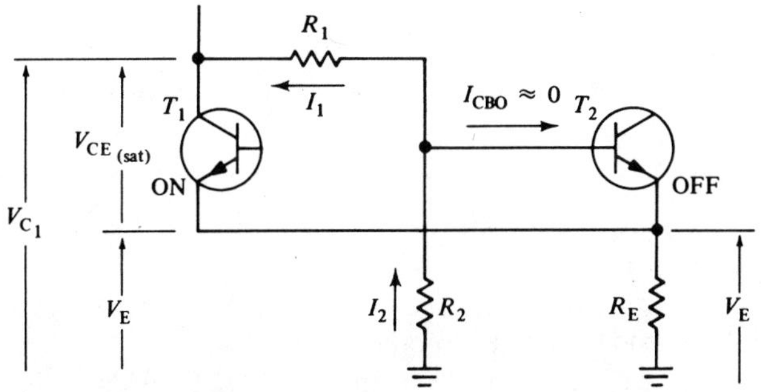

Figure 9-5 OFF Circuit

OFF circuit equation:

$$I_2 = I_1 + I_{CBO}$$

$$\frac{E_{R_2}}{R_2} = \frac{E_{R_1}}{R_1} + I_{CBO}$$

$$\frac{[V_E + V_{BE_{off}}] - 0}{R_2} = \frac{V_{C_1} - [V_E + V_{BE_{off}}]}{R_1} + I_{CBO}$$

$$\frac{(4.7 - 0.5) - 0}{R_2} = \frac{(4.7 + 0.3) - (4.7 - 0.5)}{R_1} + 0$$

$$\frac{4.2}{R_2} = \frac{5 - 4.2}{R_1} + 0$$

$$\boxed{R_2 = 5.25R_1}$$ OFF Equation

Substituting the value of R_2 determined in the OFF equation for the value of R_2 in the ON equation,

$$\frac{9.6}{0.53\text{ k} + R_1} = \frac{5.4}{5.25R_1} + 1.33\text{ mA}$$

$$9.6R_1 = (0.53\text{ k} + R_1)(1.03 + 1.33\text{ mA }R_1)$$

$$1.33\text{ mA}R_1^2 - 7.87R_1 + 0.55\text{ k} = 0$$

$$X = \frac{-b \pm \sqrt{b^2 - 4ac}}{2a}$$ Quadratic Eq.   [22]

$$R_1 = \frac{7.87 \pm \sqrt{(7.87)^2 - \cdot(4)(1.33\text{ mA})(0.55\text{ k})}}{(2)(1.33\text{ mA})}$$

$$R_1 = \frac{7.87 \pm 7.68}{0.00266}$$

$$R_1 = 5.86\text{ k}\Omega \quad \text{or} \quad 74.2\ \Omega$$

Because R_1 loads the OFF transistor, minimum loading will occur when R_1 is as large as possible. Hence, the larger of the two values should be considered.

Standard color code values—5.6 and 6.8 kΩ
Select $R_1 = 5.6$ kΩ to ensure base current enough to meet the ON circuit
requirement.

$$R_2 = 5.25R_1$$
$$R_2 = 5.25 \times 5.85\text{ k}\Omega$$
$$R_2 = 30.7\text{ k}\Omega$$

Standard color code values—27 and 33 kΩ
Select $R_2 = 27$ kΩ in order to fulfill the OFF circuit requirement by
ensuring the maximum reverse-bias voltage across the resistor
R_1.

Refer to Fig. 9-4. In many applications of this circuit the voltage drop across R_L, which is produced by I_1, may be neglected because I_1 is small; therefore, $V_{C_{off}} = V_{cc}$.

9.6 SAMPLE CIRCUIT ANALYSIS PROBLEM

Prove that the circuit shown in Fig. 9-6 will, or will not, function properly as a saturated, emitter-coupled, bistable multivibrator.

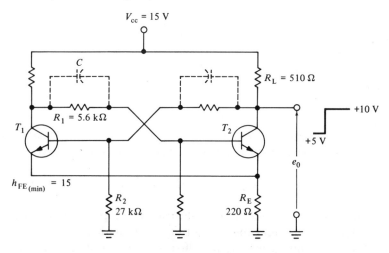

Figure 9-6

Solution:

A. Prove that the ON transistor will conduct at saturation, by determining its actual dc current gain. If this gain is less than $h_{FE_{min}}$ of the transistor used, the ON transistor will conduct at saturation. The current gain of the ON transistor may be determined by establishing the actual values of the collector and the base currents flowing in the circuit. To establish these values, replace the complex networks, at the base and at the collector of the ON transistor, with simplified but equivalent circuits. The base-biasing network of the ON transistor may be replaced by an equivalent circuit, determined by the use of Thevenin's theorem. Similarly, the collector circuit of the ON transistor may be replaced by an equivalent circuit, also determined by the application of Thevenin's theorem.

1. From the circuit shown in Fig. 9-6, determine the value of the parameters of the collector and of the base equivalent circuits by use of Thevenin's theorem.

   (a) The parameters of the equivalent circuit of the collector of the ON transistor may be determined from the circuit illustrated in Fig. 9-7.

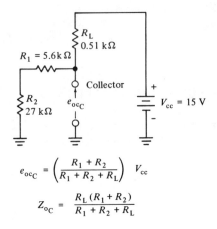

$$e_{oc_C} = \left(\frac{R_1 + R_2}{R_1 + R_2 + R_L}\right) V_{cc}$$

$$Z_{o_C} = \frac{R_L (R_1 + R_2)}{R_1 + R_2 + R_L}$$

Figure 9-7

(b)   The parameters of the equivalent circuit of the base of the ON transistor may be determined from the circuit illustrated in Fig. 9-8.

(c)   The complete equivalent ON circuit may be drawn as shown in Fig. 9-9.

2.   From the equivalent circuit, illustrated in Fig. 9-9, write Kirchhoff's voltage-input-loop equation in terms of I_B and I_C.

3.   From the equivalent circuit, illustrated in Fig. 9-9, write Kirchhoff's voltage-output loop equation in terms of I_B and I_C.

4.   Simultaneously solve the input and the output loop equations for I_B and I_C.

5.   Determine the actual value of h_{FE} from the computed values of I_C and I_B. This value must be less than the $h_{FE_{min}}$ value of the transistors used, in order to ensure that the transistor will conduct at saturation when ON.

$$e_{oc_C} = \left(\frac{5.6\ k + 27\ k}{5.6\ k + 27\ k + 0.51\ k}\right) 15 = \left(\frac{32.6\ k}{33.1\ k}\right) 15$$

$$e_{oc_C} = 14.8\ V$$

$$Z_{o_C} = \frac{0.51\ k(5.6\ k + 27\ k)}{5.6\ k + 27\ k + 0.51\ k}$$

$$Z_{o_C} = 0.492\ k\Omega$$

$$e_{oc_B} = \left(\frac{27\ k}{5.6\ k + 27\ k + 0.51\ k}\right) 15 = \left(\frac{27\ k}{33.1\ k}\right) 15$$

$$e_{oc_B} = 12.2\ V$$

$$Z_{o_B} = \frac{27\ k\ (5.6\ k + 0.51\ k)}{33.1\ k} = \frac{(27\ k)(6.1\ k)}{33.1\ k}$$

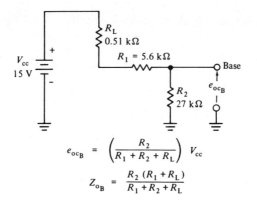

$$e_{oc_B} = \left(\frac{R_2}{R_1 + R_2 + R_L} \right) V_{cc}$$

$$Z_{o_B} = \frac{R_2 (R_1 + R_L)}{R_1 + R_2 + R_L}$$

Figure 9-8

$$Z_{o_B} = 4.97 \text{ k}\Omega$$
$$e_{oc_B} = I_B Z_{o_B} + V_{BE_{sat}} + I_B R_E + I_C R_E$$
$$12.2 = 4.97 \text{ k } I_B + 0.7 + 0.22 \text{ k } I_B + 0.22 \text{ k } I_C$$

$5.19 \text{ k } I_B + 0.22 \text{ k } I_C = 11.5$	Input Equation

$$e_{oc_C} = I_B R_E + I_C R_E + V_{CE_{sat}} + I_C Z_{oc}$$
$$14.8 = 0.22 \text{ k } I_B + 0.22 \text{ k } I_C + 0.3 + 0.492 \text{ k } I_C$$

$0.22 \text{ k } I_B + 0.712 \text{ k } I_C = 14.5$	Output Equation

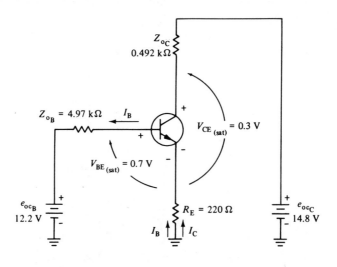

Figure 9-9

Solution for the input and output equations for I_B and I_C by determinants:

$$I_B = \frac{\begin{array}{cc} 11.5 & 0.22\text{ k} \\ 14.5 & 0.712\text{ k} \end{array}}{\begin{array}{cc} 5.19\text{ k} & 0.22\text{ k} \\ 0.22\text{ k} & 0.712\text{ k} \end{array}}$$

$$I_B = \frac{(11.5)(0.712\text{ k}) - (14.5)(0.22\text{ k})}{(5.19\text{ k})(0.712\text{ k}) - (0.22\text{ k})(0.22\text{ k})} = \frac{8.19\text{ k} - 3.19\text{ k}}{3.69\text{ k} - 0.04\text{ k}}$$

$$I_B = \frac{5\text{ k}}{3.65\text{ kk}}$$

$$I_B = 1.37\text{ mA}$$

$$I_C = \frac{\begin{array}{cc} 5.19\text{ k} & 11.5 \\ 0.22\text{ k} & 14.5 \end{array}}{\Sigma = 3.65\text{ kk}} = \frac{(5.19\text{ k})(14.5) - (11.5)(0.22\text{ k})}{3.65\text{ kk}}$$

$$I_C = \frac{75.3\text{ k} - 2.53\text{ k}}{3.65\text{ kk}} = \frac{71.7\text{ k}}{3.65\text{ kk}}$$

$$I_C = 19.6\text{ mA}$$

$$h_{FE} = \frac{I_C}{I_B} = \frac{19.6\text{ mA}}{1.37\text{ mA}} = 14.3$$

Because $h_{FE_{min}} = 15$ for the transistors, the ON transistor will conduct at the required saturation.

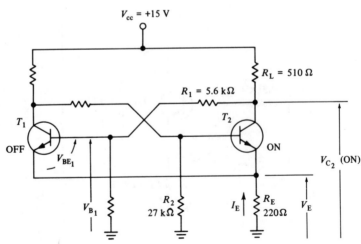

Figure 9-10

B. Prove that the OFF transistor is at cutoff.
1. Assume that transistor T_1 is OFF and that transistor T_2 is ON, as shown in Fig. 9-10. Since it has been ascertained that the ON transistor is conducting at saturation, the ON circuit currents and voltages may be used to determine the polarity and magnitude of V_{BE_1}.
2. If the base of transistor T_1 is negative, with respect to the emitter, transistor T_1 is reversed biased and hence at the required cutoff (NPN).

$$I_E = I_B + I_C = 1.37 \text{ mA} + 19.6 \text{ mA}$$
$$I_E = 20.97 \text{ mA}$$
$$V_E = I_E R_E = (20.97 \text{ mA})(0.22 \text{ k})$$
$$V_E = +4.61 \text{ V}$$
$$V_{C_{2on}} = V_E + V_{CE_{sat}}$$
$$V_{C_{2on}} = 4.61 + 0.3$$
$$V_{C_{2on}} = 4.91 \text{ V}$$

$$V_{B_1} = \left(\frac{R_2}{R_1 + R_2}\right) V_{C_{2on}} = \left(\frac{27 \text{ k}}{5.6 \text{ k} + 27 \text{ k}}\right) 4.91$$
$$V_{B_1} = +4.08 \text{ V}$$
$$V_{BE_1} = V_{B_1} - V_E = 4.08 - 4.61$$
$$V_{BE_1} = -0.53 \text{ V} \qquad \text{(reverse bias)}$$

Hence, the OFF transistor is at cutoff. Since both the ON and the OFF circuit conditions have been fulfilled, the circuit, shown in Fig. 9-6, will operate properly as a saturated emitter-coupled bistable multivibrator.

EMITTER-COUPLED BISTABLE MULTIVIBRATOR

OBJECT:

1. To design a saturated emitter-coupled bistable multivibrator
2. To analyze the operation of an emitter-coupled bistable multivibrator
3. To examine the requirements for the stable states of an emitter-coupled bistable multivibrator

MATERIALS:

2 Switching transistors (example—2N3646 silicon NPN)

1 Manufacturer's specification sheet for type of transistor used (2N3646—see Appendix B)

1 Oscilloscope, dc time-base type; frequency response dc to 450 kHz; vertical sensitivity, 100 mV/cm

1 Transistor power supply (0 to 30 V and 0 to 250 mA)

7 Resistor substitution boxes (10 Ω to 10 MΩ, 1 W)

1 Vacuum-tube voltmeter, VTVM; dc–ac type

PROCEDURE:

1. Design a saturated, emitter-coupled, bistable multivibrator circuit from the following specifications:

$$e_o = 10 \text{ V peak}$$

$$I_C = 10 \text{ mA}$$

Assume:

$$V_{cc} = 12 \text{ V}$$

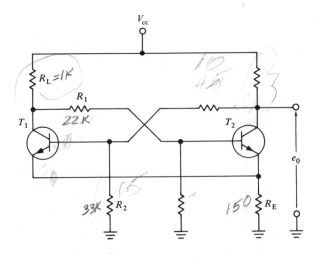

Figure 9-1X

2. Design the saturated, emitter-coupled, bistable multivibrator, illustrated in Fig. 9-1X, for the circuit conditions specified. Include all calculations. Explain why any assumptions which you may have made are valid.

3. Draw a schematic of the circuit designed in Step 2, and label it with the practical standard color code values.

4. Analyze the circuit designed in Step 2, and prove that it will operate satisfactorily for the required specifications when standard-value resistors are used. Include all calculations.

5. Connect the designed circuit. Check the operation of the circuit. Record all necessary data.

6. The stable states may be reversed by momentarily removing the forward bias from the ON transistor. This may be accomplished by momentarily grounding the base of the ON transistor. Check the operation of the circuit in this reversed stable state. Record all necessary data.

7. Compare measured values with design values, and explain any discrepancies.

QUESTIONS AND EXERCISES

1. What are the advantages of the emitter-coupled multivibrator over the collector-coupled multivibrator?

2. What are the disadvantages of the emitter-coupled bistable multivibrator compared to the collector-coupled bistable multivibrator?

3. Explain why both transistors do not conduct when voltage is initially applied to an emitter-coupled bistable multivibrator. Explain why the ON transistor conducts at saturation and why the OFF transistor remains at cutoff.

4. Which circuit parameters must be considered in order to select a value for C_E which is proper for the circuit shown in Fig. 9-1?

5. How does the emitter capacitor C_E, illustrated in Fig. 9-1, improve the operation of the emitter-coupled bistable multivibrator circuit?

6. Prove that the emitter-coupled bistable multivibrator, illustrated in Fig. 9-2X,

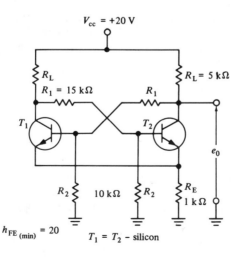

$V_{cc} = +20$ V

R_L
$R_1 = 15$ kΩ
T_1
R_1
T_2
$R_L = 5$ kΩ
e_0

R_2 10 kΩ R_2 R_E 1 kΩ

$h_{FE\,(min)} = 20$ $T_1 = T_2$ – silicon

Figure 9-2X

will or will not operate properly as a saturated flip–flop. Assume typical junction voltages.

7. In the circuit, shown in Fig. 9-2X, what is the amplitude of the output-voltage pulse when the circuit switches from one stable state to the other?

8. Prove that the emitter-coupled bistable multivibrator, illustrated in Fig. 9-3X, will or will not operate properly as a saturated flip–flop.

Assume:

$$V_{CE_{sat}} = 0 \text{ V}$$
$$I_{CBO} = 0 \text{ A}$$
$$V_{BE_{off}} = 0 \text{ V}$$
$$V_{BE_{sat}} = 0 \text{ V}$$

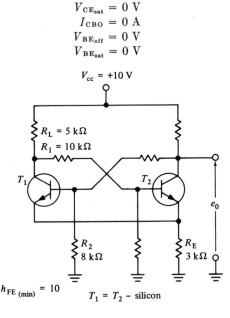

$V_{cc} = +10 \text{ V}$

$R_L = 5 \text{ k}\Omega$

$R_1 = 10 \text{ k}\Omega$

T_1 T_2

e_0

R_2
$8 \text{ k}\Omega$

R_E
$3 \text{ k}\Omega$

$h_{FE \text{ (min)}} = 10$ $T_1 = T_2$ – silicon

Figure 9-3X

9. If a 0.1 μF capacitor C_E were placed in parallel with the 1 kΩ resistor R_E, illustrated in Fig. 9-2X, what would be the longest switching interval for which the emitter capacitor might practically perform the stabilizing action?

10. Explain the operation, and justify the requirements, of a saturated, emitter-coupled, bistable multivibrator.

11. From the circuit shown in Fig. 9-1, design a saturated, emitter-coupled, bistable multivibrator with the following characteristics:

$$e_o = 15 \text{ V peak}$$
$$I_C = 10 \text{ mA}$$

Given:

Silicon NPN transistors, $h_{FE_{min}} = 30$
$$I_{CBO} = 0$$

25 V dc source

Assume:

$V_{BE_{off}} = -0.5 \text{ V}$
Standard junction voltages

12. Redesign the circuit designed in Prob. 11. The circuit must be operable under temperature conditions up to 75°C. Assume $I_{CBO} = 5$ μA at 25°C.

13. Refer to Fig. 9-1. Prove that the emitter-coupled bistable multivibrator will, or will not, operate properly when

$$R_1 = 22 \text{ k}\Omega \qquad R_2 = 180 \text{ k}\Omega \qquad R_L = 1.2 \text{ k}\Omega$$
$$R_E = 270 \text{ }\Omega \qquad V_{cc} = +20 \text{ V} \qquad e_o = 12 \text{ V}$$

Silicon NPN transistors, $h_{FE_{min}} = 20$
Assume standard junction voltages.

14. In the circuit analyzed in Prob. 13, what component or components must be changed in order for the circuit to operate properly? To what value, or values, must the component, or components, be changed?

chapter 10

TRIGGERING

Triggering is necessary.to change the stable state of a multivibrator The circuit which produces the trigger pulse is called a *trigger circuit*, and it is generally a "differentiator" waveshaping circuit. Usually, the input voltage to a trigger circuit is a pulse of short duration or a step-voltage waveform; the output is a differentiated waveform. The function of the triggering circuit is twofold: it reshapes the applied input pulse and then it directs the resultant voltage waveform to the appropriate transistor element of the multivibrator.

10.1 BASIC TYPES OF TRIGGERING

The two basic types of triggering are asymmetrical and symmetrical. Applied to computer applications, asymmetrical triggering is referred to as set–reset, and symmetrical triggering is referred to as complementary.

Asymmetrical triggering is accomplished by the application of a triggering pulse to one of the two transistors of a multivibrator. This pulse is capable of switching the multivibrator in one direction only. An applied pulse from another circuit is needed to switch the multivibrator back to the original stable state.

Symmetrical triggering is employed to reverse the existing stable state of a bistable multivibrator circuit. This is accomplished by the apparent simultaneous application of a single input-trigger pulse to both transistors of the multivibrator. Each subsequent application of a trigger pulse will reverse the stable state produced by the preceding pulse.

Triggering pulses may be used to turn ON transistors OFF or to turn OFF transistors ON. Triggering pulses are usually applied to the ON transistor, however, because the amplification of the ON transistor may be used to turn the OFF transistor ON. In so doing, the amplitude of the required trigger voltage is greatly reduced.

Triggering pulses may be applied to the collector or to the base of the transistor. In addition, they may be applied to the emitters of an emitter-coupled circuit from which the emitter capacitor has been omitted.

10.2 ASYMMETRICAL BASE TRIGGERING

Figure 10-1 illustrates one of the simplest forms of asymmetrical triggering. This circuit is designed to switch transistor T_2 from ON to OFF when a negative-trigger pulse is applied to the base of the ON (NPN) transistor. When transistor T_2 is turned OFF, it turns transistor T_1 ON. The circuit remains in this stable state until it is reversed by the action of another circuit (not shown). Subsequent negative pulses applied to transistor T_2 have no effect on the circuit.

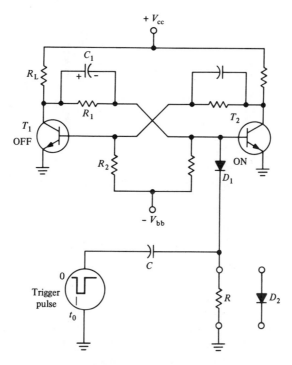

Figure 10-1 Asymmetrical Base Triggering

Refer to Fig. 10-2, which represents only the triggering part of the circuit shown in Fig. 10-1. When a negative-trigger pulse is applied at t_{+0} time, it forward biases diode D_1. Capacitor C charges through the path indicated in Fig. 10-2(b). This capacitor-charging current is the reverse base current which

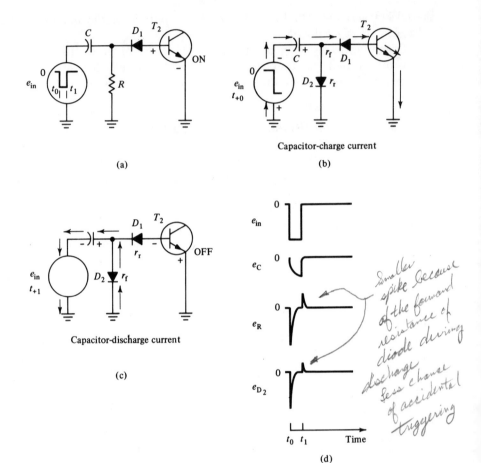

Figure 10-2

reverse biases transistor T_2, thereby turning it OFF. Capacitor C charges, as indicated in Fig. 10-2(d). The resultant voltage waveform across R is also indicated.

The resistor R must be as large as possible when the trigger pulse is applied, in order to prevent the capacitor from charging through the resistor. When the trigger pulse is removed, however, resistor R should be as small as possible, in order to provide a low resistance discharge path for capacitor C. Hence, a resistor which has two divergent values is needed. Therefore, as indicated in Fig. 10-1, frequently resistor R is replaced by diode D_2 because it automatically provides this requirement.

With the addition of diode D_2, the triggering circuit operates as a positive clipper to the normal differentiated output. In this capacity, it prevents the trailing edge of the triggering pulse from turning transistor T_2 ON again. Hence, diode D_2 is forward biased and provides a low resistance discharge path for capacitor C when the trigger pulse is removed at t_{+1} time. See Fig. 10-2(c).

If the PNP transistors were used in this circuit, a positive trigger pulse would be required and the polarity of both diodes would be reversed.

10.3   MULTIVIBRATOR VOLTAGE GAIN REQUIREMENTS

For a collector-coupled, or an emitter-coupled, bistable multivibrator to be practical, it must be designed to switch from one stable state to the other when a trigger pulse is applied.   When a multivibrator makes this transition, it traverses the active region of operation of the V_{CE}—I_C characteristic curves. This transition will only occur when the class A voltage gain of the transistor amplifier is greater than unity. Refer to Fig. 10-3. During the transition interval, a change in voltage, initiated from the base to emitter of T_1, may be

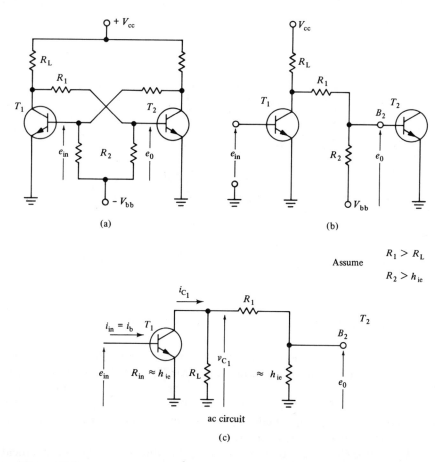

(a)

(b)

Assume $R_1 > R_L$
          $R_2 > h_{ie}$

ac circuit

(c)

Figure 10-3

thought of as an ac input voltage (e_{in}). This input voltage is amplified by transistor T_1. A portion of this amplified output voltage from T_1 appears as the resultant output voltage from base to emitter of T_2. See Fig. 10-3(b). This resultant output voltage must be greater than e_{in} or the transition from one stable state to the other will not occur.

Refer to the ac circuit shown in Fig. 10-3(c). Assume that R_1 is much larger than R_L. Practically, all the ac collector current from transistor T_1 goes through R_L. Assume that the value of R_2 is much larger than the ac input resistance of the transistor (h_{ie}). The resistance, from base to ground, equals the parallel resistance of R_2 and h_{ie} which, in turn, approximately equals h_{ie}.

$$A_V = \frac{e_o}{e_{in}}$$

$$e_o \approx \left(\frac{h_{ie}}{h_{ie} + R_1}\right) V_{C_1}$$

$$V_{C_1} \approx i_{C_1} R_L$$

$$i_{C_1} \approx i_{B_1} h_{fe}$$

$$V_{C_1} \approx i_{B_1} h_{fe} R_L$$

$$e_o \approx \left(\frac{h_{ie}}{h_{ie} + R_1}\right) i_{B_1} h_{fe} R_L$$

But

$$i_{B_1} \approx \frac{e_{in}}{h_{ie}}$$

$$e_o \approx \left(\frac{h_{ie}}{h_{ie} + R_1}\right) \frac{h_{fe} R_L e_{in}}{h_{ie}}$$

$$A_V \approx \frac{e_o}{e_{in}} \approx \frac{h_{fe} R_L}{h_{ie} + R_1} = >1$$

For the multivibrator to make the transition from one stable state to the other, the voltage gain A_V must be greater than unity.

10.4   ASYMMETRICAL COLLECTOR TRIGGERING

To trace the operation of the circuit, shown in Fig. 10-4, assume that no triggering pulse is applied and that transistor T_1 is OFF and transistor T_2 is ON. The current I_1 produces a small voltage drop across R_L as it flows through the load resistor of the OFF transistor T_1. This voltage drop is of the polarity to reverse bias diode D_1. Diode D_1 would not conduct, even if I_1 were zero, and, therefore, had zero reverse bias.

Capacitor C_1 charges to approximately V_{cc} and is of the polarity indicated in Fig. 10-4 when transistor T_2 conducts at saturation.

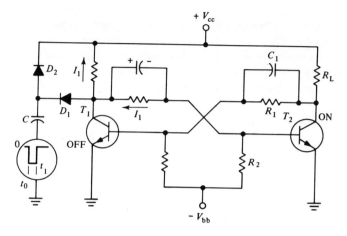

Figure 10-4 Asymmetrical Collector Triggering

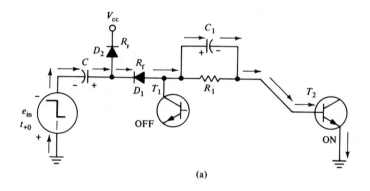

(a)

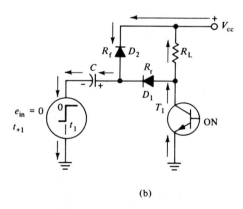

(b)

Figure 10-5 (a) Capacitor-Charge Current; (b) Capacitor-Discharge Current

When the negative-trigger pulse is applied, at t_{+0} time, it forward biases diode D_1. Acting as a voltage source, charged capacitor C_1 is effectively in series-aiding the negative-trigger pulse. Hence, capacitor C charges by the path indicated in Fig. 10-5(a). This capacitor-charging current is actually the reverse base current which flows through transistor T_2 and which turns transistor T_2 OFF.

The removal of the base storage charge may be very rapid and the duration of the triggering pulse may be extremely short because capacitor C_1 acts as a source in series with the triggering pulse.

The collector voltage of T_2 rises toward V_{cc} when transistor T_2 starts to turn OFF. This collector voltage rise is transmitted through the voltage divider R_1 and R_2 as forward bias to transistor T_1. This forward bias turns T_1 ON.

If transistor T_1 is ON and transistor T_2 is OFF, capacitor C discharges, by the path indicated in Fig. 10-5(b), when the triggering pulse is removed at t_{+1} time.

10.5   SYMMETRICAL TRIGGERING

Symmetrical triggering is used when expediency requires that each successive trigger pulse reverse the previous stable state of a multivibrator. A triggering circuit must meet two basic requirements in order to execute symmetrical triggering efficiently. It must shape the input-trigger pulse, and it must direct the trigger pulse to the ON transistor and prevent it from reaching the OFF transistor. Three basic methods of symmetrical triggering are base triggering, collector triggering, and base triggering with collector steering (hybrid).

10.6   SYMMETRICAL BASE TRIGGERING

To understand the following description of the operation of symmetrical base triggering, illustrated in Fig. 10-6, assume that transistor T_1 is OFF and that transistor T_2 is ON before the trigger pulse is applied. The respective forward- and reverse-bias polarities of the two transistors are indicated in Fig. 10-6. The negative-trigger pulse is applied at t_{+0} time. The applied pulse is of a polarity to forward bias diode D_1 and diode D_2. The reverse-bias voltage, however, which holds transistor T_1 OFF, is in series-opposing the trigger pulse; therefore, it prevents the trigger pulse from forward biasing diode D_1. The amplitude of the trigger voltage, which appears across diode D_1, must be less than the reverse bias of the OFF transistor T_1 or diode D_1 will not remain reverse biased. The value of this amplitude, therefore, limits the amplitude of the trigger pulse which may be applied to trigger the multivibrator.

The negative-trigger pulse is applied in series-aiding the forward biased transistor T_2. Hence, diode D_2 is forward biased. Capacitor C starts to charge

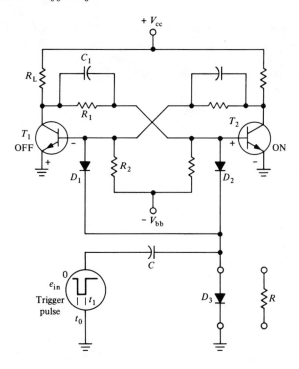

Figure 10-6   Symmetrical Base Triggering

by the path indicated in Fig. 10-7(a), and it continues to charge rapidly because the resistance of the forward biased emitter-base junction is very small. This capacitor-charging current is the reverse base-current flow, which turns the ON transistor OFF. The emitter-base junction resistance increases abruptly; thus, it effectively stops the capacitor from charging any further.

The trigger pulse is of the polarity to reverse bias diode D_3, which, therefore, is effectively open-circuited during the duration of the trigger pulse.

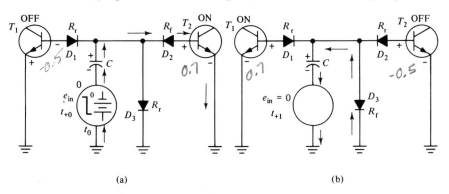

(a)   (b)

Figure 10-7   (a) Capacitor-Charging Current-Turning $T_2 -$ OFF; (b) Capacitor Discharging after Trigger Pulse Has Passed

Charged capacitor C, acting as a voltage source, is of the polarity to reverse bias diode D_1 and diode D_2. This occurs when the trigger pulse is removed at t_{+1} time. The voltage across capacitor C, however, is of the polarity to forward bias diode D_3, thereby providing a low resistance discharge path for capacitor C, as indicated in Fig. 10-7(b). Therefore, capacitor C discharges through diode D_3. Thus, the triggering cycle is completed.

Base triggering has an advantage over collector triggering; only a small amplitude of triggering pulse is required to trigger a multivibrator. This characteristic may become a disadvantage, however, because a small noise pulse may also trigger the circuit. The fact that the amplitude of the triggering pulse is critical is also a disadvantage.

10.7 SYMMETRICAL COLLECTOR TRIGGERING

To understand the following description of the operation of symmetrical, collector triggering, illustrated in Fig. 10-8, assume that transistor T_1 is OFF and that transistor T_2 is ON before the trigger pulse is applied. The collector-to-emitter voltage of the ON transistor T_2 is V_{CEsat} or almost 0 V, and the cathode of diode D_2 has a large positive voltage with reference to ground ($\approx V_{cc}$). This static voltage is of a polarity to reverse bias diode D_2. The negative-trigger pulse is applied at t_{+0} time and is of a polarity to forward bias diodes D_1 and D_2. For proper triggering action to occur, the diode in the collector circuit of the ON transistor must be reverse biased when the trigger pulse is applied. For this condition to exist, the amplitude of the trigger pulse

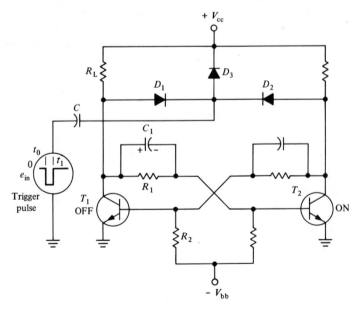

Figure 10-8 Symmetrical Collector Triggering

must be less than the static reverse-bias voltage across diode D_2. Therefore, diode D_2 will remain reverse biased as long as the amplitude of the trigger pulse is less than $[V_{cc} - V_{CEsat}]$, as indicated in Fig. 10-9.

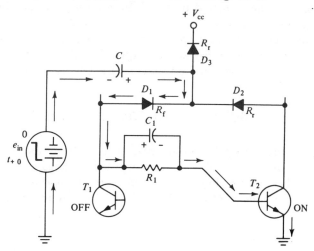

Figure 10-9   Capacitor C Charging Current-Turning T_2 — OFF

The negative-trigger pulse forward biases diode D_1. Therefore, diode D_1 acts as a closed switch, and the negative-trigger pulse is in series-aiding charged capacitor C_1. Unaltered in shape, the trigger pulse appears across the emitter-base junction of transistor T_2. Hence, capacitor C charges and capacitor C_1 discharges by the path indicated in Fig. 10-9. This reverse base-current flow reverse biases transistor T_2 and thereby turns it OFF.

During the pulse interval, the negative pulse reverse biases diode D_3 and thus prevents capacitor C from charging through D_3 and toward V_{cc}. When the trigger pulse is removed at t_{+1} time, charged capacitor C forward biases diode D_3 and discharges through it, as indicated in Fig. 10-10. When a subsequent nega-

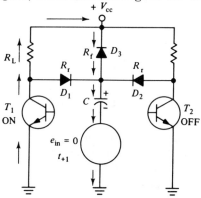

Figure 10-10   Capacitor Discharging after Trigger Pulse Has Passed

tive-trigger pulse is applied, it is directed to the ON transistor, thereby turning it OFF.

The advantage of collector triggering over base triggering is the fact that the circuit may not be triggered by small noise pulses because the amplitude of the input pulse is not critical.

A disadvantage of collector triggering is the fact that an additional stage of triggering amplification is frequently required. This added amplification is necessary because the pulse required for collector triggering is large.

10.8   BASE TRIGGERING WITH COLLECTOR STEERING

Base triggering with collector steering, a hybrid method of symmetrical triggering, is frequently employed. This method of triggering provides the advantages of the two previously discussed methods, without entailing their respective disadvantages. Refer to Fig. 10-11. This circuit incorporates two

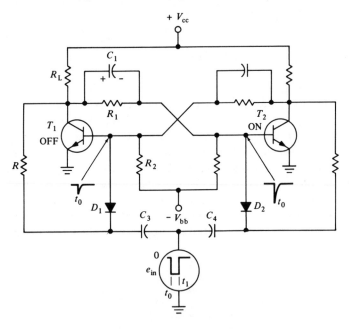

Figure 10-11   Symmetrical Base Triggering with Collector Steering

separate differentiator circuits, one for each transistor of the multivibrator. The resistors R are tied to the collectors of the transistors. Because one transistor is ON and one is OFF, the two differentiator circuits will react differently when a common trigger pulse is applied.

To understand the following description of the operation of hybrid triggering, illustrated in Fig. 10-12(a), assume that transistor T_2 is ON before the triggering pulse is applied. The collector voltage of T_1 approaches V_{cc}; hence, it reverse biases diode D_1. The collector voltage of transistor T_2 is approximately zero [$V_{CEsat} = 0.3$ V]; hence, diode D_2 is reverse biased and cannot conduct.

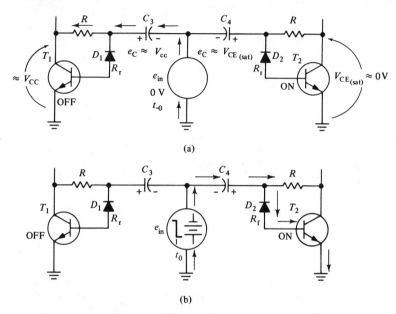

(a)

(b)

Figure 10-12   (a) Circuit Conditions at t_{-0} Time; (b) Circuit Conditions at t_{+0} Time

The negative-trigger pulse, applied at t_{+0} time, forward biases diode D_2. Hence, capacitor C_4 starts to charge by the path indicated in Fig. 10-12(b). This capacitor-charging current becomes the reverse base-current flow which turns transistor T_2 OFF.

The same negative-trigger pulse is applied simultaneously to the circuit of the OFF transistor T_2. The charged capacitor C_3 (acting as a source) is in series-opposing the trigger voltage. As long as the trigger voltage is less than the voltage across C_3, diode D_1 remains reverse biased. As long as diode D_1 is reverse biased, the trigger pulse cannot reach transistor T_1 and, therefore, cannot affect it.

10.9   SYMMETRICAL TRIGGERING—EMITTER-COUPLED MULTIVIBRATOR

Symmetrical triggering of an emitter-coupled bistable multivibrator may be accomplished by the application of a trigger pulse to the common-emitter resistor R_E, as illustrated in Fig. 10-13. To understand the following description

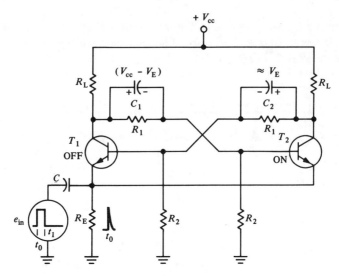

Figure 10-13 Symmetrical-Triggering Emitter-Coupled Bistable Multivibrator

of the operation of emitter-coupled symmetrical triggering, assume that transistor T_1 is OFF and that transistor T_2 is ON before the triggering pulse is applied. When the trigger pulse is applied at t_{+0} time, the differentiated waveform, which appears across R_E, is superimposed on the static dc voltage across R_E. This additional spike of voltage cuts off transistor T_2. When both transistors are cut off, the commutating capacitors C_1 and C_2 control the action of the circuit. Just before the trigger pulse is applied, at t_{-0} time, capacitor C_1 charges to $(V_{cc} - V_E)$, and capacitor C_2 charges to $\approx V_E$. The charge on both capacitors is of the polarity to reverse bias both transistors, respectively, as illustrated in Fig. 10-13. Because the charge on capacitor C_1 is larger, however, it holds transistor T_2 in deep momentary cutoff. Hence, transistor T_1 starts to conduct and thereby holds transistor T_2 OFF, even after capacitor C_1 has discharged. Thus, the stable state of the multivibrator is reversed. The value of capacitance of C_1 and C_2 must be much larger for an emitter-coupled circuit than for a collector-coupled circuit.

10.10   COMMUTATING CAPACITORS

In multivibrator circuits, the speedup capacitors C_1 are referred to as commutating capacitors because they turn or direct the base storage charge (electrons—NPN) into and out of the base. In so doing, the capacitors enable the transistor to turn ON and OFF more rapidly. The selection of a commutating capacitor, the value of which is correct for a particular circuit, is dependent upon two factors: the base storage charge of the transistor to be used in the circuit and the amount of voltage (V_{cc}) to be applied across the capacitor. The necessary base storage charge required for collector-current saturation when the pulse is first applied is obtained from the charging current of the commutating capacitor. The total charge stored in the base of a transistor is dependent upon the collector current. Some manufacturers of switching transistors list the base storage charge in a graph which depicts the value of the charge as a function of collector current. Values of base storage charge, for a typical switching transistor, range from 50 to 2000 pC for corresponding collector currents which range from 1 to 500 mA. Refer to Appendix B, manufacturer's specification sheet of 2N3903.

When a given voltage is applied across a commutating capacitor, the latter must hold a charge equal to the base storage charge of the ON transistor. Larger values of commutating capacitors will limit the upper triggering frequency of a multivibrator.

The total charge on a capacitor is equal to the product of capacitance times the potential across the capacitor. See Eq. 23.

$$Q = CE \qquad \qquad [23]$$

where
C = capacitance, farads (F)
Q = charge, coulombs (C)
E = potential across capacitor, volts (V)

10.11   DESIGN EXAMPLE

Refer to Fig. 10-1. Assume

$$V_{cc} = 10 \text{ V}$$
$$I_C = 20 \text{ mA}$$
$$T_1 = T_2 = 2\text{N}3903$$

Determine the minimum value of capacitance for capacitors C_1.

The base storage charge for a 20 mA collector current is 400 pC. See Appendix B, manufacturer's specification sheet 2N3903.

$$C_1 = \frac{Q}{E} = \frac{Q}{V_{cc}} = \frac{400 \times 10^{-12}}{10}$$
$$C_1 = 40 \text{ pF}$$

A practical first-order approximation of the necessary minimum value of capacitance may be obtained from this simple relationship.

10.12   TRIGGERING—CIRCUIT CAPACITANCE

Refer to Fig. 10-1. The total stored base charge is employed to determine the minimum value of capacitance for capacitor C. Recall that only the base storage charge necessary for collector saturation was used to determine the value of C_1. An empirical factor of 2 is commonly used to determine the proper value of C. Hence, if the sample design conditions used are the same as those used to determine the value of C_1,

$$C = \frac{2Q}{V_{cc}} = \frac{2 \times 400 \times 10^{-12}}{10}$$
$$C = 80 \text{ pF}$$

10.13 RESISTOR R—HYBRID TRIGGERING

Refer to Fig. 10-11. The value of resistor R should be large in order to minimize the loading effect on the ON transistor. If the value of resistor R is too large, however, it will establish the upper frequency limit of the triggering of the multivibrator. If the upper frequency limit of the triggering is not a problem, R is usually selected to be 100 times larger than R_L.

TRIGGERING

OBJECT:

1. To familiarize the student with the set–reset type of triggering
2. To familiarize the student with base-symmetrical triggering
3. To familiarize the student with collector-symmetrical triggering
4. To familiarize the student with hybrid-symmetrical triggering

MATERIALS:

2   Switching transistors (example—2N3903 silicon NPN)
4   Switching diodes (example—1N914 silicon)
1   Manufacturer's specification sheet for type of transistor used (2N3903—see Appendix B)
8   Resistor substitution boxes (10 Ω to 10 MΩ, 1 W)
4   Capacitor substitution boxes (0.0001 to 0.22 μF at 450 V)
2   Transistor power supplies (0 to 30 V and 0 to 250 mA)
1   Square-wave generator (20 Hz to 200 kHz; $Z_o = 600\ \Omega$)
1   Oscilloscope, dc time base, frequency response dc to 450 kHz; vertical sensitivity, 100 mV/cm

PROCEDURE:

1. Design a saturated collector-coupled bistable multivibrator with the following characteristics:

$$V_{cc} = 15\ \text{V}$$
$$V_{bb} = 5\ \text{V}$$
$$I_C = 10\ \text{mA} \qquad \text{for the ON transistor}$$

Refer to chapter 8 for design procedure. Include all calculations and explain why any assumptions which have been made are valid.

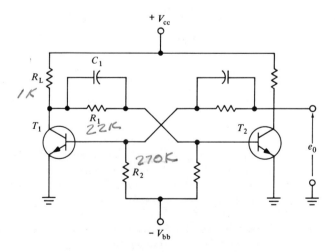

Figure 10-1X   Collector-Coupled Bistable Multivibrator

2. Set–Reset. Asymmetrical Base Triggering
    (a) Connect the circuit designed. Check the operation of the circuit.
    (b) Refer to Fig. 10-2X. Calculate C.
    (c) Connect the set–reset triggering circuit shown in Fig. 10-2X.
    (d) Alternately, connect the V_{bb} supply (acting as a negative-trigger voltage) to the trigger input of the ON transistor. Note that a momentary contact will perform the switching action. Monitor the output of the multivibrator; use an oscilloscope.
    (e) Explain the operation of the set–reset type of asymmetrical triggering.

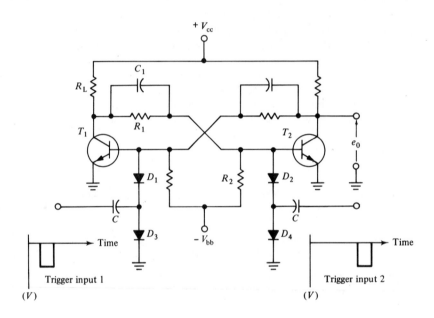

Figure 10-2X   Base Set-Reset Triggering

3. Symmetrical Base Triggering
    (a) Remove the set–reset base triggering circuit from the multivibrator circuit, and connect the symmetrical base triggering circuit, shown in Fig. 10-6.
    (b) Apply a 5000 Hz square wave of voltage as the trigger pulse. Determine the minimum value of trigger voltage necessary to trigger the multivibrator. Verify that the multivibrator is not being triggered by the trailing edge of the trigger pulse. This may be accomplished by measuring the input- and output-pulse widths. The multivibrator is triggering properly if the pulse width of the input triggering voltage is equal to half of the pulse width of the output. If the triggering pulse width is the same as the output-pulse width, the multivibrator is being triggered by the trailing edge of the trigger pulse as well as by the leading edge.

(c) With reference to ground, measure and record the amplitude and wave-shape of the waveforms at the following points in the circuit: D_3, base, and collector.

(d) Draw the following voltage waveforms to scale, and, one above the other, to the same time base: e_{in}, D_3, base, and collector. Draw a schematic of the circuit, on the same sheet of paper as that used for the waveform illustrations.

(e) Explain the operation of symmetrical base triggering. Use the waveforms for your analysis.

4. Symmetrical Collector Triggering
   (a) Remove the symmetrical base triggering circuit from the multivibrator circuit, and connect the symmetrical collector-triggering circuit, shown in Fig. 10-8.
   (b) Repeat Steps (b) through (e) of symmetrical base triggering.

5. Symmetrical Hybrid Triggering
   (a) Remove the symmetrical collector-triggering circuit from the multivibrator circuit, and connect the symmetrical hybrid-triggering circuit, shown in Fig. 10-11.
   (b) Repeat Step (b) of symmetrical base triggering.
   (c) With reference to ground, measure and record the amplitude and wave-shape of the waveforms, at the following points in the circuit: e_{in}, C_3, base, and collector.
   (d) Draw the voltage waveforms measured in Step (c) to scale and, one above the other, to the same time base.
   (e) Explain the operation of hybrid-symmetrical triggering. Use the wave-forms for your analysis.

6. Symmetrical Triggering—Emitter-Coupled Multivibrator
   (a) Connect the circuit, shown in Fig. 10-3X, or connect the emitter-coupled bistable multivibrator designed in the experiment of chapter 9 (see procedure in that experiment).

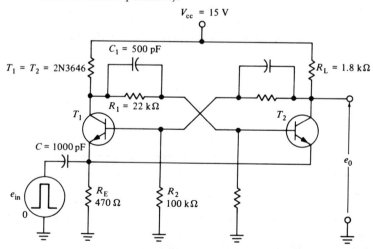

Figure 10-3X   Symmetrical-Triggering Emitter-Coupled Bistable Multivibrator

(b) Apply a 5000 Hz square wave of voltage as the trigger pulse. Determine the minimum value of trigger voltage necessary to trigger the multivibrator. Record.

(c) Verify that the multivibrator is not being triggered by the trailing edge of the trigger pulse.

(d) With reference to ground, measure and record the amplitude and waveshape of the waveforms at the following points in the circuit: e_{in}, R_E, base, and collector.

(e) Draw the waveforms observed in Step (d) to scale and, one above the other, to the same time base. Draw a schematic of the circuit on the same sheet of paper as that used for the waveform illustrations.

(f) Explain the operation of the symmetrically triggered emitter-coupled multivibrator. Use the waveforms for your analysis.

QUESTIONS AND EXERCISES

1. In general, can the amplitude of the triggering voltage be too small? Explain.
2. In general, can the amplitude of the triggering voltage be too large? Explain.
3. What is the advantage of collector triggering over base triggering? Explain.
4. What is the disadvantage of base triggering? Explain.
5. What is the disadvantage of collector triggering? Explain.
6. List two advantages of symmetrical hybrid triggering over symmetrical base triggering or over symmetrical collector triggering.
7. What is the advantage of symmetrical emitter triggering?
8. What factors determine the selection of proper commutating capacitance for a multivibrator circuit? Explain.
9. What factors determine the selection of proper capacitance for a triggering circuit? Explain.
10. List and explain two disadvantages of symmetrical emitter triggering.
11. Why is it necessary for the voltage loop gain of a multivibrator to be greater than unity? Explain.
12. Prove that the multivibrator circuit, shown in Fig. 8-6, fulfills the voltage loop gain requirements for proper switching action. This voltage loop gain requirement is expressed

$$A_V \approx \frac{h_{fe}R_L}{h_{ie} + R_1} = >1$$

Assume:

$$h_{ie} = 1 \text{ k}\Omega \qquad h_{fe} = h_{FE\,min}$$

13. Prove that the multivibrator circuit, analyzed in Prob. 9 of chapter 8, fulfills the voltage loop gain requirements for proper switching action. Use the voltage gain expression in Prob. 12 as the criteria. Assume:

$$h_{ie} = 1 \text{ k}\Omega \qquad h_{fe} = h_{FE\,min}$$

14. Prove that the multivibrator circuit, analyzed in Prob. 13 of chapter 8, fulfills the voltage loop gain requirements for proper switching action. Use the voltage gain expression in Prob. 12 as the criterion. Assume:

$$h_{ie} = 1 \text{ k}\Omega \qquad h_{fe} = h_{FE\,min}$$

chapter 11

SCHMITT TRIGGER

The Schmitt trigger circuit is a special form of the emitter-coupled, bistable, multivibrator circuit; the stable state of the former is determined by the amplitude of the input voltage, while the stable state of the latter is determined by the application of successive trigger pulses. For a given Schmitt trigger circuit, two finite values of input voltage cause the circuit to switch from one stable state to the other. Therefore, this circuit may be used to discriminate between two dc voltage levels; in this capacity, it is called a comparer circuit. It may also be used as a pulse squaring circuit because the output voltage is always a square wave (rectangular) with two distinct output-voltage levels.

11.1 DESCRIPTION OF OPERATION

To trace the operation of the Schmitt trigger circuit, refer to the schematic shown in Fig. 11-1. When no input voltage is applied to the circuit, transistor T_1 is cut off and transistor T_2 conducts at saturation. When transistor T_2 conducts at saturation, the voltage drop across R_E may be defined as V_E. Hence, the output voltage at t_1 time is $[V_{E_2} + V_{CEsat}]$. Before transistor T_1 can conduct, the input voltage must be greater than V_{E_2}. The amplitude of the input voltage required to cause transistor T_1 to conduct is called the upper trigger potential (UTP).

$$UTP = V_{E_2} + V_{BE_a}$$

V_{E_2} = voltage drop across R_E
    when transistor T_2 conducts
    at saturation

V_{BE_a} = amount of forward bias
    necessary to place a transistor
    in the active region (silicon,
    0.5 V)

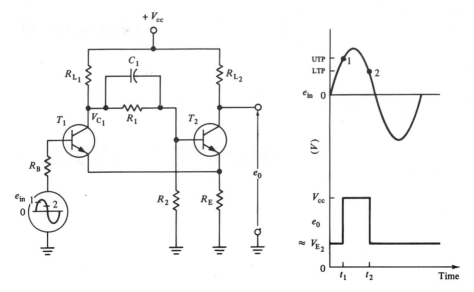

Figure 11-1   Schmitt Trigger

The collector voltage of transistor T_1 (V_{C_1}) decreases from V_{cc} as T_1 starts to conduct. This decrease in V_{C_1} decreases the forward bias of transistor T_2. During this saturation, both transistors conduct in the active region. This switching action continues until transistor T_2 is cut off. Hence, at t_{+1} time, the output voltage is V_{cc}, and transistor T_1 conducts at saturation. If the input signal has a smaller amplitude than that required to cause the transistor to operate at saturation, transistor T_1 may operate in the active region.

Transistor T_2 remains at cutoff until the input voltage drops to a predetermined value less than UTP; at this time, transistor T_2 reverses its stable state and again conducts at saturation. The amplitude of input voltage, required for transistor T_2 to conduct at saturation again, is referred to as the lower trigger potential (LTP).

The circuit will not switch back to its original stable state until the amplified voltage, appearing across R_2, is equal to the voltage across R_E when transistor T_1 conducts, (V_{E_1}).

The amplitude of the input voltage, at which this switching action occurs, is less than UTP because the voltage across R_2 is a fraction of the amplified input voltage which appears from V_{C_1} to ground.

The collector voltage of transistor T_1 (V_{C_1}) decreases from V_{cc} as T_1 starts to conduct. This decrease in V_{C_1} decreases the forward bias of transistor T_2. During this transition, both transistors conduct in the action region. This switching action will continue until transistor T_2 is cut off. Hence, at t_{+1} time, the output voltage is V_{cc}, and transistor T_1 may be at saturation or may be operating in the active region.

Transistor T_2 remains at cutoff until the input voltage drops to a predetermined value less than UTP; at this time, transistor T_2 reverses its stable state and again conducts at saturation. The input voltage, required for transistor T_2 to conduct at saturation again, is referred to as the lower trigger potential (LTP). When the input voltage decreases and again returns to the upper trigger potential, the circuit will not switch to its original stable state if the loop gain of the circuit is not greater than unity.

11.2 DETERMINATION OF UTP

To determine the upper trigger potential for the circuit, shown in Fig. 11-1, assume that no input voltage is applied; hence, transistor T_1 is OFF and transistor T_2 is ON. The base biasing network of transistor T_2 is replaced by a simplified equivalent circuit derived by the use of Thevenin's theorem. See

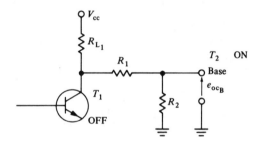

Figure 11-2

Fig. 11-2. Assume that the transistor junction voltages are negligible. Hence,

$$e_{oc_B} = \frac{R_2 V_{cc}}{R_1 + R_2 + R_{L_1}}$$

$$Z_{o_B} = \frac{R_2(R_1 + R_{L_1})}{R_1 + R_2 + R_{L_1}}$$

The base biasing circuit equivalent to that of transistor T_2 is substituted for R_1, R_2, and R_{L_1}, as illustrated in Fig. 11-3. Kirchhoff's voltage loop equation is written from the schematic shown in Fig. 11-3.

$$V_{E_2} + e_{zo_B} = e_{oc_B}$$
$$V_{E_2} = e_{oc_B} - I_{B_2} Z_{o_B}$$

Substituting for e_{oc_B} and Z_{o_B},

$$V_{E_2} = \frac{R_2 V_{cc}}{R_1 + R_2 + R_{L_1}} - I_{B_2} \left[\frac{R_2(R_1 + R_{L_1})}{R_1 + R_2 + R_{L_1}} \right]$$

But $V_{E_2} = I_{E_2} R_E$

and $I_{E_2} = (h_{FE} + 1) I_{B_2}$

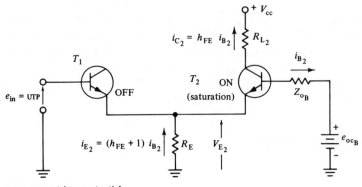

UTP = upper trigger potential

Figure 11-3

Hence $\qquad V_{E_2} = (h_{FE} + 1)I_{B_2}R_E$

$$I_{B_2} = \frac{V_{E_2}}{R_E(h_{FE} + 1)}$$

$$V_{E_2} = \frac{R_2 V_{cc} - I_{B_2}R_2(R_1 + R_{L_1})}{R_1 + R_2 + R_{L_1}}$$

$$V_{E_2} = \frac{R_2 V_{cc} - \dfrac{V_{E_2}R_2(R_1 + R_{L_1})}{R_E(h_{FE} + 1)}}{R_1 + R_2 + R_{L_1}}$$

$$V_{E_2} = \frac{R_2 R_E(h_{FE} + 1)V_{cc}}{R_E(R_1 + R_2 + R_{L_1})(h_{FE} + 1) + R_2(R_1 + R_{L_1})}$$

$$V_{E_2} = \frac{V_{cc}}{\dfrac{R_1 + R_2 + R_{L_1}}{R_2} + \dfrac{R_1 + R_{L_1}}{R_E(h_{FE} + 1)}}$$

Since the junction voltages of the transistor are neglected, transistor T_1 starts to conduct when the input voltage equals V_{E_2}. The voltage at which transistor T_1 starts to conduct is the *UTP*. Hence,

$$UTP = V_{E_2}$$

$$UTP = \frac{V_{cc}}{\dfrac{R_1 + R_2 + R_{L_1}}{R_2} + \dfrac{R_1 + R_{L_1}}{R_E(h_{FE} + 1)}} \qquad \text{(neglecting junction voltages)}$$

$$\hspace{10cm} [24]$$

11.3   DETERMINATION OF *LTP*

The potential which causes transistor T_1 to be cut off is called the lower trigger potential (*LTP*). To determine this lower trigger potential, for the circuit shown in Fig. 11-4, the collector circuit of transistor T_1 is replaced by a

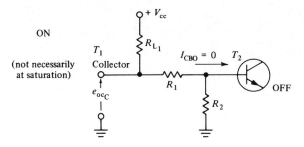

Figure 11-4

simplified equivalent circuit derived by the use of Thevenin's theorem. The transistor junction voltages are negligible. Hence,

$$e_{o c_C} = \frac{(R_1 + R_2) V_{cc}}{R_1 + R_2 + R_{L_1}}$$

$$Z_{o_C} = \frac{R_{L_1}(R_1 + R_2)}{R_1 + R_2 + R_{L_1}}$$

The original collector circuit for transistor T_1 is replaced by the simplified

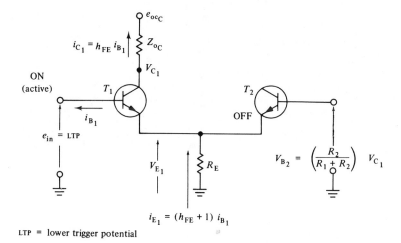

Figure 11-5

equivalent circuit, shown in Fig. 11-5. Kirchhoff's voltage loop equation is written from the schematic, shown in Fig. 11-5.

$$e_{o c_C} = V_{C_1} + E_{z o_C}$$
$$e_{o c_C} = V_{C_1} + h_{FE} I_{B_1} Z_{o_C}$$

But
$$I_{E_1} = \frac{V_{E_1}}{R_E}$$

$$I_{E_1} = (h_{FE} + 1)I_{B_1}$$

$$I_{B_1} = \frac{V_{E_1}}{R_E(h_{FE} + 1)}$$

$$e_{oc_C} = V_{C_1} + \frac{h_{FE}V_{E_1}Z_{oc}}{R_E(h_{FE} + 1)}$$

but
$$e_{oc_C} = \frac{(R_1 + R_2)V_{cc}}{R_1 + R_2 + R_{L_1}}$$

and
$$Z_{oc} = \frac{R_{L_1}(R_1 + R_2)}{R_1 + R_2 + R_{L_1}}$$

Hence:

$$\frac{(R_1 + R_2)V_{cc}}{R_1 + R_2 + R_{L_1}} = \frac{\dfrac{h_{FE}V_{E_1}R_{L_1}(R_1 + R_2)}{R_1 + R_2 + R_{L_1}}}{R_E(h_{FE} + 1)} + V_{C_1}$$

But
$$V_{E_1} = \left(\frac{R_2}{R_1 + R_2}\right)V_{C_1}$$

$$V_{C_1} = \frac{V_{E_1}(R_1 + R_2)}{R_2}$$

Substituting the value for V_{C_1},

$$\frac{(R_1 + R_2)V_{cc}}{R_1 + R_2 + R_{L_1}} = \frac{\dfrac{h_{FE}V_{E_1}R_{L_1}(R_1 + R_2)}{R_1 + R_2 + R_{L_1}}}{R_E(h_{FE} + 1)} + \frac{V_{E_1}(R_1 + R_2)}{R_2}$$

$$V_{E_1} = \frac{R_E R_2(h_{FE} + 1)V_{cc}}{R_E(h_{FE} + 1)(R_1 + R_2 + R_{L_1}) + h_{FE}R_2 R_{L_1}}$$

$$V_{E_1} = \frac{V_{cc}}{\dfrac{R_1 + R_2 + R_{L_1}}{R_2} + \dfrac{h_{FE}R_{L_1}}{R_E(h_{FE} + 1)}}$$

Since the junction voltages of the transistor are neglected, transistor T_1 is cut off when the input voltage drops to V_{E_1}. The voltage at which transistor T_1 is cut off is the *LTP*. Hence,

$$LTP = V_{E_1}$$

$$LTP = \frac{V_{cc}}{\dfrac{R_1 + R_2 + R_{L_1}}{R_2} + \dfrac{h_{FE}R_{L_1}}{R_E(h_{FE} + 1)}} \qquad \text{(neglecting junction voltages)}$$

$$[25]$$

NOTE: The value of h_{FE}, selected for use in Eqs. 24 and 25, must be the actual value of current gain in the specific Schmitt trigger circuit in question.

11.4 HYSTERESIS

The term *hysteresis,* as applied to the Schmitt trigger circuit, implies a retardation of the initiation of the switching action. This time lag, caused by the regenerative feedback, which is produced by the voltage drop across the common-emitter resistor, is evidenced by a drop in the input voltage to a value (lower trigger potential) below that at which the preceding switching action occurred (upper trigger potential). Hence, switching action is delayed until the input voltage reaches its lower trigger potential.

Voltage loop gain determines the time at which the input voltage reaches its lower trigger potential. The loop gain must be greater than unity or the circuit cannot be triggered nor, therefore, operate properly. The loop gain may be adjusted by the use of any of several methods. One method consists of the adjustment of the ratio of the value of R_1 to R_{L_1}. The relationship between the limiting values of these two resistances, which is necessary to establish unity loop gain (zero hysteresis), exists when $UTP = LTP$. Hence, if the expression for UTP (Eq. 24) is set equal to the expression for LTP (Eq. 25), the relationship between R_1 and R_{L_1} may be established.

$$UTP = \frac{V_{cc}}{\dfrac{R_1 + R_2 + R_{L_1}}{R_2} + \dfrac{R_1 + R_{L_1}}{R_E(h_{FE} + 1)}} = \frac{V_{cc}}{\dfrac{R_1 + R_2 + R_{L_1}}{R_2} + \dfrac{h_{FE}R_{L_1}}{R_E(h_{FE} + 1)}} = LTP$$

$$R_{L_1}h_{FE} = R_1 + R_{L_1}$$
$$(h_{FE} - 1)R_{L_1} = R_1$$
$$R_{L_1} = \frac{R_1}{h_{FE} - 1} = \text{unity loop gain}$$

Hence, for the circuit to trigger, the loop gain must be greater than unity or

$$R_{L_1} > \frac{R_1}{h_{FE} - 1}$$

Refer to Fig. 11-1. A resistor R_B is frequently employed in the base lead of transistor T_1, to limit the base-current flow when transistor T_1 is conducting. This resistor has no effect on the upper trigger potential because no current will flow through it until transistor T_1 starts to conduct. The lower trigger potential increases with an increase in R_B. The value of R_B will have no appreciable effect on the triggering potentials as long as $R_B < h_{FE}R_E$. Thus the value of R_B may be adjusted to produce unity gain, or zero hysteresis.

11.5 SAMPLE DESIGN PROBLEM

From the circuit, shown in Fig. 11-6, design a Schmitt trigger circuit with the following characteristics:

$$V_{cc} = 15 \text{ V}$$
$$UTP = 5 \text{ V}$$
$$I_{C_2} = 5 \text{ mA}$$
$$LTP = 3 \text{ V}$$

Given:

2 silicon NPN transistors, $h_{FEmin} = 20$
1 15 V dc source

Assume:

Ideal transistors
All junction voltages to be zero

$$I_2 = 10\% \text{ of } I_{C_2}$$
$$I_{CBO} = 0$$

Determine:

R_1, R_2, R_E, R_{L_1}, R_{L_2}, and R_B

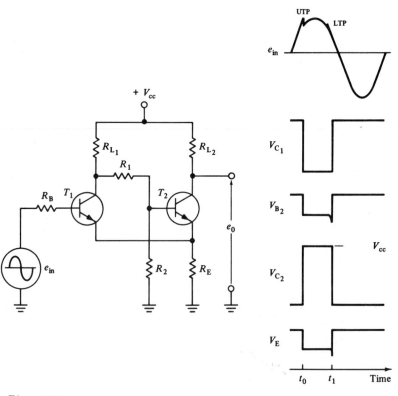

Figure 11-6

Solution:

1. By the use of Ohm's law, determine the value of $(R_{L_2} + R_E)$ when T_2 is ON at saturation and no input voltage is applied.
2. When the junction voltages are neglected, $V_{E_2} = UTP$. In the voltage divider, which consists of R_E and R_{L_2}, V_{E_2} may be expressed as a fraction of V_{cc}. Hence, R_E may be determined from this relationship.
3. The value of resistance for R_{L_2} may be determined from the results of Steps 1 and 2.
4. Since the junction voltages are neglected, $V_{E_1} = LTP$. In the voltage divider, which consists of R_E and R_{L_1}, V_{E_1} may be expressed as a fraction of V_{cc}. Hence, the value for R_{L_1} may be determined from this relationship.
5. Determine the value of resistance for R_2. Assume that the current which flows through R_2 is 10 percent of the value of the collector current which flows through T_2 when that transistor conducts at saturation. For transistor T_2 to make the transition from cutoff, at t_{-1} time, to conduction, at t_{+1} time, the emitter-base junction voltage must be zero. Since $V_{E_1} = LTP$, and $E_{R_2} = V_{E_1} = LTP$, Ohm's law may be used to determine the value of R_2.
6. Determine the value of resistance for R_1. Write the current-node equation for the circuit shown in Fig. 11-7, and solve for the value of R_1.
7. Because T_1 need not conduct at saturation, R_B is selected with a value of resistance which is less than $h_{FE}R_E$ and will, therefore, limit the base current of T_1.

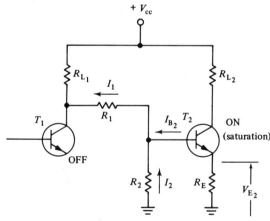

Figure 11-7

Assume:

$$I_{C_2} \approx I_{E_2}$$

$$(R_{L_2} + R_E) = \frac{V_{cc}}{I_{C_2}} = \frac{15}{5 \text{ mA}} = 3 \text{ k}\Omega$$

Approximations

$$UTP = V_{E_2} = 5 \text{ V}$$

$$V_{E_2} = \frac{R_E V_{cc}}{(R_{L_2} + R_E)}$$

$$R_E = V_{E_2} \frac{(R_{L_2} + R_E)}{V_{cc}} = \frac{5(3 \text{ k}\Omega)}{15}$$

$$R_E = 1 \text{ k}\Omega$$

$$R_{L_2} = (R_{L_2} + R_E) - R_E = 3 \text{ k}\Omega - 1 \text{ k}\Omega$$

$$R_{L_2} = 2 \text{ k}\Omega$$

$$LTP = V_{E_1} = 3 \text{ V}$$

$$V_{E_1} = \frac{R_E V_{cc}}{R_E + R_{L_1}}$$

$$R_{L_1} = \frac{R_E V_{cc}}{V_{E_1}} - R_E$$

$$R_{L_1} = \frac{(1 \text{ k}\Omega)(15)}{3} - 1 \cdot \text{k}\Omega$$

$$R_{L_1} = 5 \text{ k}\Omega - 1 \text{ k}\Omega = 4 \text{ k}\Omega \rightarrow \text{use 3.9 k}\Omega \text{ standard} \\ \text{color code value}$$

Assume:

$$I_2 = 10\% \ I_{C_2} = (10\%)(5 \text{ mA}) = 0.5 \text{ mA}$$

$$R_2 = \frac{E_{R_2}}{I_2} = \frac{V_{E_1}}{I_2} = \frac{LTP}{I_2} = \frac{3}{0.5 \text{ mA}}$$

$$R_2 = 6 \text{ k}\Omega \rightarrow \text{use 5.6 k}\Omega \\ \text{standard color code value}$$

Refer to Fig. 11-7.

$$I_1 = I_2 + I_{B_2}$$

$$\frac{E_{R_{L_1}} + E_{R_1}}{R_1 + R_{L_1}} = \frac{E_{R_2}}{R_2} + I_{B_2}$$

$$\frac{V_{cc} - UTP}{R_1 + R_{L_1}} = \frac{UTP}{R_2} + I_{B_2}$$

$$I_{B_2} = \frac{I_{C_2}}{h_{FE_{min}}} = \frac{5 \text{ mA}}{20} = 0.25 \text{ mA}$$

$$\frac{15 - 5}{3.9 \text{ k}\Omega + R_1} = \frac{5}{5.6 \text{ k}\Omega} + 0.25 \text{ mA}$$

$$\frac{10}{3.9 \text{ k}\Omega + R_1} = 0.892 \text{ mA} + 0.25 \text{ mA}$$

$$\frac{10}{3.9 \text{ k}\Omega + R_1} = 1.142 \text{ mA}$$

$$10 = 4.45 + 1.142 \text{ mA } R_1$$

$$R_1 = \frac{5.55}{1.142 \text{ mA}} = 4.87 \text{ k}\Omega \rightarrow \text{use } 4.7 \text{ k}\Omega \text{ standard color code value}$$

$$R_B < h_{FE}R_E$$

$$R_B = \frac{h_{FE}R_E}{10} = \frac{(20)(1 \text{ k}\Omega)}{10} = 2 \text{ k}\Omega$$

11.6 SAMPLE CIRCUIT ANALYSIS PROBLEM

Assume ideal transistors. Prove that the circuit shown in Fig. 11-8 will, or will not, function properly as a Schmitt trigger circuit. Determine the upper

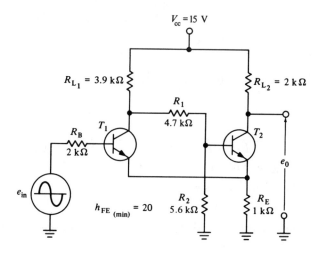

Figure 11-8

trigger potential, the lower trigger potential, and the variation of output-voltage levels.

1. Determine V_{E_2}, a fraction of V_{cc}, when T_2 is ON. Because the junction voltages are zero, $V_{E_2} = UTP$.
2. The actual base current when transistor T_2 is ON may be determined from the difference between I_1 and I_2, as illustrated in Fig. 11-7.
3. The actual collector current when transistor T_2 is ON may be determined by the use of Ohm's law.
4. The actual dc current gain, required for transistor T_2 to conduct at saturation, is equal to the ratio of the actual collector current to the actual base current. To ensure that T_2 will conduct at saturation, this current gain must be less than the $h_{FE_{min}}$ of the transistor.

5. Determine V_{E_1}. Neglect junction voltages, and express V_{E_1} as a fraction of V_{cc}.

$$V_{E_2} = \left(\frac{R_E}{R_E + R_{L_2}}\right) V_{cc} = \frac{(1 \text{ k}\Omega)(15)}{1 \text{ k}\Omega + 2 \text{ k}\Omega} = \frac{15}{3 \text{ k}\Omega}$$

$$UTP = V_{E_2} = 5 \text{ V}$$

$$I_{B_2} = \frac{V_{cc} - V_{B_2}}{R_1 + R_{L_1}} - \frac{V_{B_2}}{R_2}$$

$$V_{B_2} = UTP$$

$$I_{B_2} = \frac{15 - 5}{4.7 \text{ k}\Omega + 3.9 \text{ k}\Omega} - \frac{5}{5.6 \text{ k}\Omega} = \frac{10}{8.6 \text{ k}\Omega} - \frac{5}{5.6 \text{ k}\Omega}$$

$$I_{B_2} = 1.16 \text{ mA} - 0.89 \text{ mA} = 0.27 \text{ mA}$$

$$I_{C_2} = \frac{V_{cc}}{R_E + R_{L_2}} = \frac{15}{1 \text{ k}\Omega + 2 \text{ k}\Omega} = \frac{15}{3 \text{ k}\Omega} = 5 \text{ mA}$$

$$h_{FE} = \frac{I_{C_2}}{I_{B_2}} = \frac{5 \text{ mA}}{0.27 \text{ mA}} = 18.5 < h_{FE_{min}} = 20$$

Hence, T_2 will conduct at the required saturation.

Assume:

$$I_{E_1} \approx I_{C_1}$$

$$I_{E_1} = \frac{V_{cc}}{R_E + R_{L_1}} = \frac{15}{1 \text{ k}\Omega + 3.9 \text{ k}\Omega} = \frac{15}{4.9 \text{ k}\Omega}$$

$$I_{E_1} = 3.05 \text{ mA}$$

$$LTP = V_{E_1} = I_{E_1}R_E = (3.05 \text{ mA})(1 \text{ k}\Omega)$$

$$LTP = 3.05 \text{ V}$$

This simplified method of design and analysis is approximate, but it is adequate for most applications.

SCHMITT TRIGGER CIRCUIT

OBJECT:
1. To design a Schmitt trigger circuit
2. To analyze the operation of a Schmitt trigger circuit
3. To examine the requirements necessary to establish the transition voltages and stable states of a Schmitt trigger circuit

MATERIALS:
2 Switching transistors (example—2N3646 silicon NPN)
1 Manufacturer's specification sheet for type of transistor used (2N3646, see Appendix B)
1 Oscilloscope, dc time-base type; frequency response dc to 450 kHz; vertical sensitivity, 100 mV/cm
1 Transistor power supply (0 to 30 V and 0 to 250 mA)
1 Sine wave generator (20 Hz to 200 kHz; $Z_o = 600 \, \Omega$)
7 Resistor substitution boxes (10 Ω to 10 MΩ, 1 W)
2 Capacitor substitution boxes (0.0001 to 0.22 μF at 450 V)

PROCEDURE:
1. Design and analyze a Schmitt trigger circuit from the following specifications:

$$V_{cc} = 10 \text{ V} \qquad UTP = 4 \text{ V}$$
$$I_{C_2} = 10 \text{ mA} \qquad LTP = 3 \text{ V}$$

AVAILABLE
2 NPN silicon transistors (example—2N3646)
1 dc power supply ($V_{cc} = 10$ V)

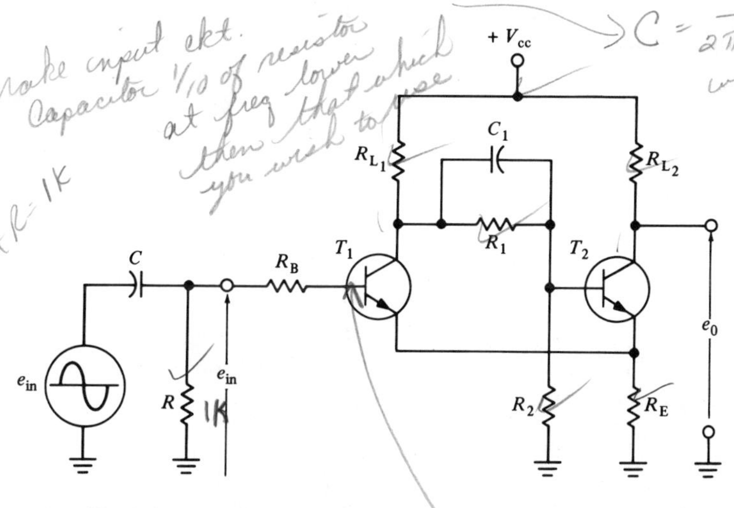

Figure 11-1X Schmitt Trigger Circuit

handwritten annotations:
make input ckt. Capacitor 1/10 of resistor at freq lower which then that which you wish to use.

$C = \dfrac{1}{2\pi f \, X_C}$ where $X_C = 100 \, \Omega$

makes voltage in phase

Input R = 1K

look at base for glish

165

2. Design the Schmitt trigger circuit, shown in Fig. 11-1X, for the circuit conditions specified. Include all calculations. Explain why any assumptions which have been made are valid.

3. Draw a schematic of the circuit designed in Step 1, and label it with the practical, standard, color coded values.

4. From the schematic drawn in Step 3, analyze the circuit designed in Step 2 to prove that it will operate satisfactorily for the required specifications when standard value resistors are used. Include all calculations.

5. Connect the circuit designed. Check the operation of the circuit. Record all necessary data. Use an input sine wave of voltage, the frequency of which is 5000 Hz and the amplitude of which is 10 V peak-to-peak.

6. Compare the measured values with the design values, and explain any discrepancies. Note that the design was based on the assumption that the junction voltages were zero.

QUESTIONS AND EXERCISES

1. Explain the operation of the Schmitt trigger circuit. Refer to Fig. 11-6. Use the voltage waveforms indicated to aid in your explanation.

2. What is hysteresis, and what is its significance in this circuit?

3. What affects the pulse width of the output voltage of a Schmitt trigger circuit? Explain.

4. Describe two applications of a Schmitt trigger circuit. Explain.

5. Explain how a change in the amplitude of the input voltage effects the output-voltage waveform.

6. Explain, in detail, why the value of the lower trigger potential is lower than that of the upper trigger potential.

7. What are the voltage loop gain requirements for a Schmitt trigger circuit?

8. For what application would a Schmitt trigger circuit, with zero hysteresis, be used? Explain.

9. From the schematic shown in Fig. 11-1 design a Schmitt trigger circuit from the following specifications:

$$V_{cc} = 10 \text{ V}, \qquad I_{C_2} = 7 \text{ mA} \qquad UTP = 3 \text{ V}, \qquad LTP = 2 \text{ V}, \qquad h_{FE_{min}} = 20$$

Assume all transistor junction voltages to be zero.

10. Determine UTP, LTP, and the amplitude of the output voltage of the circuit shown in Fig. 11-1 when

$$R_{L_1} = 1.8 \text{ k}\Omega, \qquad R_{L2} = 1 \text{ k}\Omega, \qquad R_1 = 3.3 \text{ k}\Omega, \qquad R_2 = 2.7 \text{ k}\Omega$$
$$R_B = 2.5 \text{ k}\Omega, \qquad R_E = 470 \ \Omega, \qquad V_{cc} = 10 \text{ V}$$

What is the minimum value of h_{FE} required for the transistors?

11. From the schematic shown in Fig. 11-1, design a Schmitt trigger circuit from the following specifications:

$$V_{cc} = 20 \text{ V}, \qquad UTP = 5 \text{ V}, \qquad LTP = 2 \text{ V}$$
$$I_{C_2} = 10 \text{ mA} \qquad h_{FE_{min}} = 50$$

Assume all transistor junction voltages to be zero.

12. Determine *UTP, LTP*, and the amplitude of the output voltage of the circuit shown in Fig. 11-1 when

$$R_{L_1} = 2 \text{ k}\Omega, \qquad R_{L2} = 2 \text{ k}\Omega, \qquad R_1 = 14 \text{ k}\Omega$$
$$R_2 = 10 \text{ k}\Omega, \qquad R_E = 1 \text{ k}\Omega, \qquad R_B = 4 \text{ k}\Omega$$

Assume:

$$V_{cc} = 12 \text{ V} \qquad h_{FE_{min}} = 40$$

chapter 12

MONOSTABLE MULTIVIBRATOR

The monostable multivibrator is an electronic circuit which has a stable state and a semistable state. The circuit remains in its stable state until an externally applied triggering signal switches it to a semistable state. The RC time constant within the circuit determines the interval during which the circuit remains in this semistable state. Upon completion of the time constant interval, the circuit reverts to its original stable state and remains in this state until another externally initiated triggering pulse is applied. This circuit is frequently called a one-shot multivibrator.

The monostable circuit is made up of two inverter circuits. The output of the first inverter circuit is the input to the second, and the output of the second circuit is the input to the first. The output of the first inverter circuit is RC coupled to the input of the second. The output of the second inverter circuit is resistively coupled to the first. Refer to Fig. 12-1.

The input-voltage waveshape required to trigger a monostable multivibrator is usually half of a differentiated waveform. The resultant output-voltage waveshape of the multivibrator is rectangular. See Fig. 12-1. This output-voltage pulse is frequently used to control other pulse circuits and occasionally, therefore, is referred to as a gating circuit.

The monostable multivibrator is also called a delay circuit because the negative-going output of the circuit (NPN) occurs only after the completed interval of the semistable state.

12.1 OPERATION

To follow the description of the operation of a monostable multivibrator, refer to Fig. 12-1. When no trigger pulse is applied, at t_{-1} time, transistor T_1 is OFF and transistor T_2 is ON. During this stable state, capacitor C charges to V_{cc} and is of the polarity indicated.

168

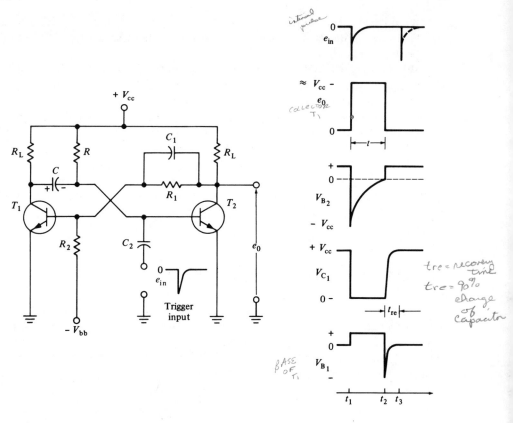

Figure 12-1 Monostable Multivibrator

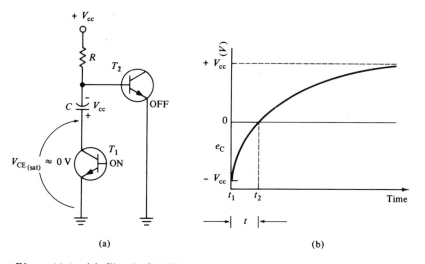

(a)   (b)

Figure 12-2 (a) Circuit Condition at t_{+1} Time; (b) Capacitor C Charge from t_{+1} to t_{-2} Time

At t_{-1} time, the voltage divider R_1 and R_2 in conjunction with the dc supply V_{bb} provides the necessary reverse bias to hold transistor T_1 at cutoff during the stable state.

When a negative-trigger pulse is applied to the base of the ON transistor at t_{+1} time, transistor T_2 is cut off. The collector voltage of T_2 rises toward V_{cc}; thus it provides the forward bias necessary to turn T_1 ON. Transistor T_1 conducts at saturation and effectively grounds the positively charged plate of capacitor C. Refer to Fig. 12-2(a). At t_{+1} time, transistor T_2 is reverse biased by an amplitude of voltage equal to V_{cc}. Capacitor C charges from $-V_{cc}$ toward $+V_{cc}$ through resistor R and transistor T_1. See Fig. 12-2(b). The voltage to which capacitor C is charged is the reverse-bias voltage for transistor T_2. Transistor T_2 is held at cutoff until capacitor C charges to 0 V. The interval during which transistor T_2 is held at cutoff determines the duration of the output-voltage pulse (t).

12.2   DURATION OF OUTPUT PULSE

Refer to Fig. 12-2. The time required for capacitor C to charge to zero volts, through resistor R, may be determined from the capacitor charge Eq. 8. To derive an equation for the output-pulse duration of a monostable multivibrator, substitute each of the general circuit parameters of the multivibrator for its corresponding symbols in Eq. 8.

$$e_C = E - (E \pm E_{co})\epsilon^{-t/RC} \qquad [8]$$
$$0 = V_{cc} - (V_{cc} + V_{cc})\epsilon^{-t/RC}$$
$$0 = V_{cc} - (2V_{cc})\epsilon^{-t/RC}$$
$$2V_{cc} = V_{cc}\epsilon^{+t/RC}$$
$$\epsilon^{t/RC} = 2$$
$$t = \frac{RC \log_{10} 2}{\log_{10} \epsilon}$$
$$t = \frac{RC(0.301)}{0.434}$$
$$t = 0.69RC$$

$$e_C = E - (E \pm E_{co})\epsilon^{-t/RC} \qquad [8]$$
$$0.9V_{cc} = V_{cc} - (V_{cc} + 0)\epsilon^{-t_{re}/R_LC}$$
$$0.9V_{cc} = V_{cc} - \frac{V_{cc}}{\epsilon^{+t_{re}/R_LC}}$$
$$\frac{V_{cc}}{\epsilon^{+t_{re}/R_LC}} = 0.1V_{cc}$$
$$\epsilon^{+t_{re}/R_LC} = 10$$
$$t_{re} = \frac{R_LC \log_{10} 10}{\log_{10} \epsilon} = \frac{R_LC(1.000)}{0.434}$$
$$t_{re} = 2.3R_LC$$

Therefore, the minimum triggering period for a monostable multivibrator (symbolized T) is

$$T = t + t_{re}$$
$$T = 0.69RC + 2.3R_LC$$
$$T = (0.69R + 2.3R_L)C$$

12.3 TRIGGERING

To turn the OFF transistor ON, in a monostable multivibrator, the trigger pulse is applied to the base of the OFF transistor T_1 and should, in this case, be positive (NPN).

At t_{+2} time, transistor T_2 is ON, and thus it effectively grounds the collector of transistor T_2. Resistor R_1 is then placed in parallel with the emitter-base junction of transistor T_1. The voltage drop across resistor R_1 is of a polarity to reverse bias transistor T_1. The amount of this reverse bias is determined by the voltage divider, which consists of V_{bb}, resistor R_1, and resistor R_2. Hence, transistor T_1 is held at cutoff by the voltage across resistor R_1. The collector voltage of transistor T_1 (V_{C_1}) will not rise to V_{cc} until capacitor C charges to V_{cc}. The charging current for capacitor C is the base current of the ON transistor T_2.

12.4 INTERVAL BETWEEN TRIGGER PULSES

Since the duration of the output pulse of the circuit is predicated upon the fact that capacitor C charges to V_{cc}, the minimum interval between trigger pulses is determined by the duration of the output pulse and by the time required to charge capacitor C to V_{cc}. Hence, the minimum interval between trigger pulses is ($t_1 - t_3$) time. Refer to Fig. 12-1.

The time required for capacitor C to charge to V_{cc} is equal to the interval between t_2 and t_3 time. This interval is called *recovery time* and is symbolized t_{re}. Practically, t_{re} is the time required for the capacitor C to charge to 90 percent of V_{cc}. The charging current for capacitor C flows through R_L and through the emitter-base resistance of the ON transistor T_2. The emitter-base resistance of the ON transistor is small; therefore, it may be neglected for practical purposes. Hence, the recovery time (t_{re}) may be expressed in terms of a time constant. It is common practice, however, to turn the ON transistor OFF; therefore, the trigger pulse is applied as indicated in Fig. 12-1.

The reverse-bias voltage applied across the emitter-base junction of transistor T_2, at t_{+1} time, is V_{cc}. If this reverse-bias voltage exceeds the emitter-base breakdown voltage of the transistor (V_{EBO}), the transistor will be destroyed. To prevent this, a diode is placed in series with transistor T_2 from

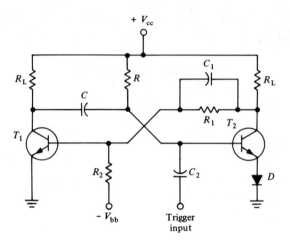

Figure 12-3

emitter to ground, as illustrated in Fig. 12-3. When capacitor C, which has charged to V_{cc}, is applied across transistor T_2 from base to ground, the voltage across the emitter-base junction is only a fraction of V_{cc}. The emitter-base junction diode, of transistor T_2, and diode D form a voltage divider for the applied reverse bias of V_{cc}. This voltage divider action prevents V_{cc} from exceeding the emitter-base-junction voltage (V_{EBO}).

12.5   SAMPLE DESIGN PROBLEM

From the circuit shown in Fig. 12-3, design a monostable multivibrator with the following characteristics:

$$e_o = 12 \text{ V peak}$$
$$I_C = 20 \text{ mA}$$
$$\text{Output-pulse duration} = 200 \ \mu\text{sec}$$

Given:

2 silicon NPN transistors, $h_{FE_{min}} = 20$
$$I_{CBO} \approx 0$$
$$V_{EBO} = 5 \text{ V}$$

1 silicon diode $PIV = 60$ V
1 12 V dc source
1 6 V dc source

Assume:

Typical values for junction voltages
$$V_{BE_{off}} = -0.5 \text{ V}$$

Determine:

$$R, R_1, R_2, \text{ and } R_L$$

Solution:

1. From the ON transistor circuit, determine R_L by use of Ohm's law.
2. Determine the minimum value of base current required for saturation.
3. Determine the value of R for the base current required in Step 2.
4. Refer to the OFF circuit shown in Fig. 12-4. Express $V_{BE_{off}}$ as a fraction of the voltage source V_{bb}. Write the OFF circuit equation; express R_1 as a fraction of R_2.
5. Refer to the ON circuit shown in Fig. 12-5. Write the ON circuit node equation in terms of R_1 and R_2.
6. Simultaneously solve the ON and OFF circuit equations for R_1 and R_2.
7. Refer to the capacitor-charge Eq. 8. Determine the value of capacitance necessary for C to charge from $-V_{cc}$ to 0 V in 200 μsec. The capacitor charges through resistor R and through voltage source V_{cc}. The capacitor has an initial charge of $-V_{cc}$ volts, and it is in series-aiding the applied voltage source V_{cc}.

$$R_L = \frac{V_{cc}}{I_C} = \frac{12}{20 \text{ mA}} = 600 \text{ } \Omega \qquad \text{color code, 620 } \Omega$$

$$I_B = \frac{I_C}{h_{FE_{min}}} = \frac{20 \text{ mA}}{20} = 1 \text{ mA}$$

$$R = \frac{V_{cc} - V_{BE_{sat}}}{I_B} = \frac{12 - 0.7}{1 \text{ mA}} = \frac{11.3}{1 \text{ mA}}$$

$$R = 11.3 \text{ k}\Omega \qquad \text{color code, 10 k}\Omega$$

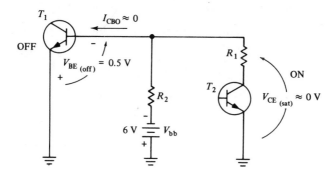

Figure 12-4   OFF Circuit

$$V_{BE_{off}} = \left(\frac{R_1}{R_1 + R_2}\right) V_{bb}$$

$$-0.5 = \left(\frac{R_1}{R_1 + R_2}\right)(-6)$$

$$R_1 + R_2 = 12R_1$$

$$R_2 = 11R_1 \qquad \text{OFF Equation}$$

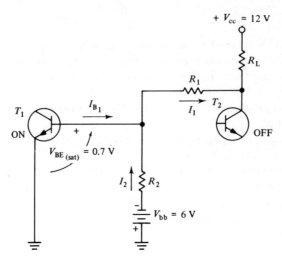

Figure 12-5   ON Circuit

$$I_1 = I_2 + I_B$$

$$\frac{V_{cc} - V_{BE_{sat}}}{R_{L_2} + R_1} = \frac{V_{BE_{sat}} - V_{bb}}{R_2} + I_B$$

$$\frac{12 - (+0.7)}{0.62 \text{ k} + R_1} = \frac{+0.7 - (-6)}{R_2} + 1 \text{ mA}$$

$$\frac{11.3}{0.62 \text{ k} + R_1} = \frac{6.7}{R_2} + 1 \text{ mA} \qquad \text{ON Equation}$$

Substitute the value of R_2, obtained from the OFF equation, for R_2, in the ON equation.

$$\frac{11.3}{0.62 \text{ k} + R_1} = \frac{6.7}{11R_1} + 1 \text{ mA}$$

$$0.011R_1^2 - 110R_1 + 4.5 \text{ k} = 0$$

Solve for value of R_1 by use of the quadratic equation

$$R_1 = \frac{-b \pm \sqrt{b^2 - 4ac}}{2a}$$

$$R_1 = \frac{+110 \pm \sqrt{(110)^2 - (4)(11 \text{ mA})(4.15 \text{ k})}}{2(0.011)}$$

$$R_1 = \frac{+110 \pm \sqrt{12{,}100 - 182}}{0.022} = \frac{+110 \pm \sqrt{11{,}900}}{0.022}$$

$$R_1 = \frac{+110 \pm 109}{0.022}$$

$$R_1 = \frac{+110 + 109}{0.022} = \frac{219}{0.022} = 9550 \ \Omega$$

$$R_1 = 10 \text{ k}\Omega \qquad \text{standard color code value}$$
$$R_2 = 11R_1 = 11(10 \text{ k}\Omega)$$
$$R_2 = 110 \text{ k}\Omega \qquad \text{use } 100 \text{ k}\Omega$$
$$\text{standard color code value}$$

The 200 μsec pulse width requirement for the circuit may be filled by the application of Eq. 26.

$$t = 0.693RC \qquad\qquad [26]$$

$$C = \frac{t}{0.693R}$$

$$C = \frac{200 \times 10^{-6}}{(0.693)(10 \times 10^{+3})}$$

$$C = 0.0289 \ \mu\text{F} \qquad \text{use } 0.03 \ \mu\text{F}$$
$$\text{standard value}$$

$$C_1 = \frac{Q}{E} = \frac{Q}{V_{cc}} = \frac{1200p}{12} = 100 \text{ pF} \qquad \text{standard value}$$

$$C_2 \approx \frac{Q}{e_{\text{in}}} = \frac{1200p}{10} = 120 \text{ pF} \qquad \text{use } 100 \text{ pF}$$

12.6   SAMPLE ANALYSIS PROBLEM

Prove that the circuit shown in Fig. 12-6 will, or will not, function properly as a monostable multivibrator. Determine the output-pulse duration. Determine the amplitude of the output voltage.

Solution:

1. Prove that transistor T_2 will conduct at saturation when no trigger pulse is applied.
   (a) Determine the actual base-current flow of transistor T_2 when ON.
   (b) Determine the actual collector-current flow of transistor T_2 when ON.
   (c) Determine the actual value of h_{FE}, from the computed values of I_C and I_B. This value must be less than the $h_{FE_{min}}$ value of the transistors used, to ensure that the transistor will conduct at saturation when ON.
2. Prove that transistor T_1 is at cutoff when transistor T_2 conducts at saturation. To accomplish this, determine the actual emitter-base voltage of transistor T_1. If the base of transistor T_1 is negative with respect to the emitter, transistor T_1 is reverse biased; hence, it is at the required cutoff (NPN).
3. Determine the actual value of h_{FE}, from the computed values of I_{C_1} and I_{B_1}. This value must be less than the $h_{FE_{min}}$ value of the transistor used, to ensure that transistor T_1 will conduct at saturation when ON.
4. The pulse width of the circuit may be determined by the use of Eq. 26.

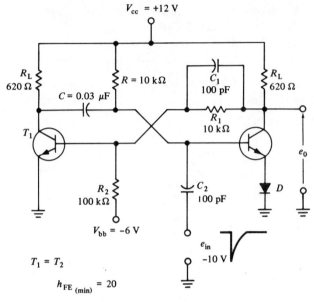

Figure 12-6

Solution:

$$I_{C_2} = \frac{V_{cc}}{R_L} = \frac{12}{620} = 19.3 \text{ mA}$$

$$I_{B_2} = \frac{V_{cc}}{R} = \frac{12}{10 \text{ k}} = 1.2 \text{ mA}$$

$$h_{FE} = \frac{I_{C_2}}{I_{B_2}} = \frac{19.3 \text{ mA}}{1.2 \text{ mA}} = 16$$

$$h_{FE_{min}} = 20$$

Since the actual value of h_{FE} is less than $h_{FE_{min}}$, transistor T_2 conducts at the required saturation when ON. Ascertain that T_1 is OFF when T_2 in ON.

$$V_{BE_{off}} = \left(\frac{R_1}{R_1 + R_2}\right) V_{bb} = \left(\frac{10 \text{ k}}{10 \text{ k} + 120 \text{ k}}\right) (-6)$$
$$V_{BE_{off}} = -0.462 \text{ V} -0.55V$$
$$100k$$

Since $V_{BE_{off}} = -0.462$ V, transistor T_1 is reverse biased; hence, it is at the required cutoff. Ascertain that T_1 is ON when T_2 is OFF.

$$I_{C_1} = \frac{V_{cc} - V_{CE_{sat}}}{R_L} = \frac{12 - 0.3}{620} = \frac{11.7}{620} = 18.9 \text{ mA}$$

Refer to Fig. 12-5.

$$I_1 = I_2 + I_{B_1}$$
$$\frac{V_{cc} - V_{BE_{sat}}}{R_1 + R_L} = \frac{V_{BE_{sat}} - V_{bb}}{R_2} + I_{B_1}$$

$$\frac{12 - 0.7}{10 \text{ k} + 0.62 \text{ k}} = \frac{+0.7 - (-6)}{120 \text{ k}} + I_{B_1}$$

$$\frac{11.3}{10.62 \text{ k}} = \frac{6.7}{120 \text{ k}} + I_{B_1}$$

$$1.06 \text{ mA} = 0.056 \text{ mA} + I_{B_1}$$

$$I_{B_1} = 1 \text{ mA}$$

$$h_{FE} = \frac{I_{C_1}}{I_{B_1}} = \frac{18.9 \text{ mA}}{1 \text{ mA}} = 18.9 \qquad 15.8$$

$$1.2 \text{ mA}$$

Since the actual value of h_{FE} is less than $h_{FE_{min}}$, transistor T_1 conducts at the required saturation when ON. Determine the output-pulse width.

$$t = 0.693 RC$$
$$t = (0.693)(10 \times 10^{+3})(0.03 \times 10^{-6})$$
$$t = 208 \text{ } \mu\text{sec}$$

In most practical circuits, the pulse width must be identical to that designed. This may not be accomplished by the use of standard-value components. The pulse width is dependent upon the value of R and C, in the circuit. Refer to Fig. 12-6. Recall that the value of R is selected to ensure base current enough to produce saturation for transistor T_2 when ON. Hence, for this sample design problem, the value of R is 10 kΩ. Resistor R may be replaced by resistor R_3 and potentiometer R_4, as illustrated in Fig. 12-7. The potentiometer R_4 is used to adjust the output-voltage pulse width.

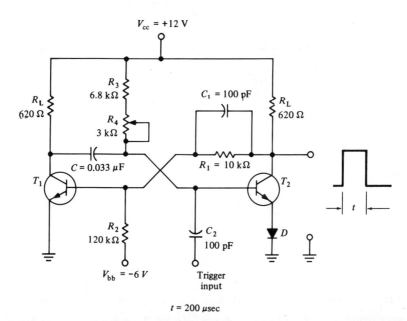

Figure 12-7

The sum of the two resistances R_3 and R_4 may not exceed the 10 kΩ resistance of the original resistor R. When the resistance of R_4 is 0 ohms, sufficient resistance remains, in this branch of the circuit, to limit the base current. As the base current increases, the transistor is driven more deeply into saturation. As this occurs, an increasing number of electrons (NPN) must be removed from the base when the transistor is reverse biased. Hence, the turn-OFF time of the transistor may be increased appreciably. Therefore, the choice of values for R_3 and R_4 must be a compromise.

12.7   DESIGN EXAMPLE

In order to obtain a pulse width of exactly 200 μsec, replace resistor R shown in Fig. 12-6 by resistor R_3 and potentiometer R_4, as illustrated in Fig. 12-7.

$$R = 10 \text{ k}\Omega$$
$$R \geq R_3 + R_4$$
$$10 \text{ k}\Omega \geq 6.8 \text{ k}\Omega + 3 \text{ k}\Omega \qquad \text{(standard values)}$$
$$10 \text{ k}\Omega > 9.8 \text{ k}\Omega$$

Hence, the resistance R may be varied from 6.8 to 9.8 kΩ. Design a time-constant circuit in which R has a mid-range value.

Hence:
$$R = 6.8 \text{ k} + \frac{9.8 \text{ k} - 6.8 \text{ k}}{2}$$

$$R = 6.8 \text{ k} + \frac{3 \text{ k}}{2}$$

$$R = 6.8 \text{ k} + 1.5 \text{ k}$$
$$R = 8.3 \text{ k}\Omega$$
$$t = 0.693RC \qquad\qquad\qquad\qquad\qquad [26]$$
$$C = \frac{t}{0.693R} = \frac{200 \times 10^{-6}}{(0.693)(8.3 \times 10^{+3})}$$
$$C = 0.035 \text{ μF} \qquad \text{use } 0.033 \text{ μF}$$
$$\text{closest standard value}$$

Check limits of pulse width.

$$t = 0.693RC$$
$$t = (0.693)(6.8 \text{ k})(0.033 \text{ μ})$$
$$t = 155 \text{ μsec} \qquad \text{(minimum)}$$
or
$$t = 0.693RC$$
$$t = (0.693)(9.8 \text{ k})(0.033 \text{ μ})$$
$$t = 224 \text{ μsec} \qquad \text{(maximum)}$$

Hence, the use of a 3 kΩ potentiometer for R_4, a 6.8 kΩ resistor for R_3, and a 0.033 μF capacitor for C_3 will provide a satisfactory solution.

MONOSTABLE MULTIVIBRATOR
(COLLECTOR-COUPLED)

OBJECT:
1. To familiarize the student with the characteristics of the collector-coupled monostable multivibrator
2. To design a collector-coupled monostable multivibrator
3. To analyze the operation of a collector-coupled monostable multivibrator

MATERIALS:
2   Switching transistors (example—2N3646 silicon NPN)

1   Manufacturer's specification sheet for type of transistor used (2N3646—see Appendix B)

1   Oscilloscope, dc time-base type; frequency response dc to 450 kHz; vertical sensitivity, 100 mV/cm

1   Square-wave generator (20 Hz to 200 kHz with $Z_o = 600\ \Omega$)

2   Transistor power supplies (0 to 30 V and 0 to 250 mA)

2   Diodes, silicon junction (example—1N914)

6   Resistor substitution boxes (10 Ω to 10 MΩ, 1 W)

4   Capacitor substitution boxes (0.0001 to 0.22 μF at 450 V)

1   Potentiometer, $R = 5$ kΩ, 1 W

PROCEDURE:
1. Design a collector-coupled monostable multivibrator from the following specifications:

$$e_o = 10 \text{ V peak}$$
$$I_{C_{ON}} = 10 \text{ mA}$$
$$t = 250\ \mu\text{sec—output-pulse width}$$

Assume:

$$V_{cc} = 10 \text{ V}$$
$$V_{bb} = -5 \text{ V}$$
$$V_{BE_{off}} = -0.5 \text{ V}$$
$$Q_T = 1200 \text{ pC}$$

2. Design the collector-coupled monostable multivibrator shown in Fig. 12-1X for the conditions specified. Include all calculations. Explain why any assumptions which have been made are valid.
3. Draw a schematic of the circuit designed in Step 2, and label it with the practical, standard, color code values.
4. From the schematic drawn in Step 3, analyze the circuit designed in Step 2; prove that it will operate satisfactorily for the required specifications when standard-value resistors are used. Include all calculations.
5. Connect the circuit designed. Check the operation of the circuit. Record all necessary data.
6. Measure and draw voltage waveforms. Compare to those illustrated in Fig. 12-1; explain any discrepancies. Compare the waveshapes, values, and

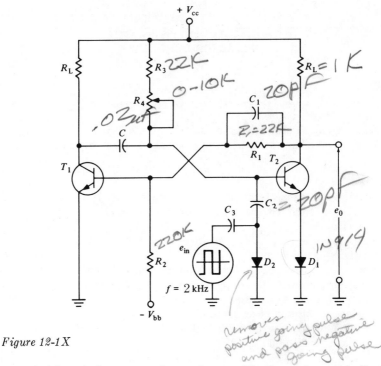

Figure 12-1X

polarities of the measured waveforms with those computed. Explain any discrepancies.

7. Vary the frequency of the triggering pulse. What is the lower triggering frequency limit? What is the upper triggering frequency limit?

QUESTIONS AND EXERCISES

1. By use of the voltage waveforms, illustrated in Fig. 12-1, explain the operation of a collector-coupled monostable multivibrator.

2. Refer to Fig. 12-1. Explain, in detail, why the collector voltage of transistor T_1 rises exponentially to V_{cc} as transistor T_1 switches from ON to OFF.

3. Refer to Fig. 12-1X. Explain why diode D_1 is, or is not, necessary.

4. What factors determine the upper triggering frequency of a monostable multivibrator? Explain.

5. Will a positive-trigger pulse trigger the circuit shown in Fig. 12-1X? Explain.

6. Would the circuit shown in Fig. 12-1X operate satisfactorily if the trigger pulse were applied to the base of transistor T_1? Explain.

7. Explain the compromise necessary to determine the resistance value of the potentiometer to be used in the circuit shown in Fig. 12-1X.

8. Draw a block diagram of the circuits necessary to delay a negative pulse for a given duration. Include the corresponding voltage waveforms in order to show the necessary development of this delayed pulse.

9. Can collector triggering be substituted for base triggering without impairing the operation of the circuit shown in Fig. 12-1X? Draw a schematic of a circuit in which this substitution has been made. Explain its operation.

10. Design a collector-coupled, monostable multivibrator from the following specifications:

$$e_o = 10 \text{ V peak}$$
$$I_{C_{ON}} = 50 \text{ mA}$$
$$t = 50 \text{ } \mu\text{sec—pulse width}$$

Assume:

$$h_{FE_{min}} = 20$$
$$V_{BE_{off}} = 0.5 \text{ V}$$
$$V_{bb} = 10 \text{ V}$$
$$Q_T = 1000 \text{ pC}$$
$$T_1 = T_2 = \text{ideal switches}$$

Show all calculations.

11. Analyze the circuit shown in Fig. 12-1, when

$V_{cc} = +15$ V,	$V_{bb} = -8$ V,	$R_L = 1$ kΩ,	$R_1 = 120$ kΩ
$R_2 = 1.8$ MΩ,	$R = 68$ kΩ,	$C = 0.0015$ μF,	$C_1 = 100$ pF

Assume $V_{BE_{off}} = -0.5$ V and $h_{FE_{min}} = 25$. Determine whether this circuit will operate properly as a monostable multivibrator.

12. Refer to the circuit shown in Fig. 12-3. Design a collector-coupled, monostable multivibrator from the following specifications:

$V_{cc} = +20$V,	$V_{bb} = -10$V,	$I_C = 15$ mA,	$t_p = 25$ μsec

Assume

$$V_{BE_{off}} = -0.5 \text{ V}, \qquad Q_T = 1500 \text{ pC} \qquad h_{FE_{min}} = 10$$

Use only standard values for resistors and capacitors.

13. Refer to the circuit designed in Prob. 12. Determine the value of resistor R_3 and that of potentiometer R_4 (as shown in Fig. 12-1X) necessary to ensure that the duration of the output pulse may be adjusted to exactly 25 μsec.

14. Refer to the circuit designed in Prob. 12. What would be the highest frequency of a square wave which might be used to trigger this circuit? Assume ideal transistors.

chapter 13

ASTABLE MULTIVIBRATOR (FREE RUNNING)

The astable multivibrator is an electronic circuit which has two semistable states. The duration of each of the two semistable states is dependent upon the two RC time constants within the multivibrator circuit. If no input signal or trigger pulse is applied, the circuit produces a square-wave (rectangular) output-voltage waveform. The circuit may be thought of as a square-wave generator or oscillator. The astable multivibrator is frequently called a free-running multivibrator.

The astable multivibrator consists of two inverter circuits. The output of the first inverter circuit is the input to the second, and the output of the second is the input to the first. The output of each inverter circuit is RC coupled to the input of the other. Refer to Fig. 13-1.

13.1 OPERATION

To follow the description of the operation of the astable multivibrator, refer to Fig. 13-1. At t_{+0} time, transistor T_2 conducts at saturation and transistor T_1 is at cutoff. The base current of transistor T_2 is the charging current for capacitor C_2. Capacitor C_2 charges to V_{cc}. The collector voltage of transistor T_1 rises exponentially to V_{cc} as capacitor C_2 charges. Resistor R_{B_2} is selected to limit the base current of transistor T_2 to the minimal value necessary for transistor T_2 to conduct at saturation when ON. Hence, transistor T_2 continues to conduct at saturation after capacitor C_2 charges to V_{cc}.

In a similar manner, capacitor C_1 charges to V_{cc} when transistor T_1 conducts at saturation. (Assume capacitor C_1 is charged to V_{cc} at t_{+0} time.) When transistor T_2 (acting as a closed switch) conducts at saturation, it effectively grounds the positively charged plate of capacitor C_1. Refer to Fig. 13-2(a).

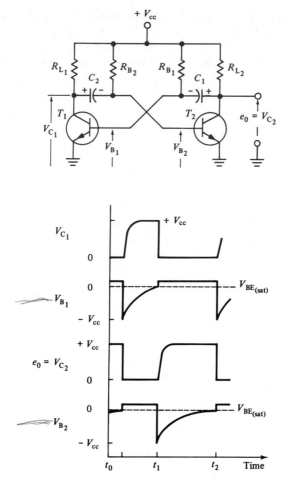

Figure 13-1   Astable Collector-Coupled Multivibrator

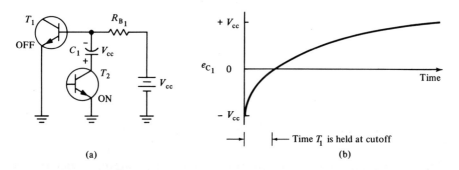

Figure 13-2   (a) Equivalent Circuit of Fig. 13-1 at t_{+0} Time; (b) Time T_1 Is Held at Cutoff

Capacitor C_1 is effectively in parallel with the emitter-base junction of transistor T_1. Since capacitor C_1 is charged to V_{cc}, the reverse bias of transistor T_1 at t_{+0} time is $-V_{cc}$. Hence, charged capacitor C_1 turns transistor T_1 OFF. Transistor T_1 remains OFF while the emitter-base junction is reverse biased. At t_{+0} time, capacitor C_1 starts to charge from $-V_{cc}$ volts toward $+V_{cc}$ volts. Refer to Fig. 13-2(b). Transistor T_1 remains OFF until t_{-1} time, at which time the bias is 0 V. At t_{+1} time, the emitter-base junction of transistor T_1 is forward biased and hence conducts.

Because transistor T_1 conducts at t_{+1} time, the positively charged plate of capacitor C_2 is effectively grounded. Hence, charged capacitor C_2 becomes the reverse bias for transistor T_2. At t_{+1} time, capacitor C_2 starts to charge from $-V_{cc}$ toward $+V_{cc}$, through the ON transistor T_1 and through R_{B_2}. Thus, charged capacitor C_2 holds transistor T_2 OFF for the duration necessary for capacitor C_2 to charge to 0 V. Transistor T_2 again starts to conduct at t_{+2} time.

During the interval between t_{+1} time and t_{-2} time, capacitor C_1 is charged to V_{cc}, by the base current of ON transistor T_1.

This process continues to repeat itself, thus producing a rectangular output-voltage waveform, as indicated in Fig. 13-1. The amplitude of the resultant output voltage is $[V_{cc} - V_{CE_{sat}}]$.

Refer to Fig. 13-1:

$$(t_1 - t_0) = t_A \text{ time}$$
and
$$(t_2 - t_1) = t_B \text{ time}$$
and
$$T = \text{one period time}$$
Hence
$$T = t_A + t_B$$
$$t = 0.69RC$$
$$t_A = 0.69R_{B_1}C_1 \qquad\qquad [26]$$
$$t_B = 0.69R_{B_2}C_2$$
$$T = 0.69R_{B_1}C_1 + 0.69R_{B_2}C_2$$
Normally,
$$R_{B_1} = R_{B_2} = R$$
Hence:
$$T = 0.69R(C_1 + C_2)$$

In a symmetrical astable multivibrator,

$$C_1 = C_2 = C$$
Hence:
$$T = 0.69R(C + C)$$
$$T = 0.69R(2C)$$
$$T = 1.38RC$$

13.2   MINIMUM PULSE DURATION

The minimum pulse width for an astable multivibrator is determined by the recovery time (t_{re}), which is the time required to charge capacitor C_1 or C_2 to V_{cc}. For practical purposes, assume that the capacitor has completely charged to V_{cc}, when, actually, it has only charged to 90% of V_{cc}. Hence, $t_{re} = 2.3R_2C$, as explained in chapter 12.

Hence:
$$t = 0.69R_{B_1}C_1$$
$$t_{re_A} = 2.3R_LC_2 \qquad \text{(recovery time for } C_2\text{)}$$

Since
$$R_{B_1} = R_{B_2} = R$$
$$R_{L_1} = R_{L_2} = R_L$$

and
$$I_C \approx \frac{V_{cc}}{R_L} \qquad V_{cc} \approx I_C R_L$$

$$I_B \approx \frac{V_{cc}}{R} \qquad V_{cc} \approx I_B R$$

$$I_C R_L = I_B R$$

$$h_{FE} = \frac{I_C}{I_B} = \frac{R}{R_L}$$

$$h_{FE_{min}} = \frac{R}{R_L}$$

$$R \leq h_{FE_{min}} R_L$$
$$t_A = 0.69R_{B_1}C_1 = 0.69RC_1$$
$$t_{re_A} = 2.3R_LC_2$$
$$\frac{t_{re_A}}{t_A} = \frac{2.3R_LC_2}{0.69RC_1} = \frac{3.33R_LC_2}{RC_1}$$

But
$$R \leq h_{FE_{min}} R_L$$
$$t_{re_A} \geq \frac{3.33R_LC_2}{h_{FE_{min}}R_LC_1} = \frac{3.33C_2}{h_{FE_{min}}C_1}$$

Similarly,
$$\frac{t_{re_B}}{t_B} \geq \frac{3.33C_1}{h_{FE_{min}}C_2} \qquad (t_{re_B} = \text{recovery time for } C_1)$$

For the special case of a symmetrical astable multivibrator, in which $t_A = t_B$, $C_1 = C_2 = C$, and $T = t_A + t_B = 2t$,

$$\frac{t_{re}}{t} \geq \frac{3.33C}{h_{FE_{min}}C} \qquad \frac{T}{2} = t$$

$$\frac{t_{re}}{t} \geq \frac{3.33}{h_{FE_{min}}}$$

$$t_{re} \geq \frac{3.33t}{h_{FE_{min}}}$$

Hence, if the transistors used in a symmetrical astable multivibrator have an $h_{FE_{min}} = 10$,

$$t_{re} = \frac{3.33t}{10} = 0.33t$$

or
$$t_{re} > 33\% t$$

13.3 SAMPLE DESIGN PROBLEM

Design an astable multivibrator from the following specifications: $e_o = 10$ V peak; output to be a positive pulse, the duration of which is exactly 20 μsec; the time between pulses to be 10 μsec; and $I_{Con} = 10$ mA. Refer to Fig. 13-1.

Available: 1 dc power supply 0 to 30 V and 0 to 250 mA
　　　　　2 NPN silicon transistors
　　　　　$h_{FE_{min}} = 20$
　　　　　$V_{EBO_{max}} = 15$ V
　　　　　$I_{CBO} \approx 0$

Assume: Standard junction voltages

$$\Delta e_o = 10 \text{ V} = V_{cc}$$

$$R_L = \frac{V_{cc} - V_{CE_{sat}}}{I_C} = \frac{10 - 0.3}{10 \text{ mA}} = \frac{9.7}{10 \text{ mA}}$$

$$R_L = 970 \ \Omega \qquad \text{use } 1 \text{ k}\Omega$$

$$I_B = \frac{I_C}{h_{FE_{min}}} = \frac{10 \text{ mA}}{20} = 0.5 \text{ mA}$$

$$R_B = R_{B_1} = R_{B_2} = \frac{V_{cc} - V_{BE_{sat}}}{I_B}$$

$$R_B = \frac{10 - 0.7}{0.5 \text{ mA}} = \frac{9.3}{0.5 \text{ mA}}$$

$$R_B = 18.6 \text{ k}\Omega \qquad \text{use } 18 \text{ k}\Omega$$

$$(t_2 - t_1) = \text{duration of output pulse} = 20 \ \mu\text{sec}$$

$$t = 0.69RC \qquad\qquad\qquad\qquad\qquad\qquad [26]$$

$$C_2 = \frac{t}{0.69R_{B_2}} = \frac{20 \times 10^{-6}}{(0.69)(18 \times 10^3)}$$

$$C_2 = \frac{20 \times 10^{-6}}{12.4 \times 10^3} = 1.61 \times 10^{-9} \text{ F}$$

$$C_2 = 0.0016 \ \mu\text{F}$$

$$C_2 = \text{use } 0.0015 \ \mu\text{F}$$

$$(t_1 - t_0) = \text{duration between output pulses} = 10 \ \mu\text{sec}$$

$$C_1 = \frac{t}{0.69R_{B_1}} = \frac{10 \times 10^{-6}}{(0.69)(18 \times 10^3)}$$

$$C_1 = 0.805 \times 10^{-9} \text{ F}$$

Use　　　　　　$C_1 = 0.00082 \ \mu\text{F}$

To adjust the duration of the output pulse to exactly 20 μsec, it is necessary to replace R_{B_2} with a fixed resistor and a potentiometer.

Therefore:

$$R_{B_2} \geq R_1 + R_2 \qquad R_1 = \text{fixed resistor}$$
$$R_2 = \text{potentiometer}$$

$$18 \text{ k}\Omega \geq 6.8 \text{ k}\Omega + 10 \text{ k}\Omega$$

$$R_{B_2} = 6.8 \text{ k} + \frac{10 \text{ k}}{2}$$

$$R_{B_2} = 6.8 \text{ k} + 5 \text{ k}$$
$$R_{B_2} = 11.8 \text{ k}\Omega$$

$$C_2 = \frac{t}{0.69 R_{B_2}} = \frac{20 \times 10^{-6}}{(0.69)(11.8 \text{ k})}$$

$$C_2 = \frac{20 \times 10^{-6}}{8.15 \times 10^3} = 2.45 \times 10^{-9}$$

$$C_2 = 0.00245 \ \mu\text{F}$$

Use
$$C_2 = 0.0022 \ \mu\text{F}$$

13.4 SAMPLE ANALYSIS PROBLEM

Given: The circuit shown in Fig. 13-3. Prove that the transistors conduct at saturation when ON.

$$I_C = \frac{V_{cc} - V_{CE_{sat}}}{R_L} = \frac{10 - 0.3}{1 \text{ k}} = \frac{9.7}{1 \text{ k}} = 9.7 \ \text{mA}$$

$$I_B = \frac{V_{cc} - V_{BE_{sat}}}{R_{B_2}} = \frac{10 - 0.7}{18 \text{ k}} = \frac{9.3}{18 \text{ k}} = 0.517 \text{ mA}$$

$$h_{FE} = \frac{I_C}{I_B} = \frac{9.7 \text{ mA}}{0.517 \text{ mA}} = 18.8 < h_{FE_{min}} = 20$$

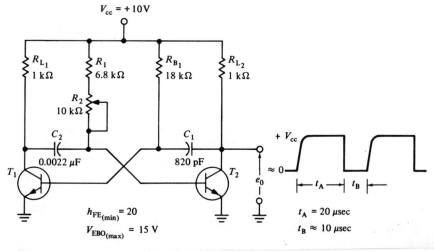

Figure 13-3

Hence, the transistors conduct at saturation when ON because the actual value of h_{FE}, in the circuit, is less than the minimal value of these transistors.

$$t = 0.69RC \qquad \text{minimum output-pulse width}$$
$$t = 0.69R_1C_2$$
$$t = (0.69)(6.8 \text{ k})(2200 \times 10^{-12})$$
$$t = 10.3 \ \mu\text{sec} \qquad \text{minimum}$$
$$t = 0.69RC_2$$
$$t = (0.69)(16.8 \text{ k})(2200 \times 10^{-12})$$
$$t = 24 \ \mu\text{sec} \qquad \text{maximum}$$

Hence, the required 20 μsec maximum may be obtained. Time duration between pulses:

$$t = 0.69R_{B_1}C_1$$
$$t = (0.69)(18 \text{ k})(820 \times 10^{-12})$$
$$t = 10.2 \ \mu\text{sec}$$

When the maximum emitter-base voltage rating of a transistor is less than the required value of V_{cc} in the circuit, diodes must be inserted in the emitter leads, as illustrated in Fig. 13-4. In turn, each diode, when reverse

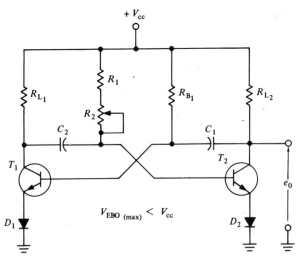

Figure 13-4   Collector-Coupled Astable Multivibrator

biased, forms a resistive voltage divider with the emitter-base junction diode. The amplitude of the output voltage is decreased 0.7 V when silicon diodes are used. This small diode-junction voltage must be taken into consideration, however, in the calculations necessary to determine the values of the circuit components, which, in turn, are necessary to establish the pulse width requirements of the circuit. Hence, Eq. 26 is not valid for this circuit because its derivation was based on the assumption that the capacitor would charge to the applied V_{cc}.

The time-constant parameters for the circuit shown in Fig. 13-4 must be determined by the use of the basic capacitor-charge Eq. 8.

13.5 DESIGN EXAMPLE PROBLEM—CIRCUIT USING EMITTER DIODES

$$V_{cc} = 10 \text{ V}$$

$$I_{C_{on}} = 10 \text{ mA}$$

$$R_L = \frac{V_{cc} - [V_{CE_{sat}} + V_D]}{I_C} = \frac{10 - (0.3 + 0.7)}{10 \text{ mA}} = \frac{10 - 1}{10 \text{ mA}}$$

$$R_L = \frac{9}{10 \text{ mA}} = 0.9 \text{ k}\Omega \qquad \text{use } 910 \ \Omega$$

$$R_B = \frac{V_{cc} - [V_D + V_{BE_{sat}}]}{I_B} \qquad\qquad I_B = \frac{I_C}{h_{FE}} = \frac{10 \text{ mA}}{20}$$

$$R_B = \frac{10 - (0.7 + 0.7)}{0.5 \text{ mA}} \qquad\qquad I_B = 0.5 \text{ mA}$$

$$R_B = \frac{10 - 1.4}{0.5 \text{ mA}} = \frac{8.6}{0.5 \text{ mA}} = 17.2 \text{ k}\Omega \qquad \text{use } 15 \text{ k}\Omega$$

$$e_C = E - (E \pm E_{co})\epsilon^{-t/RC} \qquad\qquad\qquad [8]$$

$$-1 = +10 - (+10 + 10)\epsilon^{-t/RC}$$

$$-1 = +10 - \frac{(+20)}{\epsilon^{+t/RC}}$$

$$\frac{20}{\epsilon^{+t/RC}} = +11$$

$$\epsilon^{+t/RC} = \frac{20}{11} = 1.82$$

$$\epsilon^{+t/RC} = 1.82$$

$$\frac{t}{RC} = \frac{\log_{10} 1.82}{\log_{10} \epsilon} = \frac{0.260}{0.434} = 0.6$$

$$t = 0.6RC$$

$$t_p = 20 \ \mu\text{sec}$$

$$R_B = 15 \text{ k}\Omega$$

$$R_B = R_1 + R_2$$

$$R_B = 4.7 \text{ k}\Omega + 10 \text{ k}\Omega \text{ (pot)}$$

$$V_B = 4.7 \text{ k} + \frac{10 \text{ k}}{2} = 4.7 \text{ k}\Omega + 5 \text{ k}\Omega = 9.7 \text{ k}\Omega$$

$$C = \frac{t}{0.6R} = \frac{20 \times 10^{-6}}{0.6 \times 9.7 \text{ k}} = \frac{20 \times 10^{-6}}{5.82 \text{ k}}$$

$$C = 3.43 \times 10^{-9} \text{ F} = 0.0034 \ \mu\text{F} \qquad \text{use } C = 0.0033 \ \mu\text{F}$$

$t_{re} \approx 10 \ \mu sec$

$$C = \frac{t}{0.6R} = \frac{10 \times 10^{-6}}{0.6 \times 15 \ k} = \frac{10 \times 10^{-6}}{9 \ k}$$

$$C = 1.11 \times 10^{-9} = 0.0011 \ \mu F \qquad use \ C = 0.001 \ \mu F$$

13.6   SYNCHRONIZATION

When the external application of a timing pulse, to one transistor or simultaneously to both transistors of an astable multivibrator, causes the multivibrator to switch stable states at the same frequency with which this and subsequent timing pulses are applied, synchronization takes place and the applied timing pulses and the transistor switching action are synchronized. These timing pulses, usually referred to as synchronizing pulses, may also be used as a frequency divider in the multivibrator circuit. In this application, synchronization will occur with every other application of the synchronizing pulse. Practically, this frequency division is not employed to synchronize an astable multivibrator in which the ratio of the frequency of the synchronizing pulses to the frequency of the switching action is greater than five to one.

The applied pulses may be positive or negative and may be applied to the emitter, to the base, or to the collector of one or both transistors. Usually, the frequency of the switching action of the astable multivibrator is lower than that of the synchronizing pulses; therefore, these pulses force the transistor to switch stable states slightly before the normal free-running transition time.

An example of the application of a positive-trigger pulse to the base of one of the transistors in an astable multivibrator circuit is shown in Fig. 13-5. The waveshape of the trigger pulse is frequently one-half of the differentiated-voltage waveform. Rectangular synchronizing pulses are sometimes used, however, in which case the RC circuits within the multivibrator provide some differentiating action.

To trace the following description of base synchronization, refer to Fig. 13-5. When transistor T_2 is at cutoff, capacitor C_2 charges from $-V_{cc}$ volts toward $+V_{cc}$ volts. Transistor T_2 is held at cutoff while the emitter-base junction voltage is negative. When a positive synchronizing pulse is applied to the base of transistor T_2, it modulates the base-voltage waveform of transistor T_2, as illustrated in Fig. 13-5 (V_{B_2}). When the amplitude of the synchronizing pulse is zero or positive, transistor T_2 is forward biased; hence, T_2 conducts and produces the output-voltage waveform indicated by the dotted line in Fig. 13-5.

Negative synchronizing pulses, applied to the base of transistor T_1, are amplified and inverted when transistor T_1 conducts. Hence, these initially negative-trigger pulses, applied to the base of transistor T_1, will appear as positive-amplified synchronizing pulses at the base of transistor T_2.

Capacitor C_3 is used to inject the synchronizing pulses into the multi-

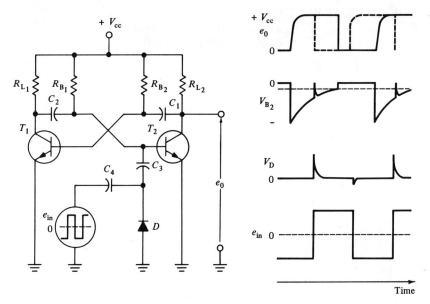

Figure 13-5 Synchronization

vibrator circuit. The value of capacitance of C_3 is not critical. To determine a value of capacitance which is practical for C_3 in the circuit shown in Fig. 13-5, follow the procedure outlined for the determination of C in the coupling circuit described in the section on symmetrical base triggering in chapter 10.

Capacitor C_4 and diode D, illustrated in Fig. 13-5, make up the wave-shaping circuit which transforms the output of the square-wave voltage source into the synchronizing-voltage waveform V_D.

ASTABLE MULTIVIBRATOR
(COLLECTOR-COUPLED)

OBJECT:

1. To familiarize the student with the characteristics of the collector-coupled astable multivibrator
2. To design a collector-coupled astable multivibrator
3. To analyze the operation of a collector-coupled astable multivibrator
4. To familiarize the student with the characteristics of synchronization

MATERIALS:

2   Switching transistors (example—2N3646 silicon NPN)
3   Diodes, silicon junction (example—1N914)
4   Capacitor substitution boxes (0.0001 to 0.22 μF at 450 V)
4   Resistor substitution boxes (10 Ω to 10 MΩ, 1 W)
1   Potentiometer, 10 kΩ, 1 W
1   Transistor power supply (0 to 30 V and 0 to 250 mA)
1   Square-wave generator (20 Hz to 200 kHz with $Z_0 = 600 \ \Omega$)
1   Oscilloscope, dc time-base type; frequency response dc to 450 kHz; vertical sensitivity, 100 mV/cm

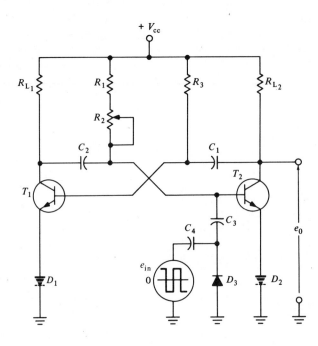

Figure 13-1X

PROCEDURE:

1. Design a collector-coupled astable multivibrator from the following specifications:

$$e_o = 8 \text{ V peak}$$
$$I_{C(on)} = 8 \text{ mA}$$
$$t = 50 \ \mu\text{sec—positive-output pulse; interval}$$
between pulses, 20 μsec

2. Design the collector-coupled astable multivibrator, shown in Fig. 13-1X, according to the conditions specified. Include all calculations. Explain why any assumptions which have been made are valid. (Do not include the synchronization circuit at this time.)

3. Draw a schematic of the circuit designed in Step 2, and label it with the practical, standard, color code values.

4. From the schematic drawn in Step 3, analyze the circuit designed in Step 2; prove that it will operate satisfactorily, according to the required specifications, when standard-value resistors and capacitors are used. Include all calculations.

5. Connect the circuit designed. Check the operation of the circuit. Record all necessary data.

6. Compare measured values with computed values, and explain any discrepancies.

7. Design the base-synchronizing circuit shown in Fig. 13-1X. Connect this synchronizing circuit to the multivibrator circuit.

8. Set the frequency and amplitude of the triggering pulse so that the output of the astable multivibrator has a positive pulse, the duration of which is 40 μsec. The triggering pulse should be applied so that the interval between positive pulses is not affected.

9. Measure and record the minimum amplitude of the synchronizing pulse which appears across diode D_3. Compute this minimum amplitude, and compare it with the measured value.

QUESTIONS AND EXERCISES

1. Explain the operation of one complete cycle of a collector-coupled astable multivibrator. Use the circuit and waveforms shown in Fig. 13-1 to aid in your explanation.

2. Explain what is meant by synchronization as applied to the astable multivibrator.

3. Explain the effect of the amplitude of a synchronizing pulse on the control of a multivibrator.

4. How do negative synchronizing pulses control an astable multivibrator when applied to the base of one of the NPN transistors?

5. Explain why the diodes which are connected from the emitters to ground may be necessary in the circuit shown in Fig. 13-1X.

6. Explain what is meant by recovery time.

7. What factors determine the minimum pulse width of an astable multivibrator? Explain.

8. In the circuit shown in Fig. 13-5, what are the synchronizing-pulse conditions necessary for frequency divisions by a factor of two?

$t_{on} = t_{off}$

9. Design a symmetrical, astable, collector-coupled multivibrator with a frequency of 2000 Hz, the pulse amplitude of which is 10 V. Assume a collector current of 10 mA. Assume that the transistors used have a $h_{FEmin} = 30$. Assume that the transistors are ideal switches.

10. Design the circuit described in Prob. 9, and assume that the transistors are not ideal switches. Assume standard junction voltages in which $V_{BEoff} = 0.5$ V.

11. Analyze the circuit shown in Fig. 13-1X when

$$V_{cc} = +20 \text{ V}, \qquad R_{L_1} = R_{L_2} = 1.8 \text{ k}\Omega, \qquad R_1 = 4.7 \text{ k}\Omega$$
$$R_2 = 5 \text{ k}\Omega, \qquad R_3 = 8.2 \text{ k}\Omega, \qquad C_1 = 0.022 \text{ }\mu\text{F}, \qquad C_2 = 0.01 \text{ }\mu\text{F}$$

Assume $h_{FEmin} = 20$. Determine the minimum and the maximum positive pulse duration.

12. Refer to the circuit shown in Fig. 13-1. Design an astable multivibrator in which the

$$prt = 1000 \text{ }\mu\text{sec}, \qquad t_p = 800 \text{ }\mu\text{sec}, \qquad e_o = +12 \text{ V pulse}, \qquad I_{C(on)} = 10 \text{ mA}$$

Assume $h_{FEmin} = 25$. Use only standard-value components.

13. Redesign the circuit, designed in Prob. 12, so that the positive pulse duration may be adjusted to exactly 800 μsec, as indicated in Fig. 13-4.

chapter 14

EXPONENTIAL RC
SWEEP CIRCUITS

Sweep circuits produce voltage or current waveshapes which increase or decrease in amplitude, at a linear rate, with reference to time. These circuits are frequently referred to as linear time-base generators. The circuit which produces a linear-voltage waveshape is called a voltage time-base generator; one which produces a linear-current waveshape is called a current time-base generator. In practice, the sweep circuit in which even a portion of voltage waveshape is linear is called a voltage time-base generator. It is with the basic forms of this circuit that this chapter deals.

Refer to Fig. 14-1. The sweep-voltage waveform is made up of two basic

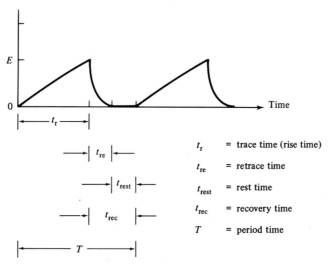

t_r = trace time (rise time)

t_{re} = retrace time

t_{rest} = rest time

t_{rec} = recovery time

T = period time

Figure 14-1 Sweep-Voltage Waveform

parts, the trace time and the recovery time. The recovery time may also be made up of two parts, the retrace time and the rest time. In many applications, the rest time is zero, in which case, the recovery time is the retrace time.

Several methods may be employed to generate linear-voltage waveshapes. The two fundamental methods, discussed in this and the following chapter, incorporate the principles upon which more complex methods are based. These two fundamental methods are

1. The charging of a capacitor through a resistor.
2. The charging of a capacitor by a constant-current source (chapter 15).

14.1 SWEEP-VOLTAGE GENERATION—CHARGING A CAPACITOR THROUGH A RESISTOR

The charging of a capacitor through a resistor produces a linear-voltage waveshape if only the initial portion of the capacitor-charging curve is used. To trace this generation of sweep voltage, refer to Fig. 14-2. Assume that capacitor C is uncharged at t_{-0} time and that switch S is in position 2. At t_{+0} time, switch S is moved to position 1; hence, capacitor C charges through resistor R_L toward $+V_{cc}$ volts. When switch S is moved to position 1, at t_{+1} time, the capacitor starts to discharge from V volts toward 0 V. The discharge path is through the closed switch S and through resistor R_{CS}. Normally,

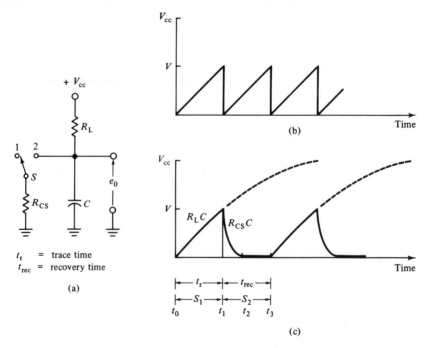

Figure 14-2 (a) Capacitor-Charging-Sweep Circuit; (b) Sawtooth-Voltage Waveform; (c) Sweep-Voltage Waveform

resistor $R_{CS} \ll R_L$; hence, the capacitor C has time to discharge completely, as shown in Fig. 14-2(c). This resultant output-voltage waveshape exhibits a high degree of linearity at the initial portion of the charge curve.

If resistor $R_{CS} \approx 0$, the capacitor can discharge in practically zero time. If the capacitor is allowed to start charging again as soon as it has discharged, the resultant sawtooth voltage waveform is generated as shown in Fig. 14-2(b). The sawtooth voltage waveform is frequently used for trace generation in electrostatically deflected cathode-ray tubes, found in oscilloscopes and some television receivers.

The degree of linearity of the resultant output voltage, which may be obtained from a circuit similar to that shown in Fig. 14-2, is dependent upon the amount of charge curve used. This degree of linearity may be expressed as a percentage of the ratio of the time the capacitor is allowed to charge to the time constant of the circuit.

Refer to Fig. 14-3. The instant the switch S is closed, the capacitor appears as a short circuit, and the initial charging current is limited by the resistance of resistor R. If this initial charging current continues until capacitor C is charged to the applied voltage, the following relationships will be valid:

$$I = \frac{Q}{t} \quad \text{and} \quad Q = Ce_C$$

$$I = \frac{Ce_C}{t}$$

But
$$I = \frac{V_{cc}}{R}$$

Hence:
$$\frac{V_{cc}}{R} = \frac{Ce_C}{t}$$

$$e_C = \frac{V_{cc}t}{RC} = \frac{V_{cc}t}{\tau} \qquad \text{(linear)}$$

When
$$t = \tau = RC$$
$$e_C = V_{cc}t$$

Capacitor C charges linearly, to the applied voltage V_{cc}, in $t = \tau = RC$ time.

In practice, capacitor C charges exponentially toward the applied V_{cc} voltage, in accordance with the capacitor-charge Eq. 5.

$$e_C = E(1 - \epsilon^{-t/RC}) \qquad [5]$$
$$e_C = V_{cc}(1 - \epsilon^{-t/RC})$$

When
$$t = \tau = RC$$
$$e_C = V_{cc} - V_{cc}\epsilon^{-1}$$
$$e_C = V_{cc} - \frac{V_{cc}}{2.718}$$

$$e_C = 0.63V_{cc}$$
$$e_C = 63\% \ V_{cc}$$

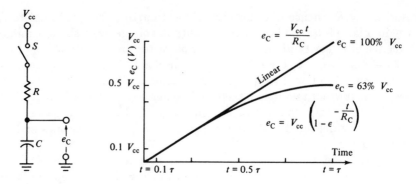

Figure 14-3

Hence, the linearity of the resultant output voltage is poor. See Fig. 14-3.

If the capacitor is only allowed to charge to one-tenth of the applied voltage, however, a relationship may be established between the time the capacitor is allowed to charge and the time constant of the circuit.

$$e_C = V(1 - \epsilon^{-t/RC})$$
$$0.1V_{cc} = V_{cc}(1 - \epsilon^{-t/RC})$$
$$\epsilon^{+t/RC} = 1.11$$

$$t = RC\frac{\log_{10} 1.11}{\log_{10} \epsilon} = \frac{0.045RC}{0.434}$$

$$t = 0.1035RC$$
$$t \approx 0.1RC \approx 0.1\tau$$

The time required for the capacitor to charge to one-tenth of the applied voltage may be expressed in terms of the RC time constant of the circuit, if the linear-charging relationship is used.

$$e_C = \frac{V_{cc}t}{RC}$$

$$0.1V_{cc} = V_{cc}\frac{t}{RC}$$

$$t = 0.1RC = 0.1\tau$$

For practical purposes, the voltage waveform, produced by the exponential charging of a capacitor through a resistor, will be linear if the time the capacitor is allowed to charge is $\leq 0.1RC$. Thus:

$$t = 0.1RC \approx 0.1035RC$$
$$\text{linear} \qquad \text{exponential}$$

The fact that the amplitude of the linear output-voltage waveform is only 10 percent of the applied voltage is a disadvantage of the capacitor-exponential-charging method of producing a linear-sweep voltage; the small linear-sweep voltage must be amplified by a second amplifier. In applications where the linearity of the sweep voltage is not critical, however, a larger portion of the capacitor-charging curve may be employed and, therefore, this method is used.

Refer to Fig. 14-2. In practical circuits, switch S is a transistor switch or inverter circuit, as illustrated in Fig. 14-4. The resistor R_{CS}, shown in Fig. 14-2, is the saturation resistance of transistor T_1 when ON, as indicated in Fig. 14-4.

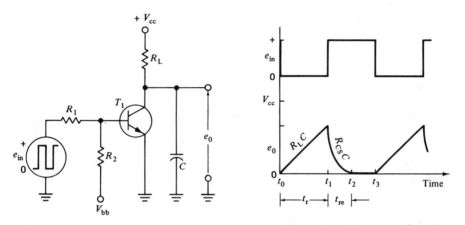

Figure 14-4   Capacitor-Charging-Sweep Circuit

14.2   OPERATION—CAPACITOR-CHARGING SWEEP CIRCUIT

To trace the operation of a capacitor-charging sweep circuit, refer to Fig. 14-4. This circuit consists of a standard inverter circuit with a capacitor C connected from collector to ground. With no input pulse applied at t_{+0} time, the transistor T_1 is OFF, and the capacitor C charges toward V_{cc} through resistor R_L. When the positive-going input pulse is applied at t_{+1} time, transistor T_1 conducts at saturation, and capacitor C discharges through the ON transistor.

The amplitude of the resultant output sweep voltage is dependent upon the interval during which the transistor is OFF because this is the time during which the capacitor is allowed to charge. The amplitude of the resultant output sweep voltage may be computed by use of the standard capacitor-charge Eq. 8.

The time necessary for the capacitor to discharge, after the input pulse is applied, may be computed by use of the standard capacitor-discharge Eq. 6. Because the capacitor discharges through transistor T_1, when conducting at saturation, it is the saturation resistance of the transistor R_{CS} which limits the discharge-current flow of the capacitor.

The standard inverter circuit is designed to produce the minimum base current necessary for collector saturation. In this standard circuit, the saturation resistance of the transistor R_{CS} may be of the order of magnitude of 25 Ω. Compared to the normal value of load resistance, 25 Ω may be considered negligible.

When the basic inverter circuit is used as a switch for a capacitor-charging sweep circuit, however, the value of the saturation resistance of the transistor is the prime factor affecting the retrace time. In most sweep applications, the retrace time should be minimized. Therefore, the saturation resistance of the transistor must also be minimized and may be of the order of magnitude of 2 Ω.

The saturation resistance of a transistor is determined by the amplitude of the base current. (See Appendix B—2N3646—collector characteristics of the saturation region of operation.) When the transistor is turned ON, in the capacitor-charging sweep circuit, the capacitor starts to discharge through the transistor. The discharge current becomes the collector-saturation current of the transistor. The amplitude of the collector-saturation current is determined by the amplitude of the base current. Practically, if the capacitor-discharge current (collector-saturation current) is five times greater than the capacitor-charging current, the saturation resistance of the transistor is minimal.

14.3   SATURATION RESISTANCE

If specification sheets are not available, the saturation resistance of the transistor may be approximately determined by use of Ohm's law if the standard junction voltage for silicon, of $V_{CE_{sat}} \approx 0.3$ V, is used.

$$R_{CS} \approx \frac{V_{CE_{sat}}}{I_{C_{sat}}}$$

where

R_{CS} = transistor-saturation resistance (Ω)
$V_{CE_{sat}}$ = collector-to-emitter voltage when transistor is at saturation (V)
$I_{C_{sat}}$ = collector current at saturation (mA)

Figure 14-5(a) illustrates a circuit, equivalent to that illustrated in Fig.

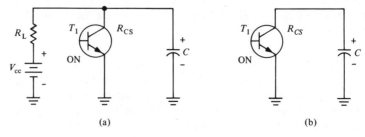

(a)   (b)

Figure 14-5   Capacitor-Discharge Circuit

14-4, at t_{+1} time. It indicates the complete discharge path existent when the transistor is ON and the capacitor C starts to discharge. The R_L branch of the discharge circuit may be neglected, as illustrated in Fig. 14-5(b) because, normally, the saturation resistance of the transistor is appreciably smaller than the load resistor of the circuit.

One of the two inverter circuits, incorporated in the design of a practical astable multivibrator, may be the same as the one shown in Fig. 14-4.

14.4 SAMPLE DESIGN PROBLEM

From the circuit shown in Fig. 14-4, design a capacitor-charging sweep circuit with the following characteristics:

$$t_r = 50 \ \mu\text{sec}$$
$$e_o = 4 \ \text{V}$$

Given:

$$1 \ \text{silicon NPN transistor}, h_{FE_{min}} = 20$$
$$I_{CBO} = 0 \ \text{A}$$

2 10 V dc sources

Assume:

Initial-charging current of 10 mA
$V_{BE_{off}} = -0.5 \ \text{V}$
e_{in} = positive pulse 10 V peak
Standard junction voltages

Solution A (Inverter designed for minimum base current)

1. Determine the value of R_L necessary for the initial capacitor-charging current to be of the value specified.
2. Using the capacitor-charge Eq. 8, determine the value of capacitance C.
3. Write the ON circuit input-node equation for the inverter circuit, in terms of R_1 and R_2.
4. Write the OFF circuit input-node equation for the inverter circuit, in terms of R_1 and R_2.
5. Simultaneously solve the ON and the OFF equations for R_1 and R_2.

$$R_L = \frac{V_{cc} - V_{CE_{sat}}}{I_C} = \frac{10 - 0.3}{10 \ \text{mA}} = \frac{9.7}{10 \ \text{mA}} = 970 \ \Omega$$

Use

$$R_L = 1 \ \text{k}\Omega$$
$$e_C = E - (E \pm E_{co})\epsilon^{-t/RC} \qquad [8]$$
$$4 = 10 - (10 - 0.3)\epsilon^{-50 \times 10^{-6}/1 \times 10^3 C}$$
$$4 = 10 - \frac{9.7}{\epsilon^{+50 \times 10^{-9}/C}}$$
$$\epsilon^{+50 \times 10^{-9}/C} = \frac{9.7}{6} = 1.615$$

$$C = \frac{50 \times 10^{-9} \log_{10} \epsilon}{\log_{10} 1.615}$$

$$C = \frac{50 \times 10^{-9} \times 0.434}{0.208}$$

$$C = 0.104 \ \mu\text{F}$$

Use $\quad\quad C = 0.1 \ \mu\text{F}$

$$I_\text{B} = \frac{I_\text{C}}{h_{\text{FE}_\text{min}}} = \frac{10 \text{ mA}}{20} = 0.5 \text{ mA}$$

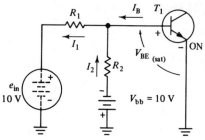

Figure 14-6   ON Input Circuit

$$I_1 = I_2 + I_\text{B}$$

$$\frac{E_{\text{R}_1}}{R_1} = \frac{E_{\text{R}_2}}{R_2} + I_\text{B}$$

$$\frac{e_\text{in} - V_{\text{BE}_\text{sat}}}{R_1} = \frac{V_{\text{BE}_\text{sat}} - V_\text{bb}}{R_2} + I_\text{B}$$

$$\frac{10 - (+0.7)}{R_1} = \frac{+0.7 - (-10)}{R_2} + 0.5 \text{ mA}$$

$$\boxed{\frac{9.3}{R_1} = \frac{10.7}{R_2} + 0.5 \text{ mA}} \quad\quad\text{ON Equation}$$

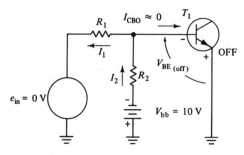

Figure 14-7   OFF Input Circuit

$$V_{\text{BE}_\text{off}} = \left(\frac{R_1}{R_1 + R_2} \right) V_\text{bb}$$

$$(-0.5) = \left(\frac{R_1}{R_1 + R_2}\right)(-10)$$

$$\boxed{R_2 = 19R_1} \qquad \text{OFF Equation}$$

$$\frac{9.3}{R_1} = \frac{10.7}{19R_1} + 0.5 \text{ mA}$$

$$\frac{9.3}{R_1} = \frac{0.563}{R_1} + 0.5 \text{ mA}$$

$$\frac{8.74}{R_1} = 0.5 \text{ mA}$$

$$R_1 = 17.48 \text{ k}\Omega$$

Use
$$R_1 = 15 \text{ k}\Omega$$
$$R_2 = 19R_1 = (19)(17.48 \text{ k}\Omega)$$
$$R_2 = 332 \text{ k}\Omega$$

Use
$$R_2 = 330 \text{ k}\Omega$$

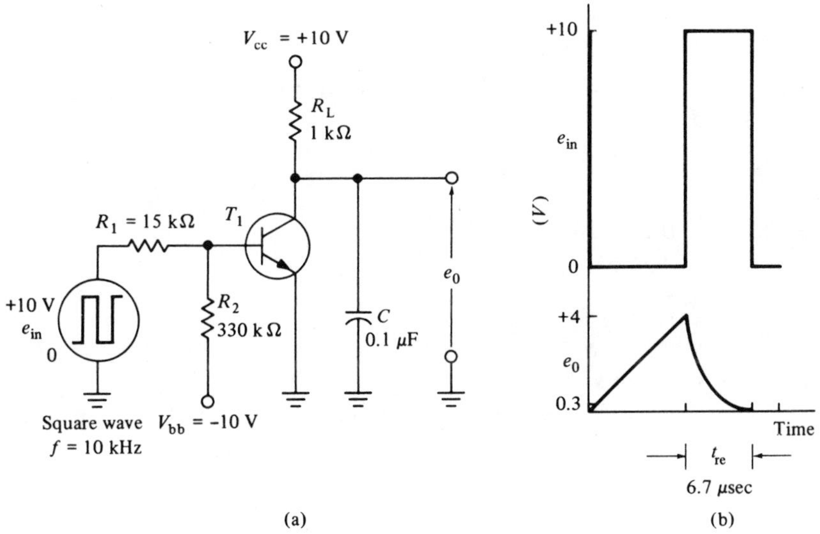

Figure 14-8 (a) Circuit Design for Minimum Base Current; (b) Capacitor-Charging-Sweep Waveforms

14.5 SAMPLE CIRCUIT ANALYSIS PROBLEM

Determine the parameters of the circuit shown in Fig. 14-8.

1. Prove that transistor T_1 is cut off when no input voltage is applied.
   (a) Refer to Fig. 14-7. Write the OFF circuit input equation.
   (b) Solve the OFF circuit input equation for $V_{BE_{off}}$. If this value is equal

to or less than 0 V, the transistor is reverse biased; hence, it is at the required OFF condition.

2. Prove that transistor T_1 will conduct at saturation when the input pulse is applied.
   (a) Refer to Fig. 14-6. Write the ON circuit input equation.
   (b) Solve the ON circuit input equation for I_B. This value of base current must be less than the value of base current determined by V_{cc}/R_L, for the transistor to be ON.

3. Determine the amplitude of the output voltage, by use of the capacitor-charge Eq. 8.

4. Determine the approximate saturation resistance of the transistor R_{CS}, by use of Eq. 25. (Assume $V_{CE_{sat}} = 0.3$ V.)

5. Determine the approximate retrace time, by use of the capacitor-discharge Eq. 6.

$$V_{BE_{off}} = E_{R_1} = \left(\frac{R_1}{R_1 + R_2}\right) V_{bb}$$

$$V_{BE_{off}} = \left(\frac{15 \text{ k}}{15 \text{ k} + 330 \text{ k}}\right)(-10)$$

$$V_{BE_{off}} = -0.435 \text{ V} \qquad \text{which is} > 0, \text{ the minimum required for cutoff}$$

$$I_1 = I_2 + I_B$$

$$\frac{e_{in} - V_{BE_{sat}}}{R_1} = \frac{V_{BE_{sat}} - V_{bb}}{R_2} + I_B$$

$$\frac{+10 - 0.7}{15 \text{ k}} = \frac{+0.7 + 10}{330 \text{ k}} + I_B$$

$$\frac{9.3}{15 \text{ k}} = \frac{10.7}{330 \text{ k}} + I_B$$

$$0.62 \text{ mA} = 0.032 \text{ mA} + I_B$$

$$I_B = 0.588 \text{ mA} > 0.5 \text{ mA} \qquad \text{minimum required for saturation;}$$
$$\text{hence } T_1 \text{ is ON}$$

When ON, C will discharge through T_1, from $+4$ V to $V_{CE_{sat}} = 0.3$ V.

$$R_{CS} \approx \frac{V_{CE_{sat}}}{I_{C_{sat}}}$$

$$I_C = h_{FE_{min}} I_B = (20)(0.588 \text{ mA}) = 11.76 \text{ mA}$$

$$R_{CS} \approx \frac{0.3}{11.76 \text{ mA}} \approx 25.5 \text{ }\Omega$$

$$e_C = E\epsilon^{-t/RC}$$

$$V_{CE_{sat}} = e_C\epsilon^{-t_{re}/R_{CS}C}$$

$$t_{re} \approx \frac{R_{CS}C \log_{10}\dfrac{e_C}{V_{CE_{sat}}}}{\log_{10} \epsilon}$$

$$t_{re} \approx \frac{(25.5)(0.1 \times 10^{-6}) \log_{10} \dfrac{4}{0.3}}{\log_{10} \epsilon}$$

$$t_{re} \approx \frac{2.55 \times 10^{-6} \log_{10} 13.8}{\log_{10} \epsilon}$$

$$t_{re} \approx \frac{(2.55 \times 10^{-6})(1.138)}{0.434} = 6.7 \ \mu\text{sec}$$

14.6 DESIGN MODIFICATION

Redesign the circuit shown in Fig. 14-8. Assume that the initial discharge current is five times greater than the initial capacitor-charging current. This will appreciably decrease the retrace time.

$$I_{C_{\text{discharge}}} = 5 I_{C_{\text{charge}}} = (5)(10 \text{ mA}) = 50 \text{ mA}$$
$$I_{C_{\text{total}}} = I_{C_{\text{discharge}}} + I_C$$
$$I_{C_{\text{total}}} = 50 \text{ mA} + 10 \text{ mA} = 60 \text{ mA}$$
$$I_B = \frac{I_{C_{\text{total}}}}{h_{FE_{\min}}} = \frac{60 \text{ mA}}{20} = 3 \text{ mA}$$

ON equation becomes

$$\frac{9.3}{R_1} = \frac{10.7}{R_2} + 3 \text{ mA}$$

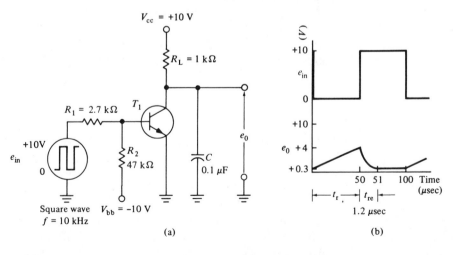

(a) (b)

Figure 14-9 (a) Circuit Design for $I_B = 6 I_{B(\min)}$; (b) Capacitor-Charging-Sweep Waveforms

OFF equation remains the same,

$$R_2 = 19R_1$$

$$\frac{9.3}{R_1} = \frac{10.7}{19R_1} + 3 \text{ mA}$$

$$R_1 = 2.91 \text{ k}\Omega$$

Use
$$R_1 = 2.7 \text{ k}\Omega$$
$$R_2 = 19R_1$$
$$R_2 = (19)(2.9 \text{ k}\Omega)$$
$$R_2 = 55.4 \text{ k}\Omega$$

Use
$$R_2 = 47 \text{ k}\Omega$$

14.7 CIRCUIT ANALYSIS OF REDESIGNED CIRCUIT

Follow procedure outlined for original circuit.

$$V_{BE_{off}} = E_{R_1} = \left(\frac{R_1}{R_1 + R_2}\right) V_{bb}$$

$$V_{BE_{off}} = \left(\frac{2.7 \text{ k}}{2.7 \text{ k} + 47 \text{ k}}\right)(-10) = -0.543 \text{ V}$$

Because -0.543 V is greater than zero, the minimum reverse bias required for cutoff, T_1 is OFF when no input pulse is applied.

$$\frac{e_{in} - V_{BE_{sat}}}{R_1} = \frac{V_{BE_{sat}} - V_{bb}}{R_2} + I_B$$

$$\frac{+10 - 0.7}{2.7 \text{ k}} = \frac{0.7 + 10}{47 \text{ k}} + I_B$$

$$I_B = 3.21 \text{ mA}$$

Since the $I_{B_{min}}$ required for saturation is 0.5 mA, transistor T_1 is ON.

$$I_C = h_{FE}I_B = (20)(3.21 \text{ mA}) = 64.2 \text{ mA}$$
$$I_{C_{discharge}} = I_{C_{total}} - I_{R_L}$$
$$I_{C_{discharge}} = 64.2 \text{ mA} - 10 \text{ mA}$$
$$I_{C_{discharge}} = 54.2 \text{ mA}$$
$$e_C = E - (E \pm E_{co})\epsilon^{-t/RC} \qquad\qquad [8]$$
$$e_C = V_{cc} - (V_{cc} - V_{CE_{sat}})\epsilon^{-t_r/R_LC}$$
$$t_r = \frac{1}{2f} = \frac{1}{2 \times 10^4} = 50 \text{ } \mu\text{sec}$$
$$e_C = 10 - (10 - 0.3)\epsilon^{-50\times10^{-6}/1\times10^3\times0.1\times10^{-6}}$$
$$e_C = 4.13 \text{ V}$$
$$\Delta e_o = e_C - V_{CE_{sat}} = 4.13 - 0.3$$
$$\Delta e_o = 3.83 \text{ V}$$
$$e_C = E\epsilon^{-t/RC}$$
$$V_{CE_{sat}} = e_C\epsilon^{-t_{re}/R_{CS}C}$$

Assume:

$$V_{CE_{sat}} \approx 0.3 \text{ V}$$

(Actually $V_{CE_{sat}}$ varies with collector current.)

$$R_{CS} \approx \frac{V_{CE_{sat}}}{I_C} \approx \frac{0.3}{64.2 \text{ mA}} \approx 4.68 \text{ }\Omega$$

$$V_{CE_{sat}} = \frac{e_C}{\epsilon^{+t_{re}/R_{CS}C}}$$

$$t_{re} \approx \frac{R_{CS}C \log_{10} \dfrac{e_C}{V_{CE_{sat}}}}{\log_{10} \epsilon}$$

$$t_{re} \approx \frac{4.68 \times 0.1 \times 10^{-6} \log_{10} \dfrac{4.13}{0.3}}{\log_{10} \epsilon}$$

$$t_{re} \approx 1.23 \text{ }\mu\text{sec}$$

Hence, retrace time is appreciably decreased. This calculated duration is only approximate because it was assumed that $V_{CE_{sat}}$ was equal to the standard-junction voltage, 0.3 V; actually, $V_{CE_{sat}}$ varies as collector current varies. Approximations, however, thus calculated, are valid for most practical circuits.

LINEAR SWEEP

$$e_C = \frac{V_{cc}t}{RC} = \frac{10 \times 50 \times 10^{-6}}{1 \times 10^3 \times 0.1 \times 10^{-6}}$$

$$e_C = 5 \text{ V}$$

$$\text{percent linearity} = \frac{e_C(\text{exponential})}{e_C(\text{linear})} \times 100$$

$$= \frac{4.13 \times 100}{5}$$

$$= 82.5 \text{ percent}$$

If an appreciable portion of the capacitor-charge curve is used, poor linearity results. This is a disadvantage of the capacitor-charging sweep circuit. If, however, only a small portion of the capacitor-charge curve is used to obtain good linearity, the amplitude of the resultant output voltage is extremely small. In addition, no appreciable current may be drawn from the circuit; hence, the output of this circuit must be connected to high input impedance.

EXPONENTIAL RC SWEEP CIRCUIT

OBJECT:
1. To familiarize the student with the characteristics of the exponential RC sweep circuit
2. To design an exponential RC sweep circuit
3. To analyze an exponential RC sweep circuit

MATERIALS:
1 Switching transistor (example—2N3646 silicon NPN)
1 Diode, silicon junction (example—1N914)
1 Capacitor substitution box (0.0001 to 0.22 μF at 450 V)
3 Resistor substitution boxes (10 Ω to 10 MΩ, 1 W)
1 Square-wave generator (20 Hz to 200 kHz, Z_0 = 600 Ω)
1 Oscilloscope, dc time-base type; frequency response dc to 450 kHz; vertical sensitivity, 100 mV/cm.
2 Transistor power supplies (0 to 30 V and 0 to 250 mA)

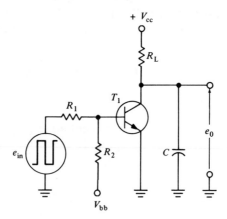

Figure 14-1X

PROCEDURE:
1. Design the capacitor-charging sweep circuit shown in Fig. 14-1X from the following specifications:

$$t_r = 40 \ \mu\text{sec}$$
$$e_o = 2 \text{ V peak}$$

Initial charging current, 5 mA

$$V_{cc} = +20 \text{ V}$$
$$V_{bb} = -10 \text{ V}$$
$$e_{in} = \text{positive pulse—20 V peak square wave}$$

2. Design the inverter part of the circuit for minimum base current and for the conditions specified. Include all calculations. Explain why any assumptions which have been made are valid.

3. Draw a schematic of the circuit designed in Step 2, and label it with the practical, standard, color code values.

4. From the schematic drawn in Step 3, analyze the circuit designed in Step 2; prove that it will operate satisfactorily, according to the required specifications, when standard-value resistors and capacitors are used. Include all calculations.

5. Connect the circuit designed. Check the operation of the circuit. Record all necessary data.

6. Compare measured values with computed values, and explain any discrepancies.

7. Redesign the circuit to minimize retrace time. Assume that the initial discharge current of the capacitor is five times greater than the initial charging current. Include all calculations.

8. Repeat Steps 3 through 6.

QUESTIONS AND EXERCISES

1. Explain the operation of the capacitor-charging sweep circuit shown in Fig. 14-1X.
2. Explain, in detail, why a capacitor charges exponentially.
3. If the amplitude of the output sweep voltage is 20 percent of V_{cc}, what is the percentage of error of the linearity of the output voltage?
4. In the circuit shown in Fig. 14-1X, does the load resistor R_L or does the base current I_B determine the collector current of the transistor? Explain.
5. What factors affect the value of the saturation resistance of a transistor (R_{CS})? Explain.
6. What factors affect the amplitude of the output voltage of the circuit shown in Fig. 14-1X?
7. What factors affect the linearity of the output voltage of the circuit shown in Fig. 14-1X?
8. Refer to Fig. 14-2X. Explain the operation of a capacitor-charging sweep circuit in which the switch is an astable multivibrator.
9. The circuit shown in Fig. 14-2X has the following component values:

$$R_L = 1 \text{ k}\Omega, \qquad R_B = 15 \text{ k}\Omega, \qquad C_2 = 0.01 \text{ } \mu\text{F}, \qquad C_1 = 0.002 \text{ } \mu\text{F}$$
$$C = 0.05 \text{ } \mu\text{F}, \qquad h_{FE_{min}} = 20, \qquad V_{cc} = 10 \text{ V}$$

Determine the amplitude and the duration of the output voltage. Determine the retrace time and the linearity of the circuit.

10. For Prob. 9, which component, in the circuit shown in Fig. 14-2X, must be changed in order to eliminate rest time? To what value should it be changed?

11. Design a capacitor-charging sweep circuit, as shown in Fig. 14-4, with the following characteristics:

$$t_r = 125 \text{ } \mu\text{sec}, \qquad e_o = 7 \text{ V}, \qquad V_{cc} = +25 \text{ V}$$
$$V_{bb} = -12 \text{ V}, \qquad e_{in} = +5 \text{ V pulse}$$

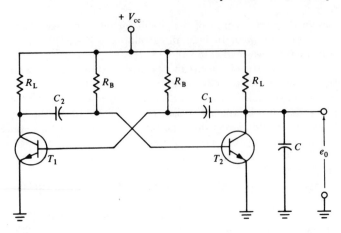

Figure 14-2X

Assume the following:

$$V_{BE_{off}} = -0.5 \text{ V}, \qquad h_{FE_{min}} = 30, \qquad \text{initial charging } I = 15 \text{ mA}$$

12. Redesign the circuit designed in Prob. 11 so that the retrace time will be minimized. Assume that the initial discharge current is five times greater than the initial capacitor-charging current.

13. Analyze the circuit shown in Fig. 14-4, when

$$V_{cc} = +15 \text{ V}, \qquad V_{bb} = -5 \text{ V}, \qquad e_{in} = +5 \text{ V pulse } e_o = 5 \text{ V}$$
$$R_L = 1 \text{ k}\Omega, \qquad R_1 = 2.4 \text{ k}\Omega, \qquad R_2 = 22 \text{ k}\Omega, \qquad C = 0.33 \text{ }\mu\text{F}$$

Assume $V_{BE_{off}} = -0.5$ V, $h_{FE_{min}} = 20$, and that the initial capacitor-discharge current is five times greater than the initial capacitor-charging current.

14. Determine the linearity of the capacitor-charging sweep circuit designed in Prob. 11.

15. Determine the linearity of the capacitor-charging sweep circuit analyzed in Prob. 13.

chapter 15

CONSTANT-CURRENT GENERATION OF LINEAR-SWEEP VOLTAGE

A linear-sweep voltage may be generated by charging a capacitor from a constant-current source. Voltage, so generated, does not have the disadvantages of that produced by an exponential-capacitor-charging circuit (chapter 14). Recall that the generation of a linear-sweep voltage by means of charging a capacitor exponentially through a resistor produces a waveshape of poor linearity if an appreciable portion of the charge curve is used. A linear waveshape will be produced if only 10 percent or less of the charge curve is used; however, the amplitude of this waveshape is relatively small when compared with the voltage source. These disadvantages may be overcome if a circuit in which the capacitor is charged by a constant-current source is employed to generate a sweep voltage. In such a circuit, the charging current of the capacitor must be maintained at a constant value, independent of partial charge on the capacitor, in order for the resultant voltage change across the capacitor to be linear with reference to time. This theoretical constant-current-source circuit may be closely approximated by one in which a transistor is employed in a grounded-base configuration. To trace the operation of this circuit, refer to Fig. 15-1.

15.1 DESCRIPTION OF OPERATION

The collector current, in this grounded-base configuration is approximately equal to the emitter current, regardless of the value of the load resistor. When the load resistor of a common-base amplifier is replaced by a capacitor, as illustrated in Fig. 15-1, the constant-collector current charges the capacitor C, linearly, to V_{cc}. The instant switch S is opened, the capacitor acts as a short circuit; hence, the collector current, which is determined by the input circuit

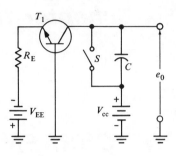 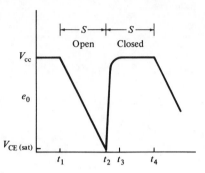

Figure 15-1

of the grounded-base amplifier, becomes the initial charging current. As the capacitor charges to V_{cc}, its effective resistance will increase from zero to ∞.

The output voltage is taken across the capacitor in series with the source V_{cc}. When switch S is closed, at t_{-1} time, the output voltage is V_{cc}. At t_{-2} time when switch S is open, at the end of the charging interval, the capacitor is charged to V_{cc} and is in series-opposing V_{cc}; hence, the resultant output voltage equals the algebraic sum of the source voltage and the voltage across the capacitor. This sum is 0 V at t_{-2} time. At t_{+2} time when the switch closes again, ideally, the capacitor discharges in zero time. Hence, the amplitude of the linear-output-sweep voltage is approximately equal to V_{cc}; it actually equals $[V_{cc} - V_{CE_{sat}}]$.

The following equations are those necessary to determine practical values of circuit components.

$$I = \frac{Q}{t} \quad \text{and} \quad Q = CE \tag{23}$$

$$I = \frac{CE}{t} \tag{27}$$

In terms of the transistor-circuit parameters,

$$I_C = \frac{V_{cc}C}{t}$$

$$I_C = h_{FB}I_E$$

where h_{FB} equals dc current gain of transistor in common-base configuration.

$$h_{FB} = \frac{h_{FE}}{h_{FE} + 1}$$

But
$$I_E = \frac{V_{EE} - V_{BE}}{R_E}$$

Neglecting V_{BE},

$$I_E \approx \frac{V_{EE}}{R_E}$$

Hence,
$$I_C = \frac{h_{FB} V_{EE}}{R_E}$$

[28]

Since
$$I_C = \frac{V_{cc} C}{t}$$

$$\frac{V_{cc} C}{t} = \frac{h_{FB} V_{EE}}{R_E}$$

Practically, switch S, shown in Fig. 15-1, is replaced by a transistor switch T_2, as illustrated in Fig. 15-2.

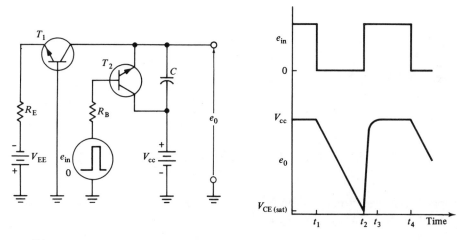

Figure 15-2

When no input is applied, transistor T_2 is OFF and operates as an open switch. At t_{+2} time, the applied positive pulse turns transistor T_2 ON, thus causing the transistor to resemble a closed switch. When transistor T_2 operates as a closed switch, charged capacitor C discharges through T_2. Operating as the switch, transistor T_2 must conduct in deep saturation, in order that the retrace time be minimum. Recall that this was also the case in the operation of the exponential-capacitor-charging circuit. Again, duplicating the operation of the exponential circuit, the initial discharge current must be five times greater than the initial charging current, in order for T_2 to conduct in deep saturation and thereby minimize retrace time.

Refer to Fig. 15-3. This schematic illustrates the circuit conditions at t_{+1} time. The emitter voltage of transistor T_2 is $+V_{cc}$ volts with reference to ground. Hence, the input voltage must be positive and in excess of the emitter voltage ($+V_{cc}$) in order to forward bias transistor T_2. Neglecting the emitter-base-junction voltage of transistor T_2, the input voltage across R_B is ($e_{in} - V_{cc}$). Therefore, to ensure proper circuit operation, the input voltage must be in excess of V_{cc}; this is a disadvantage of this circuit.

An advantage of the operation of this circuit, over that of the exponential-

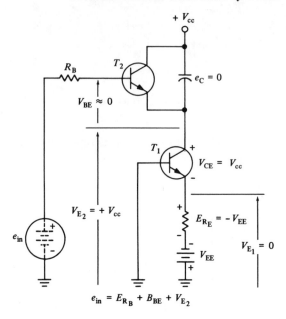

$$e_{in} = E_{R_B} + B_{BE} + V_{E_2}$$

Figure 15-3   Circuit of Fig. 15-2 at t_{+1} Time

capacitor-charge circuit, is the fact that the output voltage is linear and that it is of an amplitude almost equal to V_{cc} volts.

15.2   SAMPLE DESIGN PROBLEM

From the circuit illustrated in Fig. 15-2, design a constant-current-sweep circuit with the following characteristics:

$$t_r = 50 \ \mu\text{sec}$$
$$e_o = 10 \text{ V peak}$$

Given:

2 silicon NPN transistors, $h_{FE_{min}} = 20$
$$I_{CBO} \approx 0$$

2 10 V dc sources

Assume:

Initial charging current of 2 mA
Discharge current equals five times charging current
e_{in} equals positive pulse, 15 V peak
Standard junction voltages

Solution:

1. Determine the proper value of R_E. Use Eq. 28.
2. Determine the proper value of capacitance for C. Use Eq. 27.

3. Determine the proper value for resistor R_B. Assume that the discharge current is five times greater than the initial charging current.

$$V_{EE} = 10 \text{ V}$$

$$V_{cc} = 10 \text{ V}$$

$$I_C = \frac{h_{FB} V_{EE}}{R_E} \qquad h_{FB} = \frac{h_{FE}}{h_{FE} + 1} = \frac{20}{20 + 1}$$

$$R_E = \frac{h_{FB} V_{EE}}{I_C} \qquad h_{FB} = 0.955$$

$$R_E = \frac{(0.955)(10)}{(2)(10^{-3})} = 4770 \ \Omega$$

Use
$$R_E = 4700 \ \Omega$$

$$I_C = \frac{V_{cc} C}{t}$$

$$C = \frac{I_C t}{V_{cc}} = \frac{2 \times 10^{-3} \times 50 \times 10^{-6}}{10}$$

$$C = 10 \times 10^{-9} = 0.01 \ \mu F$$

$$I_{C_{discharge}} = 5 I_{C_{charge}} = (5)(2 \text{ mA})$$

$$I_{C_{discharge}} = 10 \text{ mA}$$

$$I_{C_{2total}} = I_{C_{discharge}} + I_C = 10 \text{ mA} + 2 \text{ mA}$$

$$I_{C_{2total}} = 12 \text{ mA}$$

$$I_{B_2} = \frac{I_{C_2}}{h_{FE_{min}}} = \frac{12 \text{ mA}}{20} = 0.6 \text{ mA}$$

$$R_B = \frac{e_{in} - V_{cc}}{I_{B_2}} = \frac{15 - 10}{0.6 \text{ mA}} = \frac{5}{0.6 \text{ mA}}$$

$$R_B = 8.33 \text{ k}\Omega$$

Use
$$R_B = 6.8 \text{ k}\Omega$$

15.3 SAMPLE CIRCUIT ANALYSIS PROBLEM

Determine whether the circuit shown in Fig. 15-4 will operate as a constant-current-sweep circuit. What are the characteristics of the circuit?

1. Determine the emitter current of transistor T_1, by the application of Ohm's law.
2. Determine the value of h_{FB} and, in turn, the collector current of transistor T_1. This collector current is equal to the constant-charging current.
3. Determine the trace time t_r for the applied input-pulse square wave.
4. By the application of Eq. 27, determine whether capacitor C has sufficient time to charge to the value of V_{cc}.
5. Determine whether transistor T_2 is ON when the input pulse is applied. This may be accomplished by determining the value of I_{B_2}.

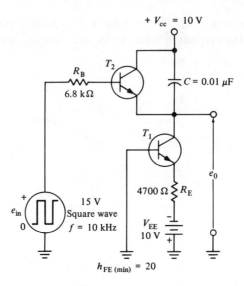

Figure 15-4 Constant-Current-Sweep Circuit

6. Determine the retrace time t_{re} by the application of the capacitor-discharge Eq. 6.

(a) Determine the value of the collector current of transistor T_2 when ON.

(b) Determine the saturation resistance of transistor T_2.

$$I_{E_1} \approx \frac{V_{EE}}{R_E} = \frac{10}{4700} = 2.13 \text{ mA}$$

$$I_{C_1} = h_{FB}I_{E_1}$$

$$h_{FB} = \frac{h_{FE}}{h_{FE} + 1} = \frac{20}{20 + 1} = 0.955$$

$$I_{C_1} = (0.955)(2.13 \text{ mA})$$

$$I_{C_1} = 2.03 \text{ mA}$$

$$T = \frac{1}{f} = \frac{1}{10 \times 10^3} = 0.1 \times 10^{-3} = 100 \ \mu\text{sec}$$

$$t_r = \frac{T}{2} = \frac{100 \ \mu\text{sec}}{2} = 50 \ \mu\text{sec}$$

$$I = \frac{CE_C}{t}$$

$$E_C = \frac{It}{C} = \frac{I_C t_r}{C} = \frac{2.03 \times 10^{-3} \times 50 \times 10^{-6}}{0.01 \times 10^{-6}}$$

$$E_C = \frac{101 \times 10^{-9}}{10^{-8}} = 10.1 \text{ V}$$

$$E_C \approx 10 \text{ V}$$

Hence, the capacitor C has sufficient time to charge to V_{cc}. Does T_2 turn ON when the input pulse is applied? Determine I_{B_2}.

$$I_{B_2} = \frac{e_{in} - V_{cc}}{R_B} = \frac{15 - 10}{6.8 \text{ k}} = 0.735 \text{ mA}$$

$$I_{C_2} = h_{FE} I_{B_2} = 20 \times 0.735 \text{ mA}$$

$$I_{C_2} = 14.7 \text{ mA}$$

For minimum retrace time,

$$I_{C_2} > 5 I_{C_1}$$
$$14.7 \text{ mA} > 5(2.03 \text{ mA})$$
$$14.7 \text{ mA} > 10.15 \text{ mA}$$

Determine retrace time (t_{re}),

$$R_{CS} \approx \frac{V_{CEsat}}{I_{Csat}} \approx \frac{0.3}{14.7 \text{ mA}}$$

$$R_{CS} \approx 20.4 \text{ }\Omega$$

$$e_C = E \epsilon^{-t/RC}$$

$$\epsilon^{+t/RC} = \frac{E}{e_C}$$

$$\frac{t}{RC} \log_{10} \epsilon = \log_{10} \frac{V_{cc}}{e_C}$$

$$t = \frac{RC \log_{10} \dfrac{V_{cc}}{e_C}}{\log_{10} \epsilon}$$

$$t_{re} = \frac{R_{CS} C \log_{10} \dfrac{V_{cc}}{e_C}}{\log_{10} \epsilon}$$

$$t_{re} \approx \frac{20 \times 0.01 \times 10^{-6} \log_{10} \dfrac{10}{0.3}}{0.434}$$

$$t_{re} = \frac{20 \times 10^{-8} \log_{10} 33.3}{0.434} = \frac{2 \times 10^{-7} \times 1.522}{0.434}$$

$$t_{re} \approx 0.7 \text{ }\mu\text{sec}$$

LABORATORY EXPERIMENT
CONSTANT-CURRENT-SWEEP CIRCUIT

OBJECT:
1. To familiarize the student with the characteristics of a constant-current-sweep circuit
2. To design a constant-current-sweep circuit
3. To analyze the operation of a constant-current-sweep circuit

MATERIALS:
2   Switching transistors (example—2N3646 silicon NPN)
1   Diode, silicon junction (example—1N914)
2   Resistor substitution boxes (10 Ω to 10 mΩ, 1 W)
1   Capacitor substitution box (0.0001 to 0.22 μF at 450 V)
2   Transistor power supplies (0 to 30 V and 0 to 250 mA)
1   Square-wave generator (20 Hz to 200 kHz with $Z_0 = 600$ Ω)
1   Oscilloscope, dc time-base type, frequency response dc to 450 kHz; vertical sensitivity, 100 mV/cm

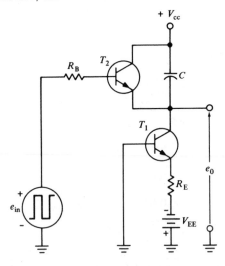

Figure 15-1X   Constant-Current-Sweep Circuit

PROCEDURE:
1. Design a constant-current-sweep circuit from the following specifications:

$$e_{in} = \text{positive pulse 15 V peak}$$
$$t_r = 45 \ \mu sec$$
$$e_o = 10 \text{ V peak}$$

Assume: $V_{EE} = 10 \text{ V}$
Initial charging current, 5 mA
Discharge current five times charging current

218

2. Design the constant-current-sweep circuit shown in Fig. 15-1X according to the conditions specified. Include all calculations. Explain why any assumptions which have been made are valid.
3. Draw a schematic of the circuit designed in Step 2, and label it with the practical standard color code values.
4. From the schematic drawn in Step 3, analyze the circuit designed in Step 2; prove that it will operate satisfactorily, according to the required specifications, when standard-value resistors and capacitor are used. Include all calculations. (Include the value for t_{re}.)
5. Connect the circuit designed. Check the operation of the circuit. Record all necessary data.
6. Compare measured values with computed values, and explain any discrepancies.

QUESTIONS AND EXERCISES

1. Explain the operation of one complete cycle of the constant-current-sweep circuit. Use the circuit and waveforms illustrated in Fig. 15-2 to aid in your explanation.
2. Explain why the amplitude of the input pulse must be greater than the amplitude of the output pulse.
3. Define R_{CS}.
4. What factors affect the value of R_{CS}?
5. What is the value of retrace time, of the circuit designed in the experiment, when the initial charging current and the initial discharge current have the same value?
6. In the circuit designed in the experiment, which component must be changed to one of a different value in order for the initial charging current and the initial discharging current to have the same value? To one of what value must this component be changed?
7. What are the advantages of the constant-current-sweep circuit when compared with the exponential-capacitor-charging circuit?
8. Refer to Fig. 15-1X. Assume that transistor T_2 is a PNP transistor. Redraw this schematic and this input-voltage waveform. Include any changes necessary for proper circuit operation.
9. Explain what is meant by a constant-current source.
10. Explain why a grounded-base amplifier closely approximates a constant-current source.
11. Design a constant-current-sweep circuit, as shown in Fig. 15-1X, with the following characteristics:

$$t_r = 175 \ \mu\text{sec}, \qquad e_o = 15 \text{ V peak}, \qquad e_{in} = +20 \text{ V pulse}, \qquad V_{EE} = 10 \text{ V}$$

Assume an initial charging current of 5 mA and an initial discharge current which is five times greater than the initial charging current. Assume $h_{FE_{min}} = 25$.

12. Determine the retrace time t_{re} of the circuit designed in Prob. 11. Assume $V_{CE_{sat}} = 0.3$ V.
13. Refer to Fig. 15-2. Analyze the circuit when

$$V_{cc} = +9 \text{ V}, \qquad V_{EE} = -10 \text{ V}, \qquad e_{in} = +12 \text{ V pulse}$$
$$R_E = 300 \ \Omega, \qquad R_B = 220 \ \Omega, \qquad C = 0.68 \ \mu\text{F}$$

Assume $h_{FE_{min}} = 15$. Determine whether this circuit would operate properly as a constant-current-sweep circuit. If it would, what would be the parameters of the output pulse?

14. What is the highest-frequency square-wave input voltage which may be applied to the circuit of Prob. 13 in order to produce an output-voltage sweep, the amplitude of which is V_{cc} volts?

chapter 16

THE UNIJUNCTION TRANSISTOR (RELAXATION OSCILLATOR)

The unijunction transistor is a special semiconductor device used in pulse and switching circuits. It is primarily used in relaxation-oscillator circuits, which, in turn, are used for timing purposes. The unijunction transistor exhibits a special characteristic, a negative resistance region. It has only one junction but is a three-element device. The three elements are base 1, base 2, and the emitter. See Fig. 16-1(a). The unijunction transistor is frequently referred to as a "unijunction" and is abbreviated "UJT."

16.1   PARAMETERS

The unijunction is made of an N-type silicon bar. Leads, connected to either end of the bar, form the two bases. A PN junction is formed along one side of the silicon bar, generally near the middle. The P-type semiconductor material, of the junction thus formed, is the emitter, as shown in Fig. 16-1(a).

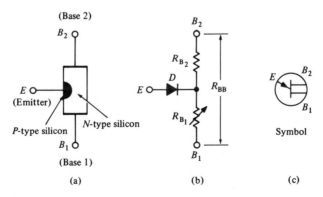

Figure 16-1   Unijunction Transistor

Refer to Fig. 16-1(b). Electrically, the unijunction resembles two resistors, in series from base 1 to base 2, and a diode, connected to the junction of the two resistors. The resistance from base 1 to base 2 (symbolized R_{BB}) is one of the major parameters of the device. The ohmic value of R_{BB} varies from one type of unijunction to another and ranges from 4 to 12 kΩ. The value of R_{BB} is both voltage and temperature dependent.

The standard symbol for the unijunction transistor is shown in Fig. 16-1(c).

Because the doping of the silicon bar is uniform, the physical location of the junction, along the bar, determines the division of the **interbase resistance**—R_{BB}.

$$R_{BB} = R_{B_1} + R_{B_2} \qquad I_E = 0 \qquad\qquad [29]$$

where

$$\begin{aligned}
R_{BB} &= \text{resistance from base 1 to} \\
&\quad \text{base 2, } (\Omega) \\
R_{B_1} &= \text{resistance from cathode of} \\
&\quad \text{junction diode to base 1} \\
&\quad \text{(with } I_E = 0\text{), } (\Omega) \\
R_{B_2} &= \text{resistance from cathode of} \\
&\quad \text{junction diode to base 2, } (\Omega)
\end{aligned}$$

To analyze the operation of the unijunction, it is necessary to know the values of R_{B_1} and R_{B_2}. Because the manufacturer does not list these values but lists instead a parameter called the intrinsic standoff ratio, symbolized η, it is necessary to solve for R_{B_1} and R_{B_2}.

$$\eta = \frac{R_{B_1}}{R_{BB}} \qquad I_E = 0 \qquad\qquad [30]$$

Practical values for the intrinsic standoff ratio range from 0.4 to 0.9, depending upon the particular unijunction in question.

16.2   UNIJUNCTION OPERATICN

To understand the following description of the operation of a unijunction, refer to Fig. 16-2. Normally, base 2 (B_2) is made positive with reference to base 1 (B_1) by voltage source V_{BB}. When $V_{EE} = 0$ V, diode D is reverse biased. Current flows from B_1 to B_2 and is limited by R_{BB}. The voltage drop produced across R_{B_1} provides the reverse-bias voltage for diode D. Since R_{B_1} and R_{B_2} form a resistive voltage divider, the reverse-bias voltage of the diode D (V_E) may be expressed

$$V_E = E_{R_{B_1}} = \left(\frac{R_{B_1}}{R_{B_1} + R_{B_2}}\right) V_{BB} = \left(\frac{R_{B_1}}{R_{BB}}\right) V_{BB}$$

But

$$\eta = \frac{R_{B_1}}{R_{BB}}$$

Hence,

$$V_E = \eta V_{BB} \qquad\qquad \text{(assuming diode } D \text{ ideal)}$$

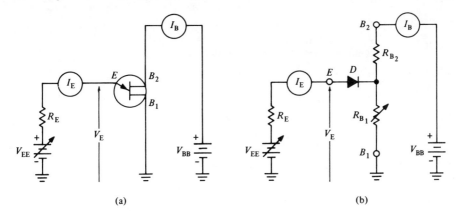

(a)   (b)

Figure 16-2

The voltage necessary to forward bias diode D is called the **peak-point voltage** and is symbolized V_P. It is equal to the intrinsic standoff ratio times the source voltage V_{BB}, when the diode junction voltage is neglected. Expressed algebraically,

$$V_P \approx \eta V_{BB} \qquad [31]$$

When diode D is forward biased by the voltage source V_{EE}, the diode operates as a closed switch. Hence, the voltage source V_{EE} injects holes, from the P-type emitter, into the N-type silicon bar. These charges carriers (holes) travel from the emitter to base 1, where they are neutralized by the electrons flowing in the external circuit. These electrons constitute the emitter-current flow. Since the resistance of semiconductor material is inversely proportional to the charge carriers (holes) in the material, the resistance of emitter-base 1 (R_{B_1}) is greatly reduced. This reduction in the value of R_{B_1} causes an increase in emitter current. Hence, the emitter-base 1 voltage V_E decreases because the resistance of R_{B_1} decreases. Thus, the emitter current increases as the emitter voltage V_E decreases; therefore, the device exhibits negative resistance.

The emitter current continues to increase and the emitter voltage continues to decrease until the rate of flow of holes into the silicon bar is greater than the rate of flow of holes which may be neutralized. Hence, the resistance of R_{B_1} starts to increase again, as the emitter current increases, and, in turn, the emitter voltage increases.

16.3   VOLT–AMPERE CHARACTERISTIC CURVES

Refer to Fig. 16-3. The emitter current which flows when $V_E = V_P$ is called the **peak-point current;** it is symbolized I_P. The peak-point current, a parameter of the device, is listed by the manufacturer. Practical balues for I_P range from 2 to 50 μA. The peak-point emitter current is both voltage and temperature dependent and varies inversely with both.

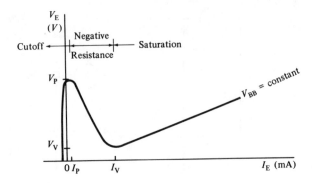

Figure 16-3 Unijunction Transistor Characteristic-Curve

The emitter current which flows when the emitter-base 1 voltage is minimum is referred to as the **valley current,** I_V. The valley current, a parameter of the device, is listed by the manufacturer. Practical values range from 1 to 10 mA.

The minimum emitter-base 1 voltage V_V is called the **valley voltage.** This parameter is not listed by the manufacturer of the device because it is very circuit dependent. Practical values of valley voltage range from 1 to 4 V.

The volt-ampere characteristic curve of a unijunction, shown in Fig. 16-3, represents a current-controlled nonlinear resistance. This characteristic curve may be approximated by straight line segments, as illustrated in Fig. 16-4.

16.4   CURRENT-CONTROLLED NONLINEAR RESISTANCE

This general volt-ampere characteristic curve represents any current-controlled nonlinear resistance: a unijunction transistor is one example. Any current-controlled nonlinear resistance device, such as a unijunction, may be used in the design of a multivibrator from any of the three classes—astable,

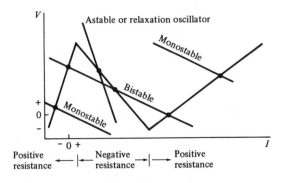

Figure 16-4   Current-Controlled Nonlinear Resistance

bistable, or monostable. The placement of the load line and the resultant Q point or points on the characteristic volt–ampere curve of the device determines the type of multivibrator designed.

The unijunction is used primarily in relaxation-oscillator circuits. The relaxation oscillator is a special case of the astable multivibrator, in which the duration of the semistable states is zero. Hence, the load line requirements for a relaxation-oscillator circuit are the same as those for the astable multivibrator.

16.5 DESCRIPTION OF OPERATION—UNIJUNCTION RELAXATION OSCILLATOR

To trace the operation of a typical relaxation-oscillator circuit which employs a unijunction, refer to Fig. 16-5. When the source voltage V_{BB} is initially applied, there is a small current flow from base 1, through the silicon

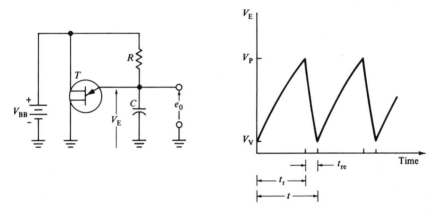

Figure 16-5   Unijunction Relaxation Oscillator

bar, to base 2. This current flow produces a voltage drop across R_{B_1}. This voltage drop is the reverse-bias voltage for the emitter-base 1 junction diode. The same time at which voltage V_{BB} is applied, capacitor C starts to charge exponentially toward V_{BB}, through resistor R. When the voltage drop across the capacitor is equal to or greater than the voltage drop across R_{B_1}, the emitter-base 1 diode is forward biased. When the diode is forward biased, the resistance of R_{B_1} decreases appreciably. This decrease in resistance of R_{B_1} provides a low resistance discharge path for capacitor C. When the capacitor has discharged, there is no further injection of holes into the silicon bar. Therefore, the resistance of R_{B_1} abruptly increases to its original high value, thus completing one cycle of operation.

For reliable relaxation oscillation to occur, it is merely necessary to place the load line, determined by resistor R, in the negative resistance region of the volt-ampere characteristic curve.

A sawtooth-voltage waveform is produced across capacitor C; thus, it becomes the output of the circuit.

The pulse width (time interval, t) of the resultant output-voltage waveform, illustrated in Fig. 16-5, is independent of the applied voltage. This characteristic is an extremely important feature of the device. The pulse duration of a standard monostable or astable multivibrator in which junction transistors are employed is dependent upon the charging of the capacitor to the applied voltage.

The equation for the time interval t of the circuit illustrated in Fig. 16-5 is derived as follows:

$$e_C = E(1 - \epsilon^{-t/RC})$$

since $V_E = e_C$ [5]

When $V_E = V_P$, the emitter-base diode is forward biased, thereby completing the charging interval of capacitor C.

$$V_P \approx \eta V_{BB}$$
$$V_E = V_{BB}(1 - \epsilon^{-t/RC})$$

But $V_E = e_C = V_P \approx \eta V_{BB}$

Hence $\eta V_{BB} = V_{BB}(1 - \epsilon^{-t/RC})$

$$\eta = 1 - \epsilon^{-t/RC}$$

$$\epsilon^{+t/RC} = \frac{1}{1 - \eta}$$

$$t = RC \, \frac{\log_{10} \dfrac{1}{1 - \eta}}{\log_{10} \epsilon}$$ [32]

Hence, the time interval t is independent of voltage. This fact makes this circuit extremely powerful when used in conjunction with one or more circuits, the operation of which is voltage dependent. The unijunction circuit may be used to force a voltage-dependent circuit to operate as if it were independent of voltage variations.

16.6   SAMPLE DESIGN PROBLEM

Design the unijunction relaxation-oscillator circuit, shown in Fig. 16-5, from the following specifications:

$$t = 200 \ \mu sec$$
$$\Delta e_o = 5 \ V$$

Given:

UJT $R_{BB} = 5 \ k\Omega$
$$\eta = 0.6$$
$$I_V = 2 \ mA$$
$$I_P = 10 \ \mu A$$

Assume:

$$V_V \approx 1 \text{ V}$$

Determine:

$$R, C, \text{ and } V_{BB}$$

Solution:

$$\Delta e_o = V_P - V_V$$
$$V_P = \Delta e_o + V_V$$
$$V_P = 5 + 1$$
$$V_P = 6 \text{ V}$$
$$V_P \approx \eta V_{BB}$$
$$V_{BB} \approx \frac{V_P}{\eta} = \frac{6}{0.6}$$
$$V_{BB} \approx 10 \text{ V}$$

For reliable relaxation oscillation to occur, the load line should intersect the V_E—I_E characteristic curve in the middle of the negative resistance region. From the parameters given by the manufacturer of a unijunction, a straight line approximation of the V_E—I_E curves may be drawn, as illustrated in Fig. 16-6. For load-line calculations, assume $V_V \approx 0$ V. The approximate minimum and maximum values for resistor R may thereby be established.

$$R_{max} = \frac{V_{BB} - V_P}{I_P} = \frac{10 - 6}{10 \ \mu A} = \frac{4}{10 \ \mu A} = 400 \text{ k}\Omega$$

$$R_{min} = \frac{V_{BB}}{I_V} \approx \frac{10}{2 \text{ mA}} \approx 5 \text{ k}\Omega$$

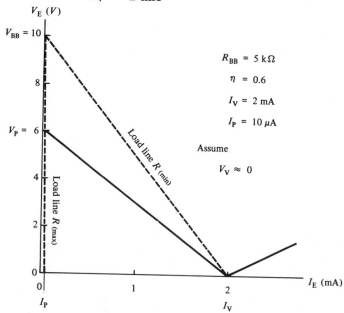

Figure 16-6

To establish a value for R, select the mean between R_{max} and R_{min}. Hence:

$$R = \sqrt{R_{max}R_{min}}$$
$$R = \sqrt{400 \times 10^3 \times 5 \times 10^3} = \sqrt{2000 \times 10^6}$$
$$R = \sqrt{20 \times 10^2 \times 10^6} = \sqrt{20 \times 10^8} = 4.45 \times 10^4$$
$$R = 44.5 \text{ k}\Omega$$

Use
$$R = 47 \text{ k}\Omega$$

$$t = RC \frac{\log_{10}\dfrac{1}{1-\eta}}{\log_{10}\epsilon} \qquad\qquad [32]$$

$$C = \frac{t \log_{10}\epsilon}{R \log_{10}\dfrac{1}{1-\eta}} = \frac{200 \times 10^{-6} \times 0.434}{4.7 \times 10^4 \log_{10}\dfrac{1}{1-0.6}}$$

$$C = \frac{200 \times 10^{-6} \times 0.434}{4.7 \times 10^4 \log_{10} 2.5} = \frac{200 \times 10^{-6} \times 0.434}{4.7 \times 10^4 \times 0.397}$$

$$C = 0.0046 \ \mu\text{F}$$

Use
$$C = 0.0047 \ \mu\text{F}$$

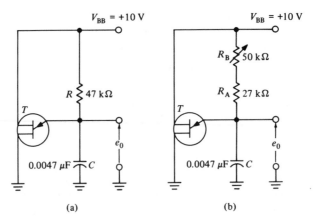

Figure 16-7

Figure 16-7 is the schematic, for the designed circuit, in which standard component values are used. To adjust the time intervals to exactly 200 μsec, resistor R must be divided into a fixed resistor R_A and a potentiometer R_B, as shown in Fig. 16-7(b). The value for resistor R_A should be less than the center design value for R (47 kΩ) and greater than the minimum value for R. Hence:

$$R_A < 47 \text{ k}\Omega \qquad \text{and} \qquad > 5 \text{ k}\Omega$$

Let
$$R_A = 27 \text{ k}\Omega$$

The value for potentiometer R_B should be selected to ensure that

$$[R_{B_{max}} + R_A] - R \approx R - R_A$$
$$[R_{B_{max}} + 27 \text{ k}\Omega] - 47 \text{ k}\Omega \approx 47 \text{ k}\Omega - 27 \text{ k}\Omega$$
$$R_{B_{max}} - 20 \text{ k}\Omega \approx 20 \text{ k}\Omega$$
$$R_{B_{max}} \approx 40 \text{ k}\Omega$$

Use
$$R_B = 50 \text{ k}\Omega \text{ potentiometer}$$

This value for R_B permits an approximately equal variation of resistance, above and below that of the center design value for R (47 kΩ).

16.7   SAMPLE ANALYSIS PROBLEM

Prove that the circuit shown in Fig. 16-7 will operate as a relaxation oscillator. Determine the characteristics of the resultant output voltage.
Solution:

1. The circuit will operate as a relaxation oscillator if the load line is in the negative resistance region of the V_E—I_E characteristic curve. This condition exists when $R_{max} > R > R_{min}$; therefore, determine the values for R_{max} and R_{min}.
2. Determine the amplitude of the output voltage.
3. By use of Fig. 16-7(b), determine the minimum and maximum interval t by the application of Eq. 32.
4. Determine the approximate capacitor-discharge interval t_{re} by use of Eq. 6—the capacitor-discharge equation. Assume $R_{B_1} \approx 100 \ \Omega$.

Given:

$$\text{UJT} \qquad R_{BB} = 5 \text{ k}\Omega$$
$$\eta = 0.6$$
$$I_V = 2 \text{ mA}$$
$$I_P = 10 \ \mu\text{A}$$

Assume:

$$V_V = 1 \text{ V}$$

$$R_{max} = \frac{V_{BB} - V_P}{I_P}$$

$$V_P \approx \eta V_{BB} \approx (0.6)(10)$$
$$V_P \approx 6 \text{ V}$$

$$R_{max} \approx \frac{10 - 6}{10 \ \mu\text{A}} \approx \frac{4}{10 \ \mu\text{A}}$$

$$R_{max} \approx 400 \text{ k}\Omega$$

$$R_{min} \approx \frac{V_{BB}}{I_V} \approx \frac{10}{2 \text{ mA}}$$

$$R_{min} \approx 5 \text{ k}\Omega$$
$$R_{max} > R > R_{min}$$
$$400 \text{ k}\Omega > 47 \text{ k}\Omega > 5 \text{ k}\Omega$$

Hence, the load line intersects the V_E—I_E characteristic curve in the negative resistance region; therefore, the circuit operates as a relaxation oscillator.

$$e_o \approx V_P - V_V$$
$$e_o \approx 6 - 1$$
$$e_o \approx 5 \text{ V}$$

$$t = RC \; \frac{\log_{10} \dfrac{1}{1 - \eta}}{\log_{10} \epsilon} \qquad\qquad [32]$$

$$t_{\min} = R_{\min} \, C \; \frac{\log_{10} \dfrac{1}{1 - \eta}}{\log_{10} \epsilon}$$

$$t_{\min} = 27 \times 10^{+3} \times 0.0047 \times 10^{-6} \; \frac{\log_{10} \dfrac{1}{1 - 0.6}}{\log_{10} \epsilon}$$

$$t_{\min} = \frac{12.7 \times 10^{-5} \times 0.397}{0.434}$$

$$t_{\min} = 116 \; \mu\text{sec}$$

$$t_{\max} = R_{\max} \, C \; \frac{\log_{10} \dfrac{1}{1 - \eta}}{\log_{10} \epsilon}$$

$$t_{\max} = 77 \times 10^3 \times 0.0047 \times 10^{-6} \; \frac{\log_{10} \dfrac{1}{1 - 0.6}}{\log_{10} \epsilon}$$

$$t_{\max} = \frac{36.2 \times 10^{-5} \times 0.397}{0.434}$$

$$t_{\max} = 331 \; \mu\text{sec}$$
$$e_C = E\epsilon^{-t/RC} \qquad\qquad [6]$$
$$V_V = V_P \epsilon^{-t_{re}/R'_{B_1} C}$$

$$t_{re} \approx R'_{B_1} C \; \frac{\log_{10} \dfrac{V_P}{V_V}}{\log_{10} \epsilon} \qquad\qquad [33]$$

$$t_{re} \approx \frac{100 \times 0.0047 \times 10^{-6} \log_{10} \dfrac{6}{1}}{\log_{10} \epsilon}$$

$$t_{re} \approx \frac{0.47 \times 10^{-6} \times 0.778}{0.434}$$

$$t_{re} \approx 0.855 \; \mu\text{sec}$$

Note: **The value of R_{B_1} is circuit dependent. A numerical value has been assigned here, merely to assist the student to understand the basic circuit action.**

16.8 DIFFERENTIATED-OUTPUT TRIGGER PULSE

The unijunction-transistor relaxation-oscillator circuit is frequently used to trigger and, hence, to time other circuits. A sharp differentiated-voltage waveform is desirable and may be obtained from the circuit, shown in Fig. 16-5, by placing a resistor R_1 between the base B_1 and ground, as shown in Fig. 16-8. To follow the detailed description of this operation, refer to Figs. 16-8 and 16-9.

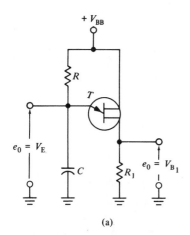
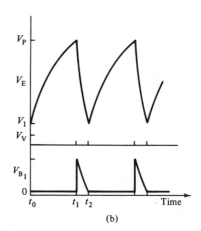

Figure 16-8

When capacitor C charges to the peak-point voltage V_P, the unijunction "fires." The emitter-base 1 resistance decreases in value to approximately 100 Ω. Refer to Fig. 16-9(a). Assume diode D to be ideal. At t_{+1} time, capacitor C charges to V_P volts and appears as a voltage source across R_{B_1} in series with R_1. This produces the characteristic differentiated-voltage waveshape, shown in Fig. 16-8(b). Hence, at t_{+1} time, the output voltage V_{B_1} is a fraction of V_P.

$$V_{B_1} \approx \left(\frac{R_1}{R_1 + R'_{B_1}} \right) V_P \Bigg|_{t_{+1}}$$

The value of peak-point voltage of a relaxation-oscillator circuit, designed to include a resistor R_1, is slightly higher than that of a comparable circuit from which R_1 has been omitted. Thus, the value of the peak-point voltage of the circuit shown in Fig. 16-8 is

$$V_P = \left(\frac{R_1 + R_{B_1}}{R_1 + R_{BB}} \right) V_{BB}$$

But

$$R_{B_1} = \eta R_{BB}$$

Hence

$$V_P = \left(\frac{R_1 + \eta R_{BB}}{R_1 + R_{BB}} \right) V_{BB} \qquad [34]$$

Refer to Fig. 16-9(a). Capacitor C, in series with R'_{B_1} and in series with diode D,

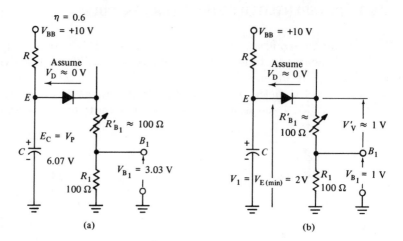

Figure 16-9   (a) Equivalent Circuit of Fig. 16-8 at t_{+1} Time; (b) Equivalent Circuit of Fig. 16-8 at t_{-2} Time

discharges through R_1 until the voltage drop across the emitter-base 1 junction is equal to the valley voltage V_V. If R'_{B_1} is equal to R_1 when the capacitor discharges to a point at which the voltage across R'_{B_1} is equal to V_V, there is a voltage drop across R_1 of V_V volts. Hence, the capacitor discharges to V_1 volts at t_{-2} time or to $2V_V$ volts, as illustrated in Fig. 16-9(b). Thus, the amplitude of the output voltage V_E is decreased when resistor R_1 is included in the circuit.

$$\Delta e_o = V_P - 2V_V \qquad \text{when } R'_{B_1} = R_1$$

Equation 32 must be modified for the interval t when, as shown in Fig. 16-8, resistor R_1 is used.

$$V_P = \left(\frac{R_1 + \eta R_{BB}}{R_1 + R_{BB}}\right) V_{BB}$$

Since
$$V_P = V_{BB}(1 - \epsilon^{-t/RC})$$

$$\left(\frac{R_1 + \eta R_{BB}}{R_1 + R_{BB}}\right) V_{BB} = V_{BB}(1 - \epsilon^{-t/RC})$$

$$\epsilon^{+t/RC} = \frac{R_1 + R_{BB}}{R_{BB}(1 - \eta)}$$

$$t = \frac{RC \log_{10}\left[\dfrac{R_1 + R_{BB}}{R_{BB}(1 - \eta)}\right]}{\log_{10} \epsilon} \qquad [35]$$

If the circuit, shown in Fig. 16-7(a), is modified by adding resistor R_1, as illustrated in Fig. 16-8, the circuit characteristics are modified.

Assume $R_1 = 100 \, \Omega$. The peak-point voltage may be determined by use of Eq. 34.

$$V_P = \left(\frac{R_1 + \eta R_{BB}}{R_1 + R_{BB}}\right) V_{BB} \qquad [34]$$

$$V_P = \left(\frac{100 + 0.6 \times 5 \text{ k}\Omega}{100 + 5 \text{ k}\Omega}\right) 10$$

$$V_P = \left(\frac{3.1 \text{ k}\Omega}{5.1 \text{ k}\Omega}\right) 10 = 6.07 \text{ V}$$

The approximate peak amplitude of the voltage pulse which appears across resistor R_1 may be determined by the use of the voltage divider equation.

$$V_{B_1} \approx \left(\frac{R_1}{R_1 + R'_{B_1}}\right) V_P$$

$$V_{B_1} \approx \left(\frac{100}{100 + 100}\right) 6.07$$

$$V_{B_1} \approx 3.03 \text{ V}$$

Assuming $R'_{B_1} \approx 100 \, \Omega$

When resistor R_1 is included in the circuit, the interval t may be determined by use of Eq. 35.

$$t = \frac{RC \log_{10}\left[\dfrac{R_1 + R_{BB}}{R_{BB}(1 - \eta)}\right]}{\log_{10} \epsilon} \qquad [35]$$

$$t = \frac{47 \times 10^3 \times 0.0047 \times 10^{-6} \log_{10}\left[\dfrac{100 + 5 \text{ k}}{5 \text{ k } (1 - 0.6)}\right]}{\log_{10} \epsilon}$$

$$t = \frac{22.1 \times 10^{-5} \log_{10} \dfrac{5.1 \text{ k}}{2 \text{ k}}}{\log_{10} \epsilon}$$

$$t = \frac{22.1 \times 10^{-5} \times 0.406}{0.434}$$

$$t = 206 \; \mu\text{sec}$$

The unijunction relaxation oscillator, an extremely simple circuit, produces either sawtooth-voltage waveforms or differentiated-voltage waveforms. Because, in this circuit, the duration of the pulse interval is independent of voltage, it may be varied within a range greater, by a factor of 10, than that imposed by voltage-dependent circuits. Therefore, this circuit is a very useful one.

An accurate and wide range square-wave generator may be built by the combination of a unijunction relaxation oscillator and an emitter-coupled

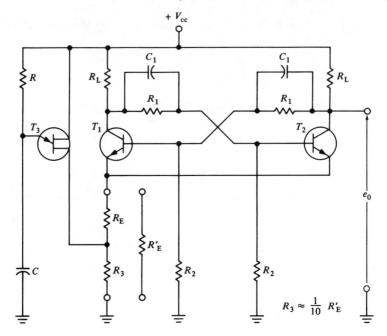

Figure 16-10

bistable multivibrator, as shown in Fig. 16-10. In this generator, each successive trigger pulse, which appears across resistor R_3, switches the stable state of the emitter-coupled bistable multivibrator. Hence, the frequency of the sawtooth-output-voltage waveform, of the unijunction relaxation oscillator, is twice that of the resultant output square wave. This circuit produces a square-wave output-voltage waveform superior to that of an astable multivibrator.

UNIJUNCTION TRANSISTOR
(RELAXATION OSCILLATOR)

OBJECT:
1. To familiarize the student with the characteristics of the unijunction transistor
2. To design a relaxation-oscillator circuit in which a unijunction transistor is employed
3. To analyze the operation of a unijunction relaxation-oscillator circuit

MATERIALS:
1   Unijunction transistor (example—2N2646 silicon UJT)
1   Manufacturer's specification sheet for type of unijunction transistor used (2N2646—see Appendix B)
2   Resistor substitution boxes (10 Ω to 10 MΩ, 1 W)
1   Capacitor substitution box (0.0001 to 0.22 μF at 450 V)
2   Transistor power supplies (0 to 30 V and 0 to 250 mA)
1   Oscilloscope, dc time-base type; frequency response dc to 450 kHz; vertical sensitivity, 100 mV/cm
1   Potentiometer, 50 kΩ, 1 W

PROCEDURE:
1. Design a unijunction-transistor relaxation-oscillator circuit from the following specifications:

$$\Delta e_o = 10 \text{ V}$$
$$t = 600 \ \mu sec$$

Circuit—Fig. 16-1X

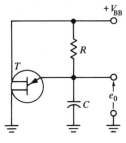

Figure 16-1X   Unijunction Relaxation Oscillator

2. Determine the value of R_{BB}, for the unijunction used, by connecting the circuit shown in Fig. 16-2X(a). For what reason were the particular values for V_{BB} and for R selected?
3. Determine the value of the intrinsic-standoff ratio, for the unijunction used, by connecting the circuit shown in Fig. 16-2X(b). Explain how and why this circuit may be used to determine η. What factors determine the value of resistor R?

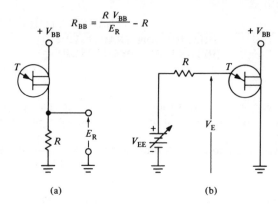

$$R_{BB} = \frac{R \, V_{BB}}{E_R} - R$$

(a)   (b)

Figure 16-2X   (a) Circuit to Determine the Value of R_{BB}; (b) Circuit to Determine the Value of η

4. Design the circuit, shown in Fig. 16-1X, from the specifications stated. Use the measured values of interbase resistance and of the intrinsic-standoff ratio.

5. Draw a schematic of the circuit designed, and label it with the practical, standard, color code values.

6. From the schematic, drawn in Step 5, analyze the circuit designed; prove that it will operate satisfactorily, for the required specifications, when components of a standard value are used. Include all calculations.

7. Connect the circuit designed. Check the operation of the circuit. Record all necessary data.

8. Replace resistor R, of the designed circuit, with a resistor R_A and a potentiometer R_B, of appropriate value, as shown in Fig. 16-7(b). For what reasons were the particular values for R_A and R_B selected?

9. Make the adjustment, in the value of resistor R_B, necessary to produce an output-pulse interval of exactly 600 μsec.

10. When the circuit is adjusted to produce a pulse interval of 600 μsec, vary the value of source voltage V_{BB} by ± 20 percent of the design value. What effect does this variation have upon the output-pulse interval? Explain.

11. Adjust the value of the source voltage to that of the original design value.

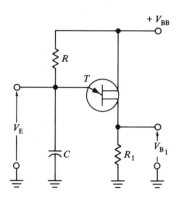

Figure 16-3X

12. Calculate the value of the minimum and of the maximum pulse interval, for the circuit designed, when resistor R_B is the only variable. Show all calculations.

13. By varying the value of resistor R_B, determine, experimentally, the maximum and minimum pulse interval of the circuit. Explain the reason for the difference between the value of the calculated and that of the measured pulse interval.

14. Refer to the schematic drawn in Step 5. To modify this circuit, insert resistor R_1 as shown in Fig. 16-3X. Let $R_1 = 47 \, \Omega$.

15. Check the operation of this modified circuit (Fig. 16-3X). Record all necessary data.

16. Determine the value of R_{B_1} when the unijunction has fired. Explain the steps necessary to determine this value.

QUESTIONS AND EXERCISES

1. Explain the operation of the basic unijunction relaxation-oscillator circuit shown in Fig. 16-5.

2. Which circuit parameters affect the amplitude of the output-sawtooth-voltage waveform of the circuit shown in Fig. 16-5? Explain.

3. Which circuit parameters affect the pulse interval of the sawtooth-output-voltage waveform of the circuit shown in Fig. 16-5? Explain.

4. May the amplitude of the sawtooth-output-voltage waveform be varied when the pulse interval is not varied? Explain.

5. Explain the significance of the negative resistance region of operation of the unijunction transistor.

6. What are the advantages of a circuit in which a unijunction is employed for timing when compared to a circuit in which junction transistors are employed for timing?

7. As illustrated in Fig. 16-10, design a square-wave generator circuit in which an emitter-coupled bistable multivibrator and a unijunction relaxation-oscillator circuit are used. The amplitude of the output square wave of voltage must be 10 V. The frequency of the output square wave must be 1000 Hz.

Given (for multivibrator):

$$I_C = 10 \text{ mA}$$
$$V_{cc} = 15 \text{ V}$$
2 silicon NPN transistors
$$h_{FE(min)} = 20$$

Refer to chapter 9 for a review of emitter-coupled multivibrator design procedure.

Given (for unijunction relaxation oscillator):

UJT
$$R_{BB} = 8 \text{ k}\Omega$$
$$\eta = 0.7$$
$$I_V = 8 \text{ mA}$$
$$I_P = 25 \, \mu\text{A}$$

Show all calculations, and explain why any assumptions which have been made are valid.

8. What are the minimum and the maximum output square-wave frequencies for the circuit designed in Prob. 7? Show all calculations.

9. Why is only one unijunction necessary in the design of an astable multivibrator, while two junction transistors are required in the design of a similar circuit? Explain.

10. Explain why a unijunction transistor is an example of a current-controlled non-linear resistance.

11. Refer to Fig. 16-5. Design a unijunction relaxation-oscillator circuit from the following specifications:

$$t = 75 \ \mu\text{sec}, \qquad \Delta e_o = 8 \ \text{V}$$

Given a unijunction with the following parameters:

$$\eta = 0.8, \qquad I_V = 9 \ \text{mA}, \qquad I_P = 10 \ \mu\text{A}$$

Assume: $\qquad\qquad\qquad V_V = 1 \ \text{V}.$

12. Refer to Fig. 16-5. Analyze the unijunction relaxation-oscillator circuit when

$$V_{BB} = +20 \ \text{V}, \qquad R = 68 \ \text{k}\Omega, \qquad C = 0.015 \ \mu\text{F}$$
$$\eta = 0.6, \qquad I_V = 6 \ \text{mA}, \qquad I_P = 4 \ \mu\text{A}$$

Assume: $\qquad\qquad\qquad V_V = 0 \ \text{V}.$

chapter 17

TUNNEL DIODE
(BISTABLE CIRCUIT)

The tunnel diode is a semiconductor junction diode which is heavily doped and which has an extremely narrow junction thickness. The tunnel diode has three important electrical characteristics: a negative resistance region, an extremely high switching speed, and a relative freedom from the effect of temperature.

Tunnel diodes may be made from any of several semiconductor materials, such as germanium, silicon, and gallium arsenide. The voltage parameters of the device are determined, primarily, by the semiconductor material from which it is made. The current capability of the diode is determined by its design, however, not by the material from which it is made. Thus, several diodes, made from the same material, may have different current capabilities.

17.1 COMPARISON OF TUNNEL DIODES WITH JUNCTION DIODES

Refer to Fig. 17-1—a step-by-step comparison between the electrical characteristics of the tunnel diode and those of the conventional junction diode, a study of which may clarify one's concept of the operation of the tunnel diode. Each step in the comparison represents the electrical characteristics of the diodes under varied applications of forward bias.

Refer to Fig. 17-1. The conventional junction diode to which no bias is applied has no forward current flow. In terms of electron energy, the barrier potential (electron energy hill) prevents the movement of electrons from the N-type semiconductor material to the P-type semiconductor material.

The P-type semiconductor material of the unbiased tunnel diode has a valence level of electron energy which is higher than the conduction level of

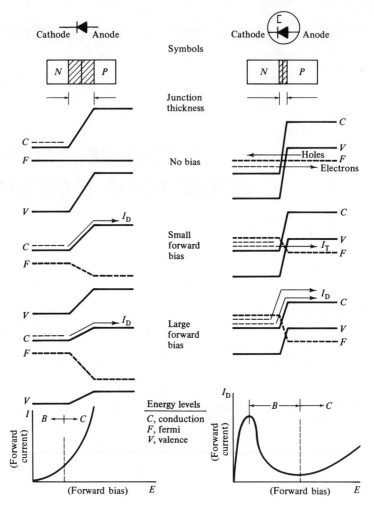

Figure 17-1   Left: Junction Diode; Right: Tunnel Diode

electron energy of the N-type semiconductor material. At a temperature of absolute zero, there is no movement of holes nor of electrons in either type of semiconductor material. Compared to absolute zero, room temperature is relatively high and represents a considerably large amount of heat energy. This thermal energy causes some of the electrons of the N-type semiconductor to pierce the junction barrier and to enter the permissible valence level of electron energy of the P-type semiconductor. Conversely, some of the holes of the P-type semiconductor material pierce the junction barrier and enter the permissible conduction level of electron energy of the N-type semiconductor. The piercing of the junction barrier is a tunneling action; hence, the "tunnel diode" is so named.

At room temperature, the net charge carrier flow (holes and electrons) of the unbiased tunnel diode is zero. This tunneling action occurs when the conduction level of electron energy of the N-type semiconductor material and the valence level of electron energy of the P-type semiconductor material overlap; thus, the charge carriers move through the depletion region to a level of energy equal to that from which they came.

Refer to Fig. 17-1. When a small forward bias is applied to a conventional junction diode, the height of the electron energy hill, at the conduction level of electron energy, is decreased slightly, thus permitting a very small flow of forward current (diffusion current I_D).

In the tunnel diode, the application of a small forward bias causes the flow of a forward tunneling current I_T. As the forward-bias voltage increases, the forward tunneling current increases. This increase in forward bias decreases the overlapping of the permissible conduction level of electron energy, of the N-type semiconductor material, with the permissible valence level of electron energy, of the P-type semiconductor material. A higher value of forward bias and, in turn, a decrease in the overlapping of permissible energy levels limits the tunneling current to a maximum value. A further increase in forward bias and, in turn, a decrease in the overlapping of permissible energy levels decreases the tunneling current. In this area of biasing, a condition exists in which the forward bias increases as the current decreases; thus, the negative resistance region of operation is established.

Refer to Fig. 17-1. When a large forward bias is applied to the conventional junction diode, the electron energy hill is decreased to a value which permits a large forward current flow.

The application of a large forward bias to a tunnel diode causes no overlapping of the permissible conduction level of electron energy, of the N-type semiconductor material, with the permissible valence level of electron energy, of the P-type semiconductor material. Hence, there can be no tunneling action. The electron energy hill, however, at the conduction level, decreases to a value which permits forward diffusion-current flow. Any further increase in forward bias causes the tunnel diode to operate as a conventional junction diode, as shown by area C in the volt–ampere characteristic curves, illustrated in Fig. 17-1.

17.2   TUNNEL-DIODE PARAMETERS

To follow the description of the parameters of a tunnel diode, refer to the volt-ampere characteristic curve shown in Fig. 17-2.

The peak-point voltage V_P is the minimum forward bias necessary to produce the first current peak before the negative resistance region is encountered. Practical values of peak-point voltage range from 50 to 200 mV.

When the tunnel diode is forward biased by the peak-point voltage, the forward current flow is the peak-point current (I_P). Practical values of peak-point current, for most tunnel diodes, range from 1 to 100 mA.

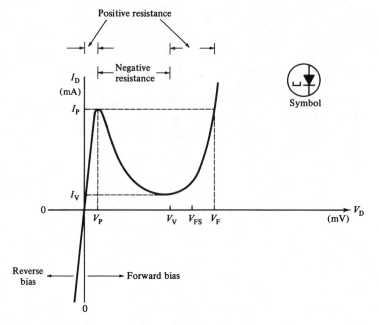

Figure 17-2 Tunnel Diode; Characteristic Curve

The forward-bias voltage which is in excess of the peak-point voltage and which produces a minimum forward current flow is called the valley-point voltage (V_V). Practical values of valley-point voltage for germanium tunnel diodes range from 300 to 400 mV.

The forward current that flows when the valley-point voltage is applied is called the valley-point current (I_V). Practical values for this current range from 0.1 to 5 mA.

The forward-bias voltage current flow of an amplitude equal to that of the peak-point current is called the forward voltage (or projected peak-point voltage) and is symbolized V_F. The amplitude of this voltage is approximately 500 mA.

17.3 LINEAR APPROXIMATIONS

These parameters are listed by the manufacturers of tunnel diodes. From them, a linear volt–ampere characteristic curve may be approximated, as illustrated in Fig. 17-3. Some manufacturers list another parameter, the forward voltage V_{FS}, as shown in Fig. 17-2 and 17-3. When this parameter is specified, it is the value of forward bias which is in excess of the valley voltage and which produces a forward-current flow, the amplitude of which is one-fourth that of the peak-point current (I_P). The value of this voltage (V_{FS}) is needed to establish the slope of the straight-line portion of the curve between V_{FS} and

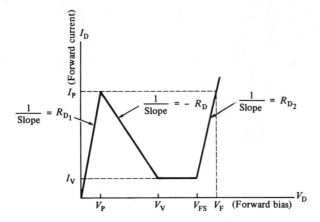

Figure 17-3 Tunnel Diode; Linear Volt–Ampere Characteristic; Curve Approximation

V_F. Therefore, when this parameter is not listed, a practical value for V_{FS} may be determined as shown:

$$V_{FS} \approx \frac{V_F - V_V}{2} + V_V \qquad [36]$$

The approximate volt–ampere characteristic curve may be used to determine the approximate resistance of the tunnel diode, in the three different areas of operation.

$$R_{D_1} \approx \frac{V_P}{I_P} \qquad [37]$$

$$-R_D \approx \frac{V_V - V_P}{I_P - I_V} \qquad [38]$$

$$R_{D_2} \approx \frac{V_F - V_{FS}}{I_P - I_V} \qquad [39]$$

17.4   VOLTAGE-CONTROLLED NONLINEAR RESISTANCE

Refer to Fig. 17-4. The tunnel diode may be employed in bistable, monostable, or astable multivibrator circuits because it operates as a voltage-controlled nonlinear resistance. The operation of the circuit in which the tunnel diode is used is determined by the electrical characteristics of the circuit, which are represented by a portion, or portions, of the characteristic curve. This chapter defines the tunnel diode as used in the bistable circuit. Chapter 18 defines the tunnel diode as used in the monostable and astable circuits.

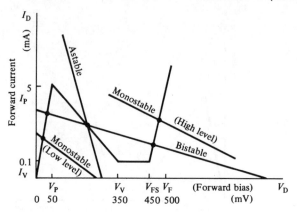

Figure 17-4   Tunnel-Diode Voltage-Controlled; Nonlinear Resistance

17.5   BISTABLE CIRCUIT

Refer to Fig. 17-5. Circuit conditions, represented by the characteristic curve in which the load line intersects the two positive resistance regions as well as the negative resistance region of operation, must exist before a tunnel diode can operate as a bistable circuit. Hence, the value of resistor R and the value of the applied voltage V_{cc} must be selected to fulfill these requirements.

With no input signal applied, the tunnel diode conducts, and the output

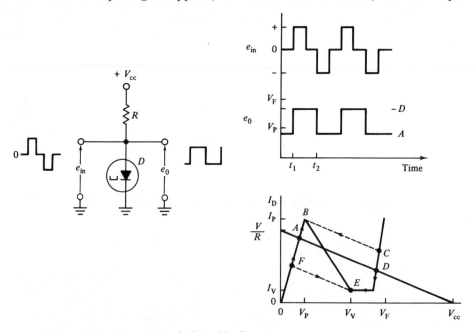

Figure 17-5   Tunnel-Diode Bistable Circuit

voltage has a value slightly less than V_P, as shown by operating point A on the characteristic curve.

The application of a positive pulse, at t_{+1} time, causes an increase in current flow, through the tunnel diode D. If the amplitude of the input voltage and, in turn, that of the current flow is greater than I_P, the tunnel diode switches to the other stable state; at this time, the output voltage is slightly less than V_F, as indicated by the operating point D, on the characteristic curve.

To switch the bistable circuit back to the original stable state (operating point A), a negative input pulse is required. The amplitude of the negative pulse must be large enough to decrease the tunnel-diode current to a value less than I_V. When this occurs, the operating point moves from point D to point E and, in turn, to point F. As the negative pulse passes, the operating point moves to the first stable state, operating point A.

17.6 SAMPLE DESIGN AND ANALYSIS PROBLEM

Design a bistable tunnel-diode circuit from the following specifications. Refer to Fig. 17-6(b).

$$e_s = \text{positive and negative pulses, 10 V p–p}$$
$$V_{cc} = 1 \text{ V}$$

Assume:

Tunnel diode—1N3715

(See Appendix B—manufacturer's specification sheet)

Determine:

$$R, R_S, C, V_{B_1}, V_{B_2}, +e_{in}, \text{ and } -e_{in}$$

Solution:

1. From the manufacturer's specification sheet, draw a straight-line volt–ampere characteristic curve approximation of the tunnel diode. See Fig. 17-6(a).
2. Arbitrarily, draw a load line which intersects the positive resistance regions and the negative resistance region of the characteristic curve. (Point A must be approximately equidistant from points B and F, for the given source voltage, $V_{cc} = 1$ V.)
3. Determine the value of R for this load line, by use of Ohm's law.
4. The minimum value of $+e_{in}$ and $-e_{in}$ may be approximated by a graphical solution, as indicated in Fig. 17-6(a).
5. The approximate values for $V_{B_1}, V_{B_2}, I_{B_1},$ and I_{B_2} may be determined graphically from the characteristic curve illustrated in Fig. 17-6(a).

The values for $+e_{in}, -e_{in}, V_{B_1}, V_{B_2}, I_{B_1},$ and I_{B_2} may also be determined by equivalent circuit analysis.

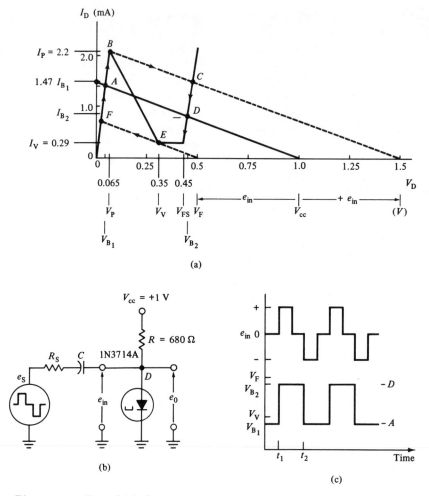

Figure 17-6   Tunnel-Diode Bistable Circuit

Equivalent Circuit Method:

6. Draw a schematic of an equivalent circuit in which circuit conditions are those necessary for operation at t_{-1} time. See Fig. 17-7. Write the loop equation, and solve for I_{B_1}. The voltage V_{B_1} is e_o for this equivalent circuit.

7. Draw a schematic of an equivalent circuit in which circuit conditions are those necessary for operation at t_{+1} time. See Fig. 17-8. Write the loop equation, and solve for $+e_{in}$.

8. Draw a schematic of an equivalent circuit in which circuit conditions are those necessary for operation at t_{-2} time. See Fig. 17-9. The output voltage e_o of this circuit is V_{B_2}.

9. Draw a schematic of an equivalent circuit in which circuit conditions are those necessary for operation at t_{+2} time. See Fig. 17-10. Write the loop equation, and solve for $-e_{in}$.

10. Refer to Fig. 17-6(b). Draw a schematic of the equivalent input circuit. See Fig. 17-11.

Solution:

$$R = \frac{V_{cc}}{I_{assumed}} = \frac{1}{1.47 \text{ mA}} = 680 \ \Omega$$

$$R_{D_1} \approx \frac{V_P}{I_P} = \frac{65 \text{ mV}}{2.2 \text{ mA}} = 29.5 \ \Omega$$

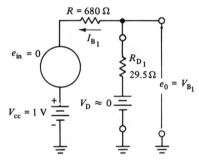

Figure 17-7 Equivalent Circuit at t_{-1} Time

$$I_{B_1} = \frac{V_{cc}}{R + R_{D_1}} = \frac{1}{680 + 29.5} = 1.41 \text{ mA}$$

$$V_{B_1} = e_o = \left(\frac{R_{D_1}}{R_{D_1} + R}\right) V$$

$$V_{B_1} = e_o = \left(\frac{29.5}{29.5 + 680}\right) (1)$$

$$V_{B_1} = 41.7 \text{ mV}$$

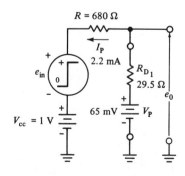

Figure 17-8 Equivalent Circuit at t_{+1} Time

$$V_{cc} + e_{in} - V_P = I_P(R_{D_1} + R)$$
$$1 + e_{in} - 0.065 = (2.2 \text{ mA})(29.5 + 680)$$
$$e_{in} + 0.935 = 1.533$$
$$+e_{in} = 0.518 \text{ V}$$

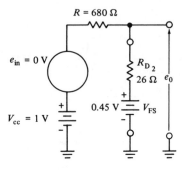

Figure 17-9   Equivalent Circuit at t_{-2} Time

$$R_{D_2} \approx \frac{V_F - V_{FS}}{I_P - I_V} = \frac{500 \text{ mV} - 450 \text{ mV}}{2.2 \text{ mA} - 0.29 \text{ mA}} = 26 \ \Omega$$
$$V_{B_2} = e_o = V_{FS} + E_{R_{D_2}}$$

$$E_{R_{D_2}} = \left(\frac{R_{D_2}}{R_{D_2} + R}\right) E$$

$$E = V_{cc} - V_{FS}$$
$$E = 1 - 0.45$$
$$E = 0.55 \text{ V}$$

$$E_{R_{D_2}} = \left(\frac{26}{26 + 680}\right)(0.55)$$
$$E_{R_{D_2}} = 36.8 \text{ mV}$$

$$V_{B_2} = V_{FS} + E_{R_{D_2}}$$
$$V_{B_2} = 0.45 + 0.0368$$
$$V_{B_2} = 0.486 \text{ V}$$

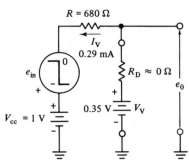

Figure 17-10   Equivalent Circuit at t_{+2} Time

$$V_{cc} - e_{in} - V_V = I_V(R_D + R)$$
$$1 - e_{in} - 0.35 = (0.29 \text{ mA})(0 + 680)$$
$$+0.65 - e_{in} = 0.197$$
$$-e_{in} = 0.46 \text{ V}$$

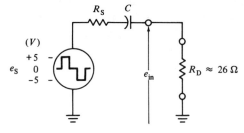

Figure 17-11 Equivalent Input Circuit

$$Z_{in} \approx R_D \approx 26 \ \Omega \qquad \text{as } Z_{in} \approx \frac{R_D R}{R_D + R} \approx R_D$$

$$-e_{in} \approx 0.46 \text{ V}$$
$$+e_{in} \approx 0.518 \text{ V}$$

Because the input voltage e_s is symmetrical, the larger of the two voltages, $-e_{in}$ and $+e_{in}$, is the controller. Hence:

$$e_{in(p-p)} = 2(+e_{in}) = 2(0.518)$$
$$e_{in} = 1.036 \text{ V} \approx 1.04 \text{ V}$$

(neglecting C)

$$e_s = E_{Rs} + e_{in}$$
$$E_{Rs} = 10 - 1.04 = 8.96 \text{ V}$$

$$E_{Rs} = \left(\frac{R_S}{R_D + R_S}\right) e_s$$

$$R_S = \frac{E_{Rs} R_D}{e_s - E_{Rs}}$$

$$R_S = \frac{(8.96)(26)}{10 - 8.96} = \frac{(8.96)(26)}{1.04}$$

$$R_S = 218 \ \Omega$$

Use
$$R_S = 180 \ \Omega \qquad \text{(standard color code value)}$$

The smaller of the two resistances of standard value, 180 and 220 Ω, should be selected to assure sufficient amplitude of signal to change the stable states.

Because capacitor C is employed to isolate the ac signal from the dc biasing circuit of the tunnel-diode switch, the value of capacitance is not critical. A lack of high-frequency components will round the leading edge of the input signal, however; therefore, the value of capacitance selected must ensure a square leading edge.

17.7   PRACTICAL BISTABLE CIRCUIT

Refer to Fig. 17-12(a). Many transistor power supplies have a lower output-voltage limit of approximately one or more volts. In many tunnel-diode circuits, the required source voltage is less than the minimum obtainable from available power supplies. Therefore, a voltage divider must be provided for a practical tunnel-diode circuit, in order to utilize a source voltage, the value of which may range from approximately 5 to 10 V.

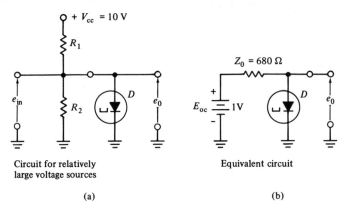

Circuit for relatively   Equivalent circuit
large voltage sources

(a)   (b)

Figure 17-12   Tunnel-Diode Bistable Circuit

Thus, to operate the sample design circuit with a 10 V source, a voltage divider (R_1 and R_2) is necessary. In order to determine the proper values for R_1 and R_2, by use of Thevenin's theorem, the effective dc source voltage and load resistance of the equivalent circuit, which was obtained by use of Thevenin's theorem, must be equal to the source voltage and load resistance of the sample design circuit. See Fig. 17-12(b).

$$Z_0 = \frac{R_1 R_2}{R_1 + R_2}$$

Hence

$$R_2 = \frac{Z_0 R_1}{R_1 - Z_0}$$

and

$$E_{oc} = \left(\frac{R_2}{R_1 + R_2}\right) V_{cc}$$

$$\frac{E_{oc}}{V_{cc}} = \frac{R_2}{R_1 + R_2}$$

Substituting the value for R_2,

$$\frac{E_{oc}}{V_{cc}} = \frac{\dfrac{Z_0 R_1}{R_1 - Z_0}}{R_1 + \dfrac{Z_0 R_1}{R_1 - Z_0}}$$

Solving this relationship for R_1,

$$R_1 = \frac{Z_0 V_{cc}}{E_{oc}} \qquad [40]$$

Since the value of R_1 has been determined, the value for R_2 may be established:

$$R_2 = \frac{Z_0 R_1}{R_1 - Z_0}$$

$$R_1 = \frac{Z_0 V_{cc}}{E_{oc}} = \frac{(680)(10)}{1}$$

$$R_1 = 6800 \ \Omega \qquad \text{(standard color code value)}$$

$$R_2 = \frac{Z_0 R_1}{R_1 - Z_0} = \frac{(6800)(680)}{6800 - 680} = \frac{(6800)(680)}{6120}$$

$$R_2 = 755 \ \Omega$$

Use $\qquad R_2 = 820 \ \Omega \qquad$ (standard color code value)

Having established the practical values for resistor R_1 and resistor R_2, calculate the actual values for E_{oc} and Z_0.

$$E_{oc} = \left(\frac{R_2}{R_1 + R_2}\right) V_{cc} = \left(\frac{820}{820 + 6800}\right) 10 = \frac{8200}{7620}$$

$$E_{oc} = 1.07 \ V$$

$$Z_0 = \frac{R_1 R_2}{R_1 + R_2} = \frac{(820)(6800)}{820 + 6800} = \frac{(820)(6800)}{7620}$$

$$Z_0 = 730 \ \Omega$$

Refer to Fig. 17-13. The effective dc source voltage for the tunnel-diode circuit is E_{oc}, and the effective load resistance for the circuit is Z_0, as shown

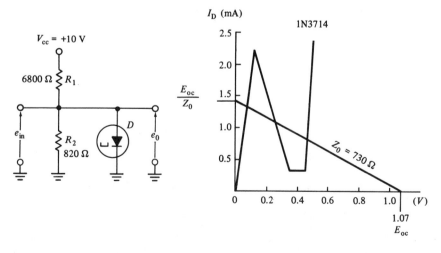

Figure 17-13

by the load line. This new circuit has been reduced to the form of the original circuit, shown in Fig. 17-6. The new circuit is equivalent to the one shown in Fig. 17-14.

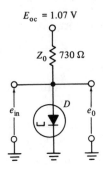

Figure 17-14   Effective Circuit of Fig. 17-13

Advantages of the tunnel diode include stability, practicability in simple circuitry, low power consumption, and high switching speed. Because no carrier storage is involved in the operation of the tunnel diode, it is, at present, the fastest switching device.

In some applications, the small amplitude of the output voltage of the tunnel diode is a disadvantage.

TUNNEL-DIODE BISTABLE CIRCUIT

OBJECT:

1. To familiarize the student with the characteristics of the tunnel diode
2. To design a tunnel-diode bistable circuit
3. To analyze the operation of a tunnel-diode bistable circuit

MATERIALS:

1 Tunnel diode (example—1N3716 germanium)

1 Manufacturer's specification sheet for tunnel diode used (1N3716—see Appendix B)

3 Resistor substitution boxes (10Ω to 10 MΩ, 1 W)

1 Capacitor substitution box (0.0001 to 0.22 μF at 450 V)

2 Transistor power supplies (0 to 30 V and 0 to 250 mA)

1 Oscilloscope, dc time-base type; frequency response dc to 450 kHz; vertical sensitivity, 100 mV/cm

1 Square-wave generator (20 Hz to 200 kHz with $Z_0 = 600$ Ω)

PROCEDURES:

1. Design a tunnel-diode bistable circuit, as shown in Fig. 17-1X, from the following specifications:

$V_{cc} = 1$ V (or lowest value obtainable from available power supply ≈ 1 V)

$e_s = \pm 5$ V

$D =$ tunnel diode (example—1N3716 germanium)

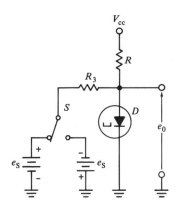

Figure 17-1X

Determine:

$$R \text{ and } R_3$$

2. By use of the data from the manufacturer's specification sheet, draw a straight-line, volt–ampere characteristic curve approximation of the tunnel diode (1N3716, see Appendix B). Label.

3. Draw a load line on the characteristic curve, constructed in Step 2, so that the current intercept is approximately equidistant from I_P and I_V.
4. From the characteristic curve and load line, establish the proper value for resistor R.
5. By use of the characteristic curve, graphically establish the maximum and the minimum values of current necessary to switch the stable states. From this data, determine the proper value for resistor R_3.
6. Connect the circuit designed. With no input voltage applied, determine the output voltage of the circuit, and determine the stable state of the circuit.
7. After the stable state of the circuit has been determined, apply an input signal, of a polarity appropriate to change the stable state. Observe and record the necessary output data. (Use dc power supply for input signal.)
8. By use of the equivalent circuit method, outlined in the sample design and analysis problem, verify the output voltage, measured in each of the stable states.

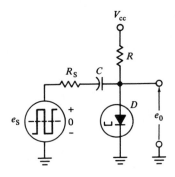

Figure 17-2X

9. Modify the connected circuit to that of Fig. 17-2X. Use a value of $C = 100$ pF. Apply an input square wave of voltage, the amplitude of which is equal to that of the dc source. Use a frequency of 20,000 Hz.
10. Observe, measure, and record the necessary data from this circuit. Compare this result with that obtained from the circuit shown in Fig. 17-1X. Explain.
11. Modify the circuit shown in Fig. 17-2X to that shown in Fig. 17-3X. Use $V_{cc} = 8$ V. Show all necessary calculations.

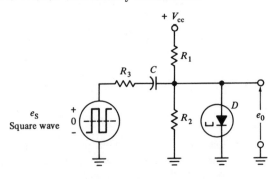

Figure 17-3X Bistable Tunnel-Diode Circuit

12. Connect the circuit designed in Step 11. Apply an input square wave, the frequency of which is 20,000 Hz and the amplitude of which is 10 V, peak to peak.

13. Observe, measure, and record the necessary data from this circuit. Include the rise time and the fall time of the resultant output voltage.

14. By use of the equivalent circuit method, outlined in the sample design and analysis problem, verify the output voltage measured in each of the stable states.

15. Compare and explain the results obtained from the three circuits—Figs. 17-1X, 17-2X, and 17-3X.

QUESTIONS AND EXERCISES

1. Explain "tunneling."
2. What is the difference between tunneling current and diffusion current?
3. Why does the tunneling current decrease as the forward bias increases?
4. What factors determine the amplitude of the output voltage of a bistable tunnel-diode circuit?
5. By use of the data obtained from the experimental results, explain how and why the measured values for rise time and fall time of the output voltage compare with the theoretical values.
6. By use of the characteristic curve and load line, explain the operation of the tunnel-diode bistable circuit.
7. Is the tunnel diode a voltage- or a current-operated device? Explain.
8. List and explain the advantages of the tunnel-diode bistable circuit, when compared with a transistor bistable circuit.
9. Explain why the negative resistance region of operation is necessary for proper operation of the bistable circuit.
10. Design and analyze a tunnel-diode bistable circuit from the following specifications:

$$V_{cc} = 5 \text{ V}$$
$$e_s = 10 \text{ V, peak–peak square wave}, f = 50,000 \text{ Hz}$$
$$D = 1\text{N}3721 \text{ (see Appendix B for data)}$$
$$\text{Circuit—Fig. 17-3X}$$

11. List and explain the disadvantages of the tunnel-diode bistable circuit, when compared with a transistor bistable circuit.
12. Draw a straight line approximation of the volt–ampere characteristic curve of a 1N3712 tunnel diode, and label. See Appendix B for manufacturer's specification sheet.
13. Refer to Fig. 17-3X. Analyze the circuit when

$$V_{cc} = +12 \text{ V}, \qquad R_1 = 24 \text{ k}\Omega, \qquad R_2 = 2.2 \text{ k}\Omega$$
$$R_3 = 68 \text{ }\Omega, \qquad D = 1\text{N}3712, \qquad e_s = 5 \text{ V peak–peak square wave}$$

14. Determine the minimum amplitude of input square-wave voltage e_s which may be used to operate the bistable circuit of Prob. 13. Explain.

chapter 18

TUNNEL-DIODE MONOSTABLE AND ASTABLE MULTIVIBRATORS

In pulse circuitry, the tunnel diode is used as the active device in monostable and in astable multivibrator circuits as well as in the bistable circuit (chapter 17). In this discussion of the tunnel-diode monostable and astable circuits, the former is defined first.

18.1 TYPES OF MONOSTABLE MULTIVIBRATORS

The tunnel-diode monostable multivibrator circuit has two modes of operation: a high level, and a low level, as shown in Fig. 18-1. The high- and low-level modes of operation are determined by the position and slope of the load

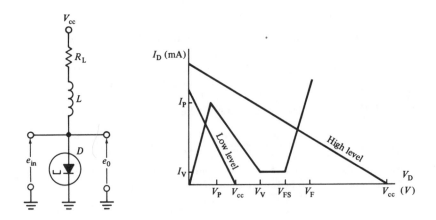

Figure 18-1 Tunnel-Diode Monostable Multivibrators

lines on the positive resistance regions of operation of the volt–ampere characteristic curve. The basic schematic for this monostable circuit is simple and differs from that of the bistable circuit by the inclusion of an inductance L.

18.2   DESCRIPTION OF OPERATION—LOW-LEVEL

To follow the description of the operation of a low-level tunnel-diode monostable multivibrator, refer to Fig. 18-2. The value of R_L and that of V_{cc} are selected to ensure that the dc operating point will be on the positive resistance region of the characteristic curve, where V_o has a value less than that of

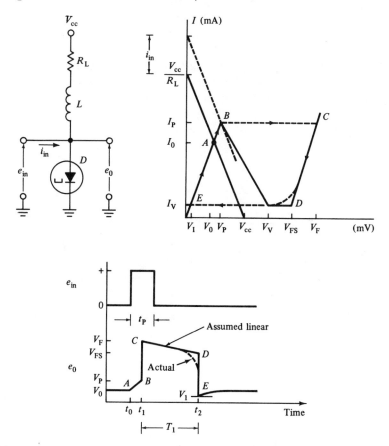

Figure 18-2   Low-Level Tunnel-Diode Monostable Multivibrator

V_P. The value of inductance is selected to ensure that L will produce an induced voltage in excess of V_F. The positive trigger pulse must be of sufficient amplitude and duration to effect an increase from I_o to I_P in the tunnel-diode current.

If no trigger pulse is applied at t_{-0} time, the steady state operating current through the tunnel diode I_o is limited by the resistance of the tunnel diode R_{D_1}, and the load resistor R_L. The application of the input trigger pulse at t_{+0} time increases the tunnel-diode current from I_o to I_P.

The delay time—the interval between the application of the trigger pulse at t_{+0} time and the resultant output pulse at t_{+1} time—is the interval required for the tunnel-diode current to increase from I_o to I_P. This increase in tunnel-diode current I_D from I_o to I_P produces an increase in voltage drop across the tunnel diode from V_o to V_P. This change in voltage across the tunnel diode causes it to act like a voltage source in series with the source voltage V_{cc} and of a polarity opposing V_{cc}. Hence, the current through inductance L and through resistor R_L decreases. The decreasing current, through L, induces a voltage across L of a polarity aiding the source voltage V_{cc}. Hence, the resultant voltage across the tunnel diode is equal to the sum of the induced voltage across L plus the source voltage. Therefore, the voltage across the tunnel diode is in excess of the source voltage V_{cc}. When the value of inductance L is correctly selected, the sum of the induced voltage across L plus the source voltage is equal to or in excess of voltage V_F. Because the tunnel-diode current may not exceed I_P and because the voltage across the tunnel diode is equal to or in excess of the forward voltage V_F, the operating point on the characteristic curve shifts instantly from point B to point C.

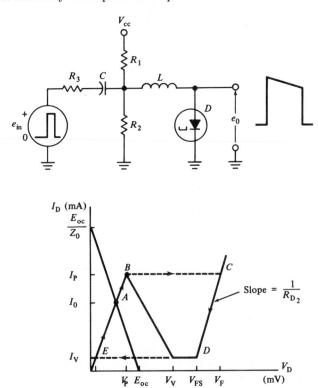

Figure 18-3 Low-Level Tunnel-Diode Monostable Multivibrator

At t_{+1} time, the current through the tunnel diode I_P starts to decrease through R_L, L, and R_{D_2} toward zero. The load line and the operating point move from point C to point D. The decreasing current and subsequent collapsing magnetic field around inductance L produce an induced voltage which acts like a source and which is of a polarity aiding the source V_{cc}. The time required for the current to decrease from I_P to I_V is established by the RL time-constant circuit. This interval is equal to the duration of the output pulse T_1. The decrease in voltage (V_F to V_{FS}) across the tunnel diode causes an increase in current through the inductance. This increase in current induces a voltage across the inductance of a polarity opposing V_{cc}. Hence, the effective voltage across the tunnel diode is less than the source voltage V_{cc}. This decrease in voltage causes the operating point to shift from point D to point E, as shown in Fig. 18-2. The tunnel-diode current increases until the stable operating point is established at point A. The resultant output voltage remains at a value equal that of V_o, until a subsequent trigger pulse is applied; thus, the cycle is completed.

Most practical tunnel-diode monostable multivibrator circuits are designed to use a greater source voltage V_{cc} than that used in the circuit shown in Fig. 18-2. Hence, the practical circuit shown in Fig. 18-3 is designed to incorporate a voltage divider. The value of source voltage E_{oc} and the effective value of load resistance Z_0 may be determined by the use of an equivalent circuit which is determined by the use of an equivalent circuit which is determined by the application of Thevenin's theorem.

18.3   INDUCTANCE L

The approximate value of inductance L may be determined by use of the basic equation which defines inductance.

$$E_{av} = L\frac{\Delta I}{\Delta t}$$

$$\frac{(V_F + 0)}{2} \approx L\frac{(I_P - I_V)}{T_1}$$

$$L \approx \frac{V_F T_1}{2(I_P - I_V)}$$

[41]

where
$$T_1 = \text{duration of output pulse (sec)}$$
$$L = \text{inductance (H)}$$

18.4   DURATION OF OUTPUT PULSE—LOW-LEVEL

The duration of the output pulse T_1 may be determined by use of the basic inductance discharge equation:

$$e_L = E_{L}\epsilon^{-Rt/L} \qquad \text{(I decreasing)} \qquad [42]$$

where E_L = initial voltage across the inductance L (V)
                   e_L = final voltage across the inductance L, at
                            the end of time interval t (V)

$$\tau = \frac{L}{R} \quad \text{(sec)}$$ [43]

The initial voltage across the inductance L may be established by solving the loop equation of an equivalent circuit for E_L. From the circuit shown in Fig. 18-3 and from the time-base diagram shown in Fig. 18-2, draw an equivalent circuit in which circuit conditions are those present in this circuit at t_{+1}

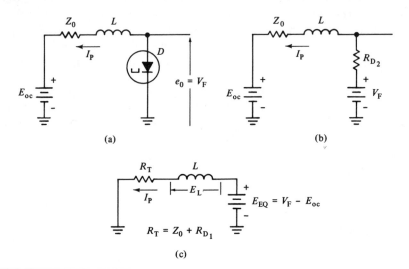

Figure 18-4   Equivalent Circuit of Schematic of Figs. 18-2 and 18-3 at t_{+1} Time

time, as illustrated in Fig. 18-4. By use of the simplified equivalent circuit shown in Fig. 18-4(c), write the loop equation and solve for E_L.

$$E_L = E_{EQ} + E_{Z_0}$$
$$E_L = V_F - E_{oc} + I_P Z_0$$

The final voltage across the inductance L may be established by solving the loop equation of an equivalent circuit for e_L. From the circuit shown in Fig. 18-3 and from the time-base diagram shown in Fig. 18-2, draw an equivalent circuit in which circuit conditions are those present in this circuit at t_{+2} time, as illustrated in Fig. 18-5.

By use of the simplified equivalent circuit, shown in Fig. 18-5(c), write the loop equation and solve for e_L.

$$e_L = e_{EQ} + E_{Z_0}$$
$$e_L = V_{FS} - E_{oc} + I_V Z_0$$

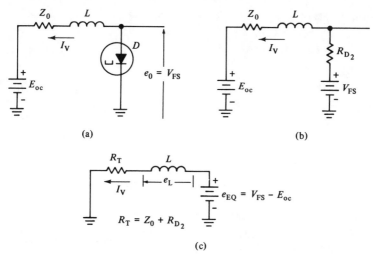

Figure 18-5 Equivalent Circuit of Schematic of Figs. 18-2 and 18-3 at t_{+2} Time

Substitute these values for E_L and for e_L, in Eq. 42, and solve for t.

$$e_L = E_L \epsilon^{-Rt/L} \qquad \text{(I decreasing)} \qquad [42]$$

$$(V_{FS} - E_{oc} + I_V Z_0) \approx (V_F - E_{oc} + I_P Z_0)\epsilon^{-Rt/L}$$

$$\epsilon^{+Rt/L} \approx \frac{V_F - E_{oc} + I_P Z_0}{V_{FS} - E_{oc} + I_V Z_0}$$

$$T_1 \approx t \approx \frac{L \log_{10} \dfrac{V_F - E_{oc} + I_P Z_0}{V_{FS} - E_{oc} + I_V Z_0}}{R \log_{10} \epsilon} \qquad [44]$$

where
$$R = Z_0 + R_{D_2}$$

and
$$R_{D_2} \approx \frac{V_F - V_{FS}}{I_P - I_V}$$

$T_1 = $ duration of output pulse (sec) as illustrated in Fig. 18-2

18.5 SAMPLE DESIGN AND ANALYSIS PROBLEM: LOW-LEVEL TUNNEL-DIODE MONOSTABLE MULTIVIBRATOR

Design a low-level tunnel-diode monostable multivibrator circuit from the following specifications:

$$T_1 \approx 5 \ \mu sec$$

Given:

Circuit shown in Fig. 18-3

$$V_{cc} = 4 \text{ V}$$
$$D = 1N3715 \qquad \text{(see Appendix B)}$$
$$e_{in} = 10 \text{ V positive pulse}, \qquad t_p = 3 \ \mu sec$$

Assume:

$$E_{oc} = 0.15 \text{ V}$$
$$I_{max} = 3 \text{ mA}$$

Solution:

1. Determine the value of Z_0 necessary for this circuit by use of Ohm's law.
2. Determine the value of resistance of R_1 by use of Eq. 40.
3. Determine the value of R_2 by use of the values determined for R_1 and Z_0.
4. Select practical color code values of resistance for resistors R_1 and R_2.
5. By use of the selected color code values of R_1 and R_2, determine the actual values of E_{oc} and Z_0.
6. Determine the value of inductance L by use of Eq. 41. Select the closest standard value for L.
7. Determine the resultant output-pulse width T_1 by use of Eq. 44.
8. Determine the value of R_3 by use of Ohm's law.
9. Assume that $C = 0.001 \ \mu\text{F}$.

$$Z_0 = \frac{E_{oc}}{I_{max}} = \frac{0.15}{3 \text{ mA}} = 50 \ \Omega \qquad [40]$$

$$R_1 = \frac{V_{cc}Z_0}{E_{oc}} = \frac{50 \times 4}{0.15} = 1330 \ \Omega$$

Use
$$R_1 = 1.2 \text{ k}\Omega$$

$$R_2 = \frac{R_1 Z_0}{R_1 - Z_0} = \frac{1330 \times 50}{1330 - 50} = 52 \ \Omega$$

Use
$$R_2 = 47 \ \Omega$$

$$E_{oc} = \frac{R_2 V_{cc}}{R_1 + R_2} = \frac{47 \times 4}{1200 + 47} = 0.15 \text{ V} \qquad [41]$$

$$Z_0 = \frac{R_1 R_2}{R_1 + R_2} = \frac{47 \times 1200}{47 + 1200} = 45 \ \Omega$$

$$L \approx \frac{V_F T_1}{2(I_P - I_V)} \approx \frac{(0.5)(5 \times 10^{-6})}{2(2.2 \text{ mA} - 0.29 \text{ mA})}$$

$$L \approx 650 \ \mu\text{H}$$

Use
$$L = 750 \ \mu\text{H}$$

$$T_1 \approx \frac{L \log_{10} \dfrac{V_F - E_{oc} + I_P R_T}{V_{FS} - E_{oc} + I_V R_T}}{R_T \log_{10} \epsilon} \qquad [44]$$

$$R_T = Z_0 + R_{D_2}$$

$$R_{D_2} \approx \frac{V_F - V_{FS}}{I_P - I_V} \approx \frac{0.5 - 0.45}{2.2 \text{ mA} - 0.29 \text{ mA}} \approx 26 \ \Omega$$

$$R_T = Z_0 + R_{D_2} = 45 + 26 = 71 \ \Omega$$

$$T_1 \approx \frac{750 \times 10^{-6} \log_{10} \dfrac{0.5 - 0.15 + (2.2 \text{ mA})(71)}{0.45 - 0.15 + (0.29 \text{ mA})(71)}}{71 \log_{10} 2.718}$$

$$T_1 \approx \frac{10.5 \times 10^{-6} \log_{10} \dfrac{0.506}{0.32}}{0.434}$$

$$T_1 \approx \frac{10.5 \times 10^{-6} \log_{10} 1.58}{0.434} \approx \frac{10.5 \times 10^{-6} \times 0.198}{0.434}$$

$$T_1 \approx 4.88 \ \mu\text{sec}$$

Since
$$e_{\text{in}} \gg E_{\text{oc}}$$
$$10 \gg 0.15$$

$$R_3 \approx \frac{e_{\text{in}}}{I_P} \approx \frac{10}{2.2 \ \text{mA}} \approx 4.54 \ \text{k}\Omega$$

Use
$$R_3 = 4.7 \ \text{k}\Omega$$

18.6   HIGH-LEVEL TUNNEL-DIODE MONOSTABLE MULTIVIBRATOR

To follow the description of the operation of a basic high-level tunnel-diode monostable multivibrator, refer to Fig. 18-6. The selection of applicable values for R_L and V_{cc} is necessary to ensure that the dc operating point on the characteristic curve is in the positive resistance region and that V_o has a value between that of V_{FS} and V_{F}. Point A on the load line is the dc operating point. The value of inductance L is established by the use of Eq. 41. The closest standard value is selected for L. The negative trigger pulse must be of sufficient amplitude and duration to effect a decrease from I_o to I_V in the tunnel-diode current. The delay time—the interval between the application of the trigger pulse at t_{+0} time and the resultant output pulse at t_{+1} time—is the interval required for the tunnel-diode current to decrease from I_o to I_V. Because the applied trigger pulse is negative, it opposes V_{cc}. Therefore, the tunnel-diode current is limited by I_V, and the voltage across the tunnel diode is approximately equal to zero; hence, the operating point switches from point B to point C.

At t_{+1} time, the tunnel-diode current increases from I_V toward I_{max} but is prevented from exceeding I_P by the tunnel diode. This increase is mathematically defined by the standard inductance charge Eq. 45 which is comparable with the capacitor-charge Eq. 5.

$$e_C = E(1 - \epsilon^{-t/RC}) \qquad [5]$$
$$i_L = I(1 - \epsilon^{-Rt/L}) \qquad [45]$$

where

i_L = instantaneous value of current through the inductance L at time t (A)
I = final value of current through the inductance L when $t = \infty$ (A)
R = value of resistance (Ω), through which the current flows in a series RL circuit
L = value of inductance (H)
t = time (sec); current is allowed to increase

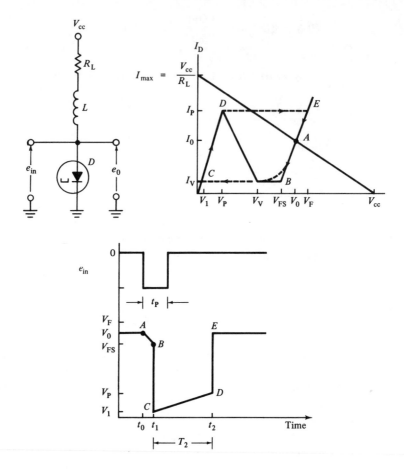

Figure 18-6   High-Level Monostable Tunnel-Diode Multivibrator

18.7   DURATION OF OUTPUT PULSE—HIGH-LEVEL

Refer to Fig. 18-6. In this circuit, the pulse duration T_2 is produced by the current, increasing from I_V to I_P, through the series RL circuit. This interval may be determined by use of the inductance charge Eq. 45, modified to include an initial value of current other than zero. This modified Eq. 46 has the same form as that of the capacitor-charge equation in which an initial charge on the capacitor has been accounted for. Recall Eq. 8.

$$e_C = E - (E \pm E_{co})\epsilon^{-t/RC} \qquad\qquad [8]$$

Thus, Eq. 45, modified to include an initial value of current other than zero, is

$$i_L = I - (I \pm I_{co})\epsilon^{-Rt/L} \qquad\qquad [46]$$

When substitutions are made in Eq. 46, appropriate for the circuit shown in Fig. 18-6,

$$I_P = \frac{V_{cc}}{R_T} - \left(\frac{V_{cc}}{R_T} - I_V\right) \epsilon^{-R_T T_2/L}$$

Solving for T_2,

$$T_2 \approx \frac{L \log_{10} \dfrac{V_{cc} - I_V R_T}{V_{cc} - I_P R_T}}{R_T \log_{10} \epsilon} \qquad [47]$$

The practical high-level tunnel-diode monostable multivibrator circuit, which is comparable to the low-level circuit, employs a higher source voltage V_{cc} than that employed in the circuit shown in Fig. 18-6. Hence, the practical circuit, shown in Fig. 18-7, is designed to incorporate a voltage divider. The

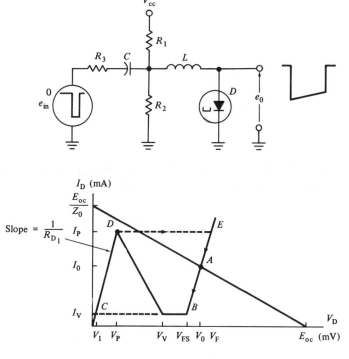

Figure 18-7   High-Level Tunnel-Diode Monostable Multivibrator

value of source voltage E_{oc} and the effective value of load resistance Z_0 may be determined by use of an equivalent circuit which is determined by the application of Thevenin's theorem, as illustrated in Fig. 18-8.

The equivalent circuit shown in Fig. 18-8 is of the same form as that of the basic circuit shown in Fig. 18-6. Therefore, the equation for the duration of the output pulse T_2, Eq. 47, is modified to incorporate the voltage divider

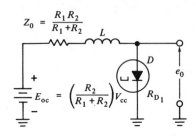

Figure 18-8   Equivalent Circuit of Fig. 18-7 Excluding Trigger Input

of the practical circuit, shown in Fig. 18-7. This is accomplished by the substitution of the value of E_{oc} for V_{cc} and that of Z_0 for R_L.

$$T_2 \approx \frac{L \log_{10} \dfrac{E_{oc} - I_V R_T}{E_{oc} - I_P R_T}}{R_T \log_{10} \epsilon}$$

[48]

where $R_T = Z_0 + R_{D_1}$

and $R_{D_1} \approx \dfrac{V_P}{I_P}$

18.8   SAMPLE DESIGN AND ANALYSIS PROBLEM: HIGH-LEVEL TUNNEL-DIODE MONOSTABLE MULTIVIBRATOR

Design a high-level tunnel-diode monostable multivibrator circuit from the following specifications:

$$T_2 \approx 5 \ \mu sec$$

Given:

Circuit shown in Fig. 18-7

$V_{cc} = 6$ V

$D = $ 1N3715   (see Appendix)

$e_{in} = 10$ V negative pulse, $t_p = 3 \ \mu sec$

Assume:

$E_{oc} = 0.5$ V

$I_{max} = 5$ mA

Solution:

1.   Determine the value of Z_0 necessary for the circuit by use of Ohm's law.
2.   Determine the value of resistance of R_1 by use of Eq. 40.

3. Determine the value of R_2 by use of the values determined for R_1 and Z_0.
4. Select practical color code values of resistance for resistors R_1 and R_2.
5. By use of the selected color code values of R_1 and R_2, determine the actual values of E_{oc} and Z_0.
6. Determine the value of inductance L by use of Eq. 41. Select the closest standard value for L.
7. Determine the resultant output-pulse width T_2 by use of Eq. 48.
8. Assume that the value of C is 0.001 μF.
9. Determine the value of resistor R_3 by use of Ohm's law.

$$Z_0 = \frac{E_{oc}}{I_{max}} = \frac{0.5}{5 \text{ mA}} = 100 \ \Omega$$

$$R_1 = \frac{Z_0 V_{cc}}{E_{oc}} = \frac{100 \times 6}{0.5} = 1200 \ \Omega \qquad [40]$$

$$R_2 = \frac{R_1 Z_0}{R_1 - Z_0} = \frac{1200 \times 100}{1200 - 100} = 109 \ \Omega$$

Use
$$R_1 = 1200 \ \Omega$$
$$R_2 = 120 \ \Omega$$

$$E_{oc} = \frac{R_2 V_{cc}}{R_1 + R_2} = \frac{120 \times 6}{1200 + 120} = 0.545 \text{ V}$$

$$Z_0 = \frac{R_1 R_2}{R_1 + R_2} = \frac{1200 \times 120}{1200 + 120} = 109 \ \Omega$$

$$L \approx \frac{V_F T_2}{2(I_P - I_V)} \approx \frac{(0.5)(5 \times 10^{-6})}{2(2.2 \text{ mA} - 0.29 \text{ mA})} \qquad [41]$$

$$L \approx 650 \ \mu\text{H}$$

Use
$$L = 750 \ \mu\text{H}$$

$$R_{D_1} \approx \frac{V_P}{I_P} \approx \frac{65 \text{ mV}}{2.2 \text{ mA}} \approx 29.5 \ \Omega \approx 30 \ \Omega$$

$$R_T = Z_0 + R_{D_1} = 109 + 30 = 139 \ \Omega$$

$$T_2 \approx \frac{L \log_{10} \dfrac{E_{oc} - I_V R_T}{E_{oc} - I_P R_T}}{R_T \log_{10} \epsilon} \qquad [48]$$

$$T_2 \approx \frac{750 \times 10^{-6} \log_{10} \dfrac{0.545 - (0.29 \text{ mA})(139)}{0.545 - (2.2 \text{ mA})(139)}}{139 \log_{10} 2.718}$$

$$T_2 \approx \frac{5.4 \times 10^{-6} \log_{10} \dfrac{0.505}{0.24}}{\log_{10} 2.718}$$

$$T_2 \approx \frac{5.4 \times 10^{-6} \log_{10} 2.1}{\log_{10} 2.718} \approx \frac{5.4 \times 10^{-6} \times 0.322}{0.434}$$

$$T_2 \approx 4 \ \mu\text{sec}$$

Since $e_{in} \gg E_{oc}$

$\qquad\qquad\qquad 10 \gg 0.545$

$$R_3 \approx \frac{e_{in}}{I_P} \approx \frac{10}{2.2 \text{ mA}} \approx 4.54 \text{ k}\Omega$$

Use $R_3 = 4.7 \text{ k}\Omega$

18.9   TUNNEL-DIODE ASTABLE MULTIVIBRATOR

The basic circuit of the astable multivibrator is the same as the basic circuit of the monostable multivibrator. Refer to Fig. 18-9. The value of load resistance R_L and that of the source voltage V_{cc} determines the difference

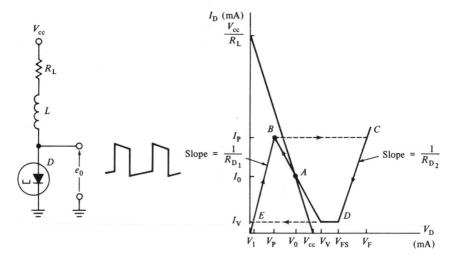

Figure 18-9   Tunnel-Diode Astable Multivibrator

between the two circuits. In the astable multivibrator circuit, the load line must intersect only the negative resistance portion of the volt–ampere characteristic curve of the tunnel diode. The proper value of load resistance and source voltage ensure this circuit condition. The negative resistance region of the volt–ampere characteristic curve is the unstable operating portion.

To trace the operation of this circuit, refer to Fig. 18-9. At the initial application of source voltage V_{cc}, the current increases. When the operating point is on the unstable negative resistance portion of the volt–ampere characteristic curve, the current does not stabilize at I_o but increaes to I_P because the tunnel diode acts as a voltage source in series-aiding the voltage source V_{cc}.

The increasing current produces an increasing magnetic field around the inductance L. At this time, a voltage source is no longer effectively provided by the tunnel diode because the operating point is no longer in the negative resistance region; hence, the current begins to decrease. When this occurs, the magnetic field around the inductor starts to collapse, and the inductor acts as a voltage source, aiding the source voltage V_{cc}. Since the current cannot exceed I_P, the operating point is forced to shift to a higher voltage operating point, point C.

The collapsing magnetic field around the inductance L and, in turn, the decreasing current cause the operating point to shift from point C to point D, as shown in Fig. 18-9. The interval required for the operating point to shift from point C to point D is determined by the inductive time-constant circuit in the same manner that the pulse duration T_1 is determined in the low-level monostable multivibrator circuit. Hence, the positive pulse duration of the resultant output-voltage waveform T_1 is determined by the use of Eq. 44.

The inductor again acts as a voltage source, as the tunnel-diode current decreases, and the operating point shifts from point C to point D. At this time, however, the inductor, acting as a voltage source, is of a polarity opposing the voltage source V_{cc}. Hence, the effective source voltage is less than the source V_{cc}. Thus, the operating point is forced to shift from point D to point E, as shown in Fig. 18-9.

As the tunnel-diode current starts to increase, the operating point moves from point E to point B. The interval required for this current increase is determined in the same manner that the pulse interval T_2 was determined in the high-level monostable circuit. Thus T_2 establishes the rest interval between the positive pulse of the resultant output-voltage waveform.

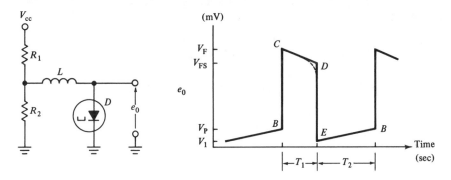

Figure 18-10　Tunnel-Diode Astable Multivibrator

Most practical tunnel-diode astable multivibrator circuits are designed to use a greater source voltage V_{cc} than that used in the circuit shown in Fig. 18-9. Hence, the practical circuit, shown in Fig. 18-10, is designed to incorporate a voltage divider. The value of source voltage E_{oc} and the effective value of load resistance Z_0 may be established by the use of an equivalent circuit, which is determined by the application of Thevenin's theorem.

18.10 SAMPLE DESIGN AND ANALYSIS PROBLEM: TUNNEL-DIODE ASTABLE MULTIVIBRATOR

Design a tunnel-diode astable multivibrator circuit from the following specifications:

$$T_1 \approx 5 \ \mu\text{sec}$$

Given:

Circuit shown in Fig. 18-10
$$V_{cc} = 5 \text{ V}$$
$$D = 1\text{N}3715 \qquad \text{(see Appendix B)}$$

Assume:

$$E_{oc} = 0.25 \text{ V}$$
$$I_{max} = 5 \text{ mA}$$

Solution:

1. Determine the value of Z_0 necessary for this circuit by use of Ohm's law.
2. Determine the value of resistance R_1 by use of Eq. 40.
3. Determine the value of R_2 by use of the values determined for R_1 and Z_0.
4. Select practical color code values for resistors R_1 and R_2.
5. By use of the selected color code values of R_1 and R_2, determine the actual values of E_{oc} and Z_0.
6. Determine the approximate value of inductance L by use of Eq. 41. Select the closest standard value for L.
7. Determine the positive output-pulse width T_1 by use of Eq. 44.
8. Determine the rest interval between positive pulses, T_2, by use of Equation 48.

$$Z_0 = \frac{E_{oc}}{I_{max}} = \frac{0.25}{5 \text{ mA}} = 50 \ \Omega$$

$$R_1 = \frac{Z_0 V_{cc}}{E_{oc}} = \frac{50 \times 5}{0.25} = 1000 \ \Omega$$

$$R_2 = \frac{Z_0 R_1}{R_1 - Z_0} = \frac{50 \times 1000}{1000 - 50} = 52.7 \ \Omega$$

Use $R_1 = 1000 \ \Omega$

$R_2 = 47 \ \Omega$

$$E_{oc} = \frac{R_2 V_{cc}}{R_1 + R_2} = \frac{47 \times 5}{47 + 1000} = 0.23 \text{ V}$$

$$Z_0 = \frac{R_1 R_2}{R_1 + R_2} = \frac{47 \times 1000}{1000 + 47} = 45.5 \ \Omega$$

$$R_{D_2} \approx \frac{V_F - V_{FS}}{I_P - I_V} \approx \frac{0.5 - 0.45}{2.2 \text{ mA} - 0.29 \text{ mA}} \approx 26 \ \Omega$$

$$L \approx \frac{V_F T_1}{2(I_P - I_V)} \approx \frac{0.5 \times 5 \times 10^{-6}}{2(2.2 \text{ mA} - 0.29 \text{ mA})} \approx 650 \ \mu\text{H}$$

Use $L = 750 \ \mu\text{H}$

Assume:

$$R_L = 7.5 \ \Omega$$

$$R_T = Z_0 + R_L + R_{D_2}$$

$$R_T = 45.5 + 7.5 + 26 = 79 \ \Omega$$

$$T_1 \approx \frac{L \log_{10} \dfrac{V_F - E_{oc} + I_P R_T}{V_{FS} - E_{oc} + I_V R_T}}{R_T \log_{10} \epsilon}$$

$$T_1 \approx \frac{750 \times 10^{-6} \log_{10} \dfrac{0.5 - 0.23 + (2.2 \ \text{mA})(79)}{0.45 - 0.23 + (0.29 \ \text{mA})(79)}}{79 \log_{10} 2.718}$$

$$T_1 \approx \frac{9.5 \times 10^{-6} \log_{10} \dfrac{0.444}{0.243}}{\log_{10} 2.718} \approx \frac{9.5 \times 10^{-6} \log_{10} 1.83}{\log_{10} 2.718}$$

$$T_1 \approx \frac{9.5 \times 10^{-6} \times 0.26}{0.434}$$

$$T_1 \approx 5.7 \ \mu\text{sec}$$

$$R_T = Z_0 + R_L + R_{D_1}$$

$$R_T = 45.5 + 7.5 + 30 = 83 \ \Omega$$

$$T_2 \approx \frac{L \log_{10} \dfrac{E_{oc} - I_V R_T}{E_{oc} - I_P R_T}}{R_T \log_{10} \epsilon}$$

$$T_2 \approx \frac{750 \times 10^{-6} \log_{10} \dfrac{0.23 - (0.29 \ \text{mA})(83)}{0.23 - (2.2 \ \text{mA})(83)}}{83 \log_{10} 2.718}$$

$$T_2 \approx \frac{9.05 \times 10^{-6} \log_{10} \dfrac{0.206}{0.047}}{\log_{10} 2.718} \approx \frac{9.05 \times 10^{-6} \times 0.641}{0.434}$$

$$T_2 \approx 13.4 \ \mu\text{sec}$$

LABORATORY EXPERIMENT
TUNNEL-DIODE MONOSTABLE AND ASTABLE MULTIVIBRATOR

OBJECT:
1. To familiarize the student with the characteristics of a tunnel-diode monostable and astable multivibrator.
2. To design and analyze a low-level tunnel-diode monstable multivibrator.
3. To design and analyze a high-level tunnel-diode monostable multivibrator.
4. To design and analyze a tunnel-diode astable multivibrator.

MATERIALS:
1 Tunnel diode—(example—1N3716 germanium)
1 Manufacturer's specification sheet for tunnel diode used (see Appendix B for 1N3716)
1 1 mH inductor—100 mA
1 Silicon junction diode (example—1N914)
3 Resistor substitution boxes (10 Ω to 10 MΩ, 1 W)
1 Capacitor substitution box (0.0001 to 0.22 μF at 450 V)
1 Square-wave generator (20 Hz to 200 kHz; Z_0 = 600 Ω)
1 Oscilloscope, dc time base; frequency response dc to 450 kHz; vertical sensitivity, 100 mV/cm
1 Transistor power supply (0 to 30 V and 0 to 250 mA)

PROCEDURE:
PART A   Low-level monostable multivibrator
1. Design a low-level monostable tunnel-diode multivibrator circuit, as shown in Fig. 18-1X, from the following specifications:

$$V_{cc} = 5 \text{ V}$$
$$L = 1 \text{ mH}$$
$$e_{in} = +10 \text{ V pulse—square wave}$$

Determine:

$$R_1, R_2, R_3, C, T_1, \text{ and } f \text{ (input trigger)}$$

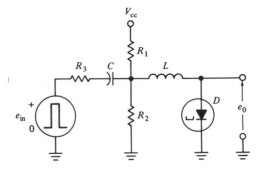

Figure 18-1X   Low-Level Tunnel-Diode Monostable Multivibrator

2. By use of the data from the manufacturer's specification sheet, draw a straight-line approximation of the volt–ampere characteristic curve for the tunnel diode used. Label (1N3716—see Appendix B).

3. From the volt–ampere curve, drawn in Step 1, determine the approximate positive resistance R_{D_1} and R_{D_2}. Show all calculations.

4. Graphically, draw a load line on the volt–ampere characteristic curve drawn in Step 2. In order for the value of quiescent current to equal approximately three-fourths the value of I_P, select the operating point so that the load line intersects the positive resistance region R_{D_1}. The voltage intercept of the load line approximately equals the value of E_{oc}. The current intercept of the load line establishes the value of maximum current. From this value of maximum current and the approximate value of E_{oc}, the approximate value of Z_0 may be determined.

5. By use of the approximate values of E_{oc} and Z_0, established in Step 4, calculate the approximate values for R_1 and R_2.

6. Select standard color code values of resistance for resistors R_1 and R_2 from the approximate calculated values.

7. By use of the standard color code values of resistance for resistors R_1 and R_2, calculate the values for E_{oc} and Z_0.

8. By use of the values of E_{oc} and Z_0, calculated in Step 5, draw the actual load line on the characteristic curve drawn in Step 2. Check the location of this load line to ensure that the requirements for a low-level monostable multivibrator are fulfilled. If the load line is satisfactory, continue with Step 9; if it is not satisfactory, start with Step 4 and select a different initial load line.

9. Calculate the duration of the output pulse T_1.

10. Determine the frequency limits for the square-wave triggering input. Show all calculations.

11. Determine the value of resistor R_3 and of capacitor C. Show all calculations.

12. Connect the circuit designed. Observe, measure, and record all necessary data.

13. Draw a schematic of the circuit designed, and label it. On the same sheet of paper, draw a graph of the input- and output-voltage waveforms, one above the other, and to the same time base.

14. Compare the calculated results with the measured results, and explain the reason for any discrepancies.

PART B   High-level monostable multivibrator

1. Design a high-level monostable tunnel-diode multivibrator circuit, as shown in Fig. 18-2X, from the following specifications:

$$V_{cc} = 5 \text{ V}$$
$$L = 1 \text{ mH}$$
$$e_{in} = -10 \text{ V pulse—square wave}$$

Determine:

$$R_1, R_2, R_3, C, T_2, \text{ and } f \text{ (input trigger)}$$

2. Graphically, draw a load line on the volt–ampere characteristic curve previously drawn in Part A, Step 2. In order for the value of quiescent current to equal approximately three-fourths the value of I_P, select the operating point so that it intersects the positive resistance region R_{D_2}. The voltage intercept of the load line approximately equals the value of E_{oc}. The current

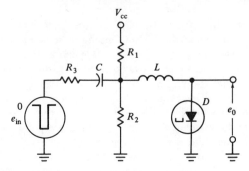

Figure 18-2X   High-Level Tunnel-Diode Monostable Multivibrator

intercept of the load line establishes the value of maximum current. From this value of maximum current and the approximate value of E_{oc}, the approximate value of Z_0 may be determined.

3. By use of the approximate values of E_{oc} and Z_0 established in Step 2, calculate the approximate values for R_1 and R_2.

4. Select standard color code values of resistance for resistors R_1 and R_2, from the approximate calculated values.

5. By use of the standard color code values of resistance for resistors R_1 and R_2, calculate the values for E_{oc} and Z_0.

6. By use of the values of E_{oc} and Z_0, calculated in Step 4, draw the actual load line on the characteristic curve drawn in Step 2. Check the location of this load line to ensure that the requirements for a high-level monostable multivibrator are fulfilled. If the load line is satisfactory, continue with Step 7; if it is not satisfactory, start with Step 2 and select a different initial load line.

7. Calculate the duration of the output pulse T_2.

8. Determine the frequency limits for the square-wave triggering input. Show all calculations.

9. Determine the value of resistor R_3 and that of capacitor C. Show all calculations.

10. Connect the circuit designed. Observe, measure, and record all necessary data.

11. Draw a schematic of the circuit designed, and label it. On the same sheet of paper, draw a graph of the input- and output-voltage waveform, one above the other, and to the same time base.

12. Compare the calculated results with the measured results, and explain the reason for any discrepancies.

PART C   Astable tunnel-diode multivibrator

1. Design an astable tunnel-diode multivibrator circuit, as shown in Fig. 18-3X, from the following specifications:

$$V_{cc} = 5 \text{ V}$$
$$L = 1 \text{ mH}$$
$$D = \text{tunnel diode (1N3716—example)}$$

Determine:

$$R_1, R_2, T_1, \text{ and } T_2$$

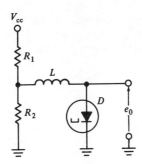

Figure 18-3X Tunnel-Diode Astable Multivibrator

2. Graphically, draw a load line on the volt–ampere characteristic curve previously drawn in Part A, Step 2. Select the operating point so that it intersects the negative resistance region of the curve approximately in the center and in no other region. The voltage intercept of the load line approximately equals the value of E_{oc}. The current intercept of the load line establishes the value of maximum current. From this value of maximum current and the approximate value of E_{oc}, the approximate value of Z_0 may be determined.

3. By use of the approximate values of E_{oc} and Z_0, established in Step 2, calculate the approximate values for R_1 and R_2.

4. Select standard color code values of resistance for resistors R_1 and R_2 from the approximate calculated values.

5. By use of the standard color code values of resistance for resistors R_1 and R_2, calculate the values for E_{oc} and Z_0.

6. By use of the values of E_{oc} and Z_0, calculated in Step 5, draw the actual load line on the characteristic curve drawn in Step 2. Check the location of this load line to ensure that the requirements for an astable multivibrator are fulfilled. If the load line is satisfactory, continue with Step 5; if it is not satisfactory, start with Step 1 and select a different initial load line.

7. Calculate the duration of the output pulse T_1.

8. Calculate the duration of the interval between positive pulses, T_2.

9. Connect the circuit designed. Observe, measure, and record all necessary data.

10. Compare the calculated results with the measured results, and explain the reason for any discrepancies.

QUESTIONS AND EXERCISES

1. Explain the operation of a low-level tunnel-diode monostable multivibrator circuit.

2. Explain the operation of a high-level tunnel-diode monostable multivibrator circuit.

3. Explain the operation of an astable tunnel-diode multivibrator circuit.

4. Why is it necessary to include an inductance in a monostable or an astable tunnel-diode multivibrator circuit?

5. Why is it necessary to use only a single tunnel-diode in the design of a multivibrator circuit but necessary to use two junction transistors in the design of a comparable circuit?

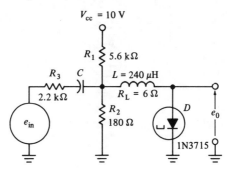

Figure 18-4X

6. Name the circuit shown in Fig. 18-4X. Determine the amplitude and pulse duration of the output-voltage waveform. Determine the amplitude and polarity of the input-voltage waveform. (Assume the pulse duration of the input trigger voltage to be half that of the resultant output pulse of the circuit.) Draw the input- and output-voltage waveforms to scale, one above the other, and to the same time base. Show all calculations.

7. What are the advantages of tunnel diodes compared with junction transistors?

8. What are the disadvantages of tunnel diodes compared with junction transistors?

9. Refer to Fig. 18-2. Explain the reason for the time delay between the application of the trigger pulse and the start of the output pulse.

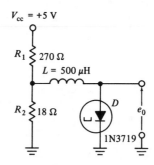

Figure 18-5X

10. Determine whether the circuit shown in Fig. 18-5X will operate properly as an astable multivibrator. Determine the amplitude of the resultant output voltage and the duration of the pulse. Show all calculations.

11. Design a low-level monostable multivibrator circuit, as shown in Fig. 18-3, with the following characteristics:

$V_{cc} = +10\,\text{V}, \qquad T_1 = 15\,\mu\text{sec}, \qquad e_{in} = 10\,\text{V positive pulse}, \qquad D = 1\text{N}3712$

Assume: $\qquad E_{oc} = 0.1\,\text{V} \qquad$ and $\qquad I_{max} = 1.5\,\text{mA}$

12. Analyze the high-level monostable multivibrator circuit, shown in Fig. 18-7, when

$$V_{cc} = +12\,\text{V}, \qquad R_1 = 5.6\,\text{k}\Omega, \qquad R_2 = 330\,\Omega, \qquad R_3 = 10\,\text{k}\Omega$$
$$L = 5\,\text{mH}, \qquad D = 1\text{N}3712, \qquad e_{\text{in}} = 10\,\text{V negative pulse}$$

13. Refer to Fig. 18-10. Analyze the tunnel-diode astable multivibrator circuit when

$$V_{cc} = +9\,\text{V}, \qquad R_1 = 2.7\,\text{k}\Omega, \qquad R_2 = 68\,\Omega, \qquad L = 470\,\mu\text{H}, \qquad D = 1\text{N}3712$$

Assume: $R_L = 0$

chapter 19

LOGIC GATES

A gating circuit is a switching circuit which controls the ON and the OFF states of another switching circuit, such as the bistable multivibrator. There are two types of gating circuits—the sampling gate and the logic gate. The sampling gate is a circuit in which an exact reproduction of the signal input voltage is permitted to appear, as an output signal, for a specified interval—sample of time. This circuit is also referred to as a linear gate, a transmission gate, or a time-selection circuit. This circuit is discussed in chapter 20. The logic gate is a circuit in which the input control voltages determine the interval during which the output-pulse voltage waveshape appears; the waveshape of the output pulse may have no resemblance to that of any of the input pulses. The sampling gate and the logic gate may use either diodes or transistors as active elements.

Most of the circuits used in digital systems are comprised of a gating circuit in conjunction with a switching circuit. Therefore, these circuits are used in systems such as digital computers, data processing systems, control systems, digital-communications systems, frequency counters, and digital voltmeters. A brief definition of the computer, for which most of these digital circuits are manufactured, is essential.

An electronic computer is capable of adding the decimal numbers 0, 1, 2, 3, 4, 5, 6, 7, 8, and 9. Because successive addition is equivalent to multiplication and because successive subtraction is equivalent to division, by "adding," the computer can also subtract, multiply, and divide. A diode or transistor, when used as a switch, may be either ON or OFF. Because ON or OFF may represent any two opposites, the binary number system is used instead of the decimal number system. In the computer logic circuit, the ON and OFF states of the semiconductor device are defined as 1 and 0.

The circuit conditions (ON and OFF), represented by 1 and 0, may be represented by any assumed values in place of these arbitrarily selected symbols; for example, 1 may be defined as a positive square-wave pulse of voltage and 0 as the absence of a positive square-wave pulse of voltage. This example of positive logic is illustrated in Fig. 19-1(a). The absence of a pulse, 0, need not be 0 V with reference to ground, as illustrated in Fig. 19-1(b).

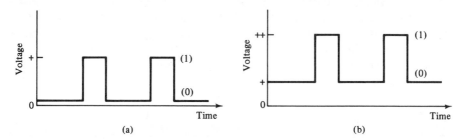

Figure 19-1   (a), (b) Positive Logic

Digital systems are designed to incorporate sets of different defined values for 1 and 0. The following examples illustrate a few of the possibilities: (1) a digital system in which 1 represents the absence of a pulse and in which 0 represents a negative pulse with reference to ground, (2) a system in which 1 represents the negative pulse with reference to ground and 0 represents the absence of a pulse.

The circuitry of a digital system must be geared to the assumed conditions represented by 1 and 0.

19.1 LOGIC GATES

Logic gates are divided into two classes—passive and active. A circuit which employs a nonamplifying electronic device which acts like a switch is called a passive gate; the most common example of this type of device is the semiconductor junction diode. A passive logic gate which employs one or more junction diodes is referred to as a diode logic (DL) gate. A circuit which employs an amplifying electronic device which acts like a switch is called an active logic gate; the most common example of this type of device is the semiconductor junction transistor.

Because a circuit which employs a transistor as a switch requires different junction voltages for proper operation, it necessarily must include resistors, diodes, or additional transistors for proper operation. Hence, active logic circuits which employ a transistor as a switch are classified according to the type of electronic device which is used in conjunction with the transistor. For example, an active logic gate which uses resistors and a transistor is called resistor-transistor logic (RTL). The most common active logic circuits are (1) resistor-transistor logic (RTL), (2) diode-transistor logic (DTL), (3) direct-coupled-transistor logic (DCTL), (4) current-mode logic (CML), and (5) transistor-transistor logic (TTL).

The three basic logic gates are the OR gate circuit, the AND gate circuit, and the NOT gate circuit.

Active logic circuits which employ transistors may be designed to operate in any one of the three basic amplifier configurations: common collector, common base, or common emitter. Since logic gates are generally required to deliver power to succeeding stages, the common-emitter or common-collector configurations are normally used. The transistor-transistor logic circuit, an exception, is designed to operate in the common-base configuration.

		Basic logic gates				
		OR	AND	NOT (inverter)	NOR (inverted OR)	NAND (Inverted AND)
Passive		DL	DL			
Active	Saturated mode	RTL	RTL	RTL	RTL	RTL
		DTL	DTL	DTL	DTL	DTL
		DCTL	DCTL	DCTL	DCTL	DCTL
		TTL	TTL	TTL	TTL	TTL
	Active mode	CML	CML	CML	CML	CML

DL	Diode logic		DCTL	Direct-coupled-transistor logic
RTL	Resistor-transistor logic		TTL	Transistor-transistor logic
DTL	Diode-transistor logic		CML	Current-mode logic

Figure 19-2 Logic Gates

An inspection of the Logic Gate Chart (Fig. 19-2) discloses the relationships among the most common types of circuit configurations used to produce the basic types of the logic gate circuits—the OR gate, the AND gate, the NOT gate, the NOR gate, and the NAND gate—and designates the type of coupling for each, with the exception of current-mode logic.

19.2   OR GATE

An OR gate is a logic circuit which produces an output pulse when one or more than one input pulse is applied; the polarity of the input is the same as that of the output. The simplest form of OR gate is the diode logic (DL) circuit.

To trace the operation of a positive logic, two-input OR gate, refer to Fig. 19-3(a). At t_{-1} time, when no input pulses are present, the diodes are zero biased and there is no output pulse. At t_{+1} time, there is an input pulse at A and no input pulse at B; hence, diode D_1 is forward biased; therefore, it con-

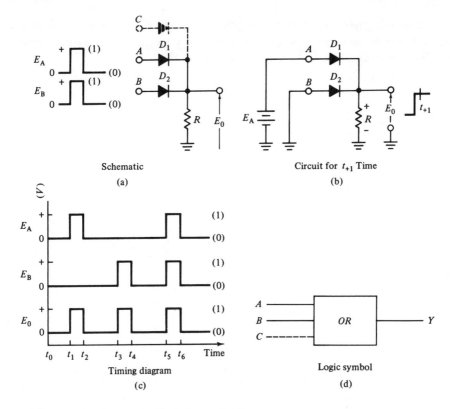

Figure 19-3 OR Gate Diode Logic (DL)

ducts and produces an output-voltage pulse, as shown in Fig. 19-3(b) and (c). The output-voltage pulse, produced across resistor R, reverse biases diode D_2.

At t_{+5} time, when both input pulses appear simultaneously, both diodes are forward biased; hence, they conduct and produce an output pulse, as shown in the timing diagram—Fig. 19-3(c).

The amplitude of the output pulse is equal to the differences between the amplitude of the input pulse and the voltage drop across the conducting diode. If silicon diodes are used, the voltage drop across the diode is approximately equal to 0.7 V. When the output of one diode gate is the input to another diode gate, the loss in output voltage becomes additive across the diodes. Hence, in a practical diode logic circuit, more than four successive gates decrease the amplitude of the output voltage to one which equals the noise level of the circuit; thus, the circuit is rendered unusable.

The resistor-transistor logic (RTL) circuit is employed to overcome the disadvantages of diode logic. See Fig. 19-4(a). In this circuit, the transistor T_1 is a common-collector amplifier (switch) which operates in the saturated mode. The common-collector configuration produces no phase inversion; therefore, it fulfills the requirements for the OR gate.

The input base resistors R_A and R_B are selected to ensure sufficient base current which, in turn, ensures saturation. In terms of input signal isolation, the base resistance R_A and R_B must be large.

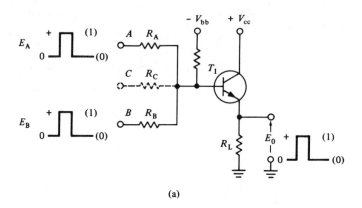

(a)

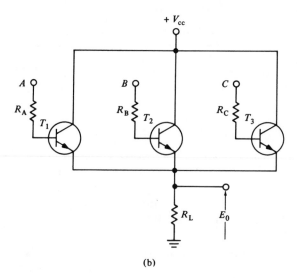

(b)

Figure 19-4   OR Gate Resistor-Transistor Logic (RTL)

When the input base resistance is large, the base current is small. When the base current is small, it is capable of switching only small amounts of collector current; hence, the value of R_L must be large. A large value of load resistance has two disadvantages: (1) It exhibits low power gain, and (2) the output voltage is temperature dependent. Since OR gates are frequently required to drive many other circuits simultaneously, the power gain requirement of the circuit is important. When the transistor is at cutoff, the leakage current may be of the same order of magnitude as that of the signal current.

Because leakage current is very temperature dependent, it may seriously affect the zero level of output voltage.

Figure 19-4(b) illustrates another form of (RTL) OR gate. By use of separate transistors for each input, signal-input isolation is achieved. This circuit requires large load resistance, however; therefore, it has the same disadvantages as those of the circuit shown in Fig. 19-4(a).

To overcome the disadvantages of resistor-transistor logic, the diode-transistor logic (DTL) circuit is employed. See Fig. 19-5. In this circuit, the

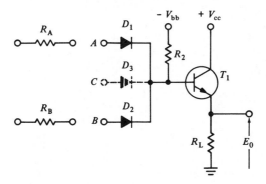

Figure 19-5   OR Gate Diode-Transistor Logic (DTL)

required signal isolation is provided by the diodes because a diode exhibits high reverse resistance when reverse biased. The diodes have very low resistances when forward biased by input pulses; hence, base currents capable of switching large values of collector current are possible. Since large values of collector current are possible, the value of load resistance R_L may be small; therefore, the circuit is no longer temperature dependent and exhibits large power gain, capable of driving many other circuits.

The value of base current must be sufficiently limited to prevent overdrive of the transistors. This is accomplished by the placement of the resistors in series with the diodes; it is necessary because the forward resistance of the diode is very small. See Figs. 19-5 and 19-15.

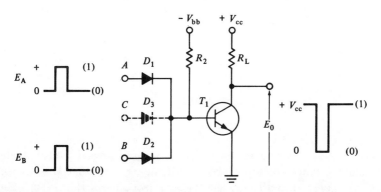

Figure 19-6   NOR Gate Diode-Transistor Logic (DTL)

An OR gate circuit, the output of which is inverted, is called a NOR gate. Hence, if a common-emitter switch (inverter-NOT circuit) employs multiple inputs, as shown in Fig. 19-6, it is a NOR gate.

19.3   AND GATES

An AND gate is a logic circuit which exhibits no phase inversion and one which produces an output pulse, only when all the input pulses are applied simultaneously. The AND gate is occasionally referred to as a "coincidence" circuit because all the input pulses must coincide in order to produce an output pulse. The schematic of the diode logic (DL) circuit, shown in Fig. 19-7, illustrates the simplest form of AND gate.

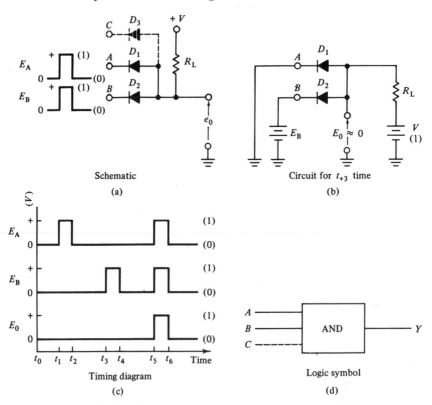

Figure 19-7   AND Gate Diode Logic (DL)

To trace the operation of a positive-logic, two-input AND gate, refer to Fig. 19-7(a). At t_{-1} time when no input pulses are present, both diodes are forward biased by the voltage source V. Ideally, the voltage drop across the diodes (E_{out}) is zero. Practically, the voltage drop across the silicon diodes is approximately equal to 0.7 V; hence, the output voltage is approximately equal to 0.7 V. In diode logic AND gates, the amplitude of the input pulses must be greater than the amplitude of the voltage source V, as shown in Fig. 19-7(a).

At t_{+1} time, there is an input pulse at A but there is no input pulse at B; hence, the input pulse at A reverse biases diode D_1. Thus, diode D_1 operates as an open switch, in parallel with diode D_2 which is still forward biased by the voltage source V. Hence, the output voltage is 0.

At t_{+5} time, all input pulses are present simultaneously; hence, each input pulse reverse biases its respective diode. Thus, all diodes are reverse biased simultaneously, and the resultant output voltage is equal to the voltage source V.

The disadvantage of this circuit is the fact that the amplitude of the input pulses must be greater than that of the output pulse. To overcome this disadvantage, the resistor-transistor logic (RTL) circuit is employed.

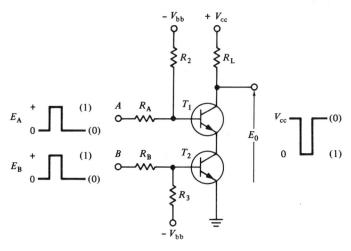

Figure 19-8   NAND Gate Resistor-Transistor Logic (RTL)

To trace the operation of the resistor-transistor logic circuit, refer to Fig. 19-8. In this circuit the output pulse is inverted; hence, the circuit is not an AND gate but it is a NAND gate. When no input pulses are applied, both transistors are held at cutoff by the voltage source V_{bb}. The output voltage is equal to V_{cc} because both transistors operate as open switches. When one input is applied, its respective transistor is forward biased and the other transistor remains at cutoff. The transistors, acting as switches, are in series; hence, if either switch is open, there is no current flow through either of them. Thus, the output voltage is equal to V_{cc}. When both input pulses are applied simultaneously, both transistors are forward biased and conduct at saturation. Hence, the output voltage is approximately equal to zero. The output voltage is $2V_{CEsat} = 2(0.3)$ V $= 0.6$ V for silicon.

If the output of the NAND gate, shown in Fig. 19-8, is the input to an inverter circuit (NOT gate), the resultant circuit is an AND gate.

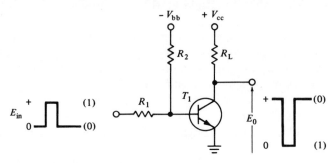

Figure 19-9   NOT Gate (Inverter) Resistor-Transistor Logic (RTL)

19.4   NOT GATE

A NOT gate is a logic circuit which produces an output pulse when no input pulse is applied. The circuit is an inverter circuit but it is referred to as a NOT gate in the study of logic circuits.

The schematic of a practical resistor-transistor logic (RTL) AND gate is shown in Fig. 19-10. The circuit is a combination of the NAND gate, shown

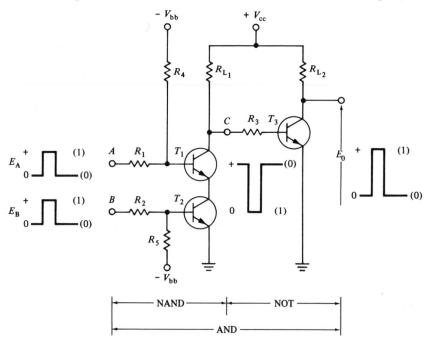

Figure 19-10   AND Gate Resistor-Transistor Logic (RTL)

in Fig. 19-8, and the NOT gate, shown in Fig. 19-9. The NOT gate is a modification of the basic inverter circuit. Since the output of the NAND gate is $+V_{cc}$ when no input pulses are applied or when one input pulse is applied, transistor T_3 normally conducts at saturation. The output of the circuit E_o is normally zero. When both input pulses are applied simultaneously to the NAND gate, the output of the NAND gate is 0 V; thus, the high forward bias to transistor T_3 is removed. Hence, transistor T_3 is cutoff and produces a positive output pulse of an amplitude of $+V_{cc}$.

19.5 CURRENT-MODE LOGIC (CML)

The current-mode logic circuit is a basic gating circuit in which a constant-current source is switched from one transistor to another. Because the current source is constant, the operation of the switched transistors is independent of base-current drive. The constant-current source prevents the switched transistors from conducting at saturation; thus, higher switching speeds are provided. Switching speeds of the order of 1 nanosecond or less are attainable in a current-mode logic circuit.

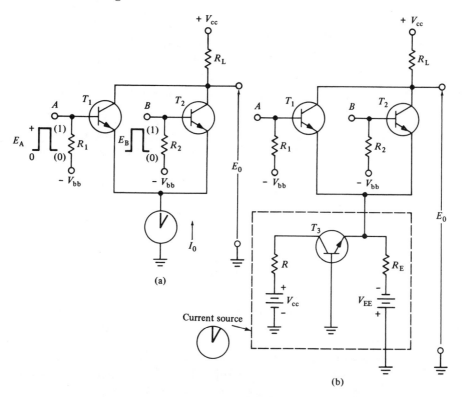

Figure 19-11   NOR Gate Current-Mode Logic (CML)

Figure 19-11(a) illustrates a two-input NOR gate in which current-mode logic is used. The common-base amplifier, shown in Fig. 19-11(b), is used to produce the necessary constant-current source.

An advantage of the current-mode logic circuit is the fact that it has high noise immunity when used with large signal swings. The relatively large power requirement of this system of logic is a principal disadvantage, particularly when it is used in integrated circuits. Current-mode logic is primarily used when extreme speed and noise immunity are needed.

19.6   DIRECT-COUPLED-TRANSISTOR LOGIC (DCTL)

Direct-coupled-transistor logic circuits are those which use a minimum number of passive elements, as illustrated in the schematic of the OR gate shown in Fig. 19-12. Direct-coupled-transistor logic circuits are normally

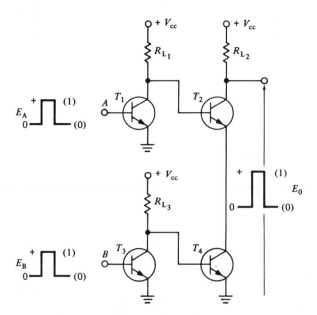

Figure 19-12   OR Gate Direct-Coupled-Transistor-Logic (DCTL)

designed to operate in the saturated mode, although they may be designed to operate in the active mode. Because it is easier to produce a transistor within an integrated circuit than it is to produce a resistor within the same circuit, direct-coupled-transistor logic lends itself to integrated circuitry.

Refer to Fig. 19-12. To ensure proper operation of this circuit, the value of V_{CEsat} across transistor T_1 must be low enough to turn transistor T_2 OFF when the input pulse E_A is applied.

19.7   TRANSISTOR-TRANSISTOR LOGIC (TTL)

Transistor-transistor logic is an outgrowth of integrated circuitry. Refer to Fig. 19-13. In this schematic of a NAND gate circuit, transistors are connected in parallel in a common-base configuration and are connected to a

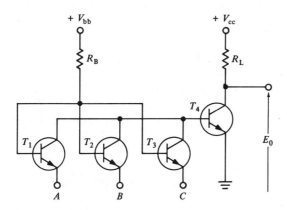

Figure 19-13   NAND Gate Transistor-Transistor Logic (TTL)

transistor in a common-emitter configuration. Observe that the bases of transistors T_1, T_2, and T_3 are connected and that the collectors of the three transistors are also connected. By means of the techniques used in integrated circuitry, a transistor in which multiple emitters are connected to a common base may be constructed, as shown in Fig. 19-14.

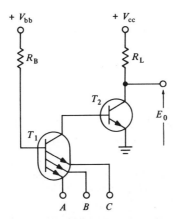

Figure 19-14   NAND Gate Integrated Transistor-Transistor Gate

19.8   SAMPLE DESIGN PROBLEM

From the circuit shown in Fig. 19-15, design a two-input diode-transistor OR gate with the following characteristics:

$$\text{Output-pulse amplitude} = +10 \text{ V}$$
$$\text{Input-gate voltages} = +15 \text{ V}$$
$$I_E = 20 \text{ mA}$$

Given:

$$V_{cc} = +10 \text{ V}$$
$$V_{bb} = -10 \text{ V}$$
$$D_1 = D_2 = \text{silicon junction diodes}$$
$$T_1 = \text{NPN silicon transistor, } h_{FE\min} = 20$$
$$I_{CBO} \approx 0$$

Assume:

$$V_{D_{on}} = 0.7 \text{ V}$$
$$V_{BE_{off}} = -1 \text{ V}$$
$$\text{Standard junction voltages}$$

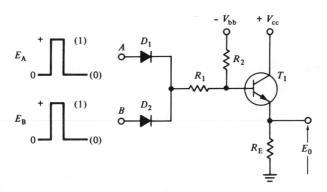

Figure 19-15   2-Input DTL OR Gate

Solution:

1. Determine the value of R_E by use of Ohm's law.
2. Determine the value of base current required for the transistor to conduct at saturation.
3. From the complete schematic shown in Fig. 19-15, draw a schematic of the ON circuit as shown in Fig. 19-16. Assume that one input-gate voltage is applied.
4. From the ON circuit shown in Fig. 19-16, write the ON circuit node equation.

5. From the complete schematic shown in Fig. 19-15, draw a schematic of the OFF circuit as shown in Fig. 19-17. Assume that no input-gate voltage is applied.
6. From the OFF circuit shown in Fig. 19-17, write the OFF circuit node equation.
7. Solve the ON and the OFF circuit equations simultaneously for R_1 and R_2.

$$R_E = \frac{V_E}{I_E} = \frac{V_{cc} - V_{CE_{sat}}}{I_E} = \frac{10 - 0.3}{20 \text{ mA}} = \frac{9.7}{20 \text{ mA}}$$

$$R_E = 485 \ \Omega$$

Use $\qquad R_E = 470 \ \Omega$

$$I_E = \frac{V_E}{R_E} = \frac{9.7}{470} = 20.6 \text{ mA}$$

$$I_E = I_B + I_C = I_B + h_{FE}I_B = (h_{FE} + 1)I_B$$

$$I_B = \frac{I_E}{h_{FE} + 1} = \frac{20.6 \text{ mA}}{20 + 1} = 0.98 \text{ mA} \approx 1 \text{ mA}$$

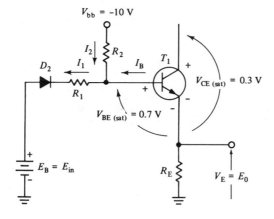

Figure 19-16 ON Circuit

$$I_1 = I_2 + I_B$$

$$\frac{(E_{in} - V_D) - [V_{BE_{sat}} + V_E]}{R_1} = \frac{[V_{BE_{sat}} + V_E] - V_{bb}}{R_2} + I_B$$

$$\frac{(15 - 0.7) - (0.7 + 9.7)}{R_1} = \frac{(+0.7 + 9.7) - (-10)}{R_2} + 1 \text{ mA}$$

$$\frac{3.9}{R_1} = \frac{20.4}{R_2} + 1 \text{ mA}$$

$$\boxed{\frac{1}{R_1} = \frac{5.23}{R_2} + 0.257 \text{ mA}} \qquad\qquad \text{ON Equation}$$

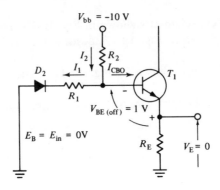

Figure 19-17   OFF Circuit

$$I_2 = I_1 + I_{CBO}$$

$$\frac{[V_E + V_{BE_{off}}] - V_{bb}}{R_2} = \frac{(E_{in} - V_D) - [V_E + V_{BE_{off}}]}{R_1} + I_{CBO}$$

$$\frac{(0 - 1) - (-10)}{R_2} = \frac{(0 - 0.7) - (0 - 1)}{R_1} + 0$$

$$\boxed{R_2 = 30R_1} \qquad \text{OFF Equation}$$

$$\frac{1}{R_1} = \frac{5.23}{R_2} + 0.257 \text{ mA}$$

$$\frac{1}{R_1} = \frac{5.23}{30R_1} + 0.257 \text{ mA}$$

$$\frac{1}{R_1} = \frac{0.174}{R_1} + 0.257 \text{ mA}$$

$$R_1 = 3.22 \text{ k}\Omega$$
$$R_2 = 30R_1 = (30)(3.22 \text{ k}\Omega)$$
$$R_2 = 96.6 \text{ k}\Omega$$

Use
$$R_1 = 2.7 \text{ k}\Omega$$
$$R_2 = 82 \text{ k}\Omega$$

19.9   SAMPLE DESIGN PROBLEM

From the circuit shown in Fig. 19-18, design a two-input resistor-transistor AND gate with the following characteristics:

$$\text{Output-pulse amplitude} = +10 \text{ V}$$
$$\text{Input-gate voltages} = +10 \text{ V}$$
$$I_{C_1} = I_{C_2} = I_{C_3} = 20 \text{ mA}$$

Given:

$$V_{cc} = +10 \text{ V}$$
$$V_{bb} = -10 \text{ V}$$
$$T_1 = T_2 = T_3 = \text{NPN silicon transistor}$$
$$h_{FE_{min}} = 20$$
$$I_{CBO} \approx 0$$

Assume:

$$V_{BE_{off}} = -1 \text{ V}$$
Standard junction voltages
$$R_{L_1} = R_{L_2}$$
$$R_1 = R_2$$
$$R_3 = R_4$$

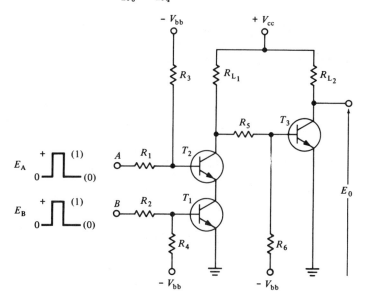

Figure 19-18   2-Input RTL AND Gate

Solution:

1. Determine the value of R_{L_2}. Assume that no input-gating pulses are present and that $E_A = E_B = 0$; hence, transistors T_1 and T_2 are OFF and transistor T_3 is conducting at saturation.

2. Refer to Fig. 19-18. Draw a schematic, equivalent to that portion of this complete circuit in which transistor T_3 is the active element and for conditions that exist when T_3 is ON. See Fig. 19-19(a).

3. From the ON circuit shown in Fig. 19-19(a), write the ON circuit node equation.

4. Refer to Fig. 19-18. Draw a schematic, equivalent to that portion of this complete circuit in which transistor T_3 is the active element and for conditions that exist when T_3 is OFF. See Fig. 19-19(b).

5. From the OFF circuit shown in Fig. 19-19(b), write the OFF circuit node equation.
6. Solve the ON and the OFF circuit equations simultaneously for R_5 and R_6.
7. Refer to Fig. 19-18. Draw a schematic, equivalent to that portion of this complete circuit in which transistor T_1 is the active element and for conditions that exist when T_1 is ON. See Fig. 19-20(a). Assume that both input-gating pulses are present.
8. From the ON circuit shown in Fig. 19-20(a), write the ON circuit node equation.
9. Refer to Fig. 19-18. Draw a schematic, equivalent to that portion of this complete circuit in which transistor T_1 is the active element and for conditions that exist when T_1 is OFF. See Fig. 19-20(b).
10. From the OFF circuit shown in Fig. 19-20(b), write the OFF circuit node equation.
11. Solve the ON and the OFF circuit equations simultaneously for R_2 and R_4.

Assume:

$$T_1 = T_2 = \text{OFF}$$

$$T_3 = \text{ON}$$

$$R_{L_2} = \frac{V_{cc} - V_{CE_{sat}}}{I_{C_3}} = \frac{10 - 0.3}{20 \text{ mA}} = \frac{9.7}{20 \text{ mA}}$$

$$R_{L_2} = 485 \ \Omega$$

Use

$$R_{L_1} = R_{L_2} = 470 \ \Omega$$

$$I_{B_3} = \frac{I_{C_3}}{h_{FE}} = \frac{20 \text{ mA}}{20} = 1 \text{ mA}$$

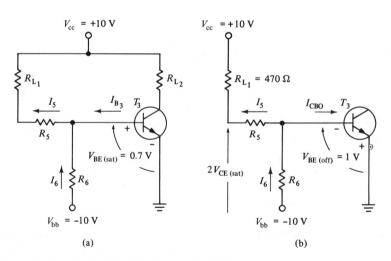

Figure 19-19   (a) T_3 ON Circuit; (b) T_3 OFF Circuit

$$I_5 = I_6 + I_{B_2}$$

$$\frac{V_{cc} - V_{BE_{sat}}}{R_5 + R_{L_1}} = \frac{V_{BE_{sat}} - V_{bb}}{R_6} + I_{B_2}$$

$$\boxed{\frac{10 - 0.7}{R_5 + 470} = \frac{+0.7 + 10}{R_6} + 1 \text{ mA}}$$ ON Equation

$$I_6 = I_5 + I_{CBO}$$

$$\frac{V_{BE_{off}} - V_{bb}}{R_6} = \frac{2V_{CE_{sat}} - V_{BE_{off}}}{R_5} + I_{CBO}$$

$$\frac{-1 + 10}{R_6} = \frac{2(0.3) - (-1)}{R_5} + 0$$

$$\frac{9}{R_6} = \frac{1.6}{R_5}$$

$$\boxed{R_6 = 5.62R_5}$$

$$\frac{9.3}{R_5 + 470} = \frac{10.7}{5.62R_5} + 1 \text{ mA}$$

$$0 = 0.00562R_5^2 - 39R_5 + 6 \text{ k}$$

$$X = \frac{-b \pm \sqrt{b^2 - 4ac}}{2a}$$

$$R_5 = \frac{+39 \pm \sqrt{(39)^2 - (4)(5.62 \text{ mA})(6 \text{ k})}}{(2)(0.00562)}$$

$$R_5 = \frac{+39 \pm \sqrt{1521 - 134.8}}{0.01124}$$

$$R_5 = \frac{+39 \pm 37.2}{0.01124}$$

$$R_5 = \frac{+39 - 37.2}{0.01124} = 160 \ \Omega$$

This solution, for one of the two values possible for resistor R_5, is not acceptable as it would permit excessive base-current flow.

$$R_5 = \frac{+39 + 37.2}{0.01124} = 6770 \ \Omega$$

$$R_6 = 5.62R_5 = 5.62 \times 6.77 \text{ k}\Omega$$

$$R_6 = 38 \text{ k}\Omega$$

Use
$$R_5 = 6.8 \text{ k}\Omega$$
$$R_6 = 33 \text{ k}\Omega$$

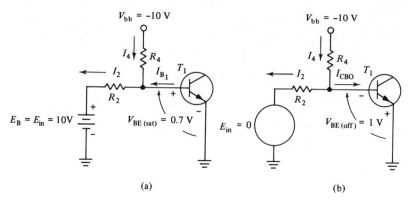

(a)   (b)

Figure 19-20   (a) T_1 ON Circuit; (b) T_1 OFF Circuit

Assume that both gating pulses are present.

$$I_2 = I_4 + I_{B_1}$$

$$\frac{E_B - V_{BE_{sat}}}{R_2} = \frac{V_{BE_{sat}} - V_{bb}}{R_4} + I_{B_1}$$

$$\frac{+10 - 0.7}{R_2} = \frac{+0.7 + 10}{R_4} + 1 \text{ mA}$$

$$\frac{9.3}{R_2} = \frac{10.7}{R_4} + 1 \text{ mA}$$

$$\boxed{\frac{1}{R_2} = \frac{1.15}{R_4} + 0.1075 \text{ mA}}$$ ON Equation

$$I_5 = I_2 + I_{CBO}$$

$$\frac{V_{BE_{off}} - V_{bb}}{R_4} = \frac{E_B - V_{BE_{off}}}{R_2} + I_{CBO}$$

$$\frac{(-1) - (-10)}{R_4} = \frac{0 + 1}{R_2} + 0$$

$$\boxed{R_4 = 9R_2}$$ OFF Equation

$$\frac{1}{R_2} = \frac{1.15}{9R_2} + 0.1075 \text{ mA}$$

$$\frac{1}{R_2} = \frac{0.128}{R_2} + 0.1075 \text{ mA}$$

$$R_2 = \frac{0.875}{0.1075 \text{ mA}} = 8.14 \text{ k}\Omega$$

$$R_4 = 9R_2 = (9)(8.14 \text{ k}\Omega)$$

$$R_4 = 73.2 \text{ k}\Omega$$

Use
$$R_2 = R_1 = 6.8 \text{ k}\Omega$$
$$R_4 = R_3 = 68 \text{ k}\Omega$$

The number of inputs which can be connected to a logic circuit is referred to as "fan-in," and the number of parallel loads which can be driven from one output of a logic circuit is referred to as "fan-out." Therefore, in the design of any logic circuit, it is expedient to provide for a large "fan-in" and a large "fan-out."

LABORATORY EXPERIMENT

LOGIC GATES

OBJECT:

1. To familiarize the student with the characteristics of the basic logic gate
2. To design and analyze a two-input diode-transistor logic OR gate
3. To design and analyze a two-input resistor-transistor logic AND gate

MATERIALS:

3 Switching transistors (example—2N3646 silicon NPN)
1 Manufacturer's specification sheet for transistor used (2N3646—see Appendix B)
2 Diodes, silicon junction (example—1N914)
8 Resistor substitution boxes (10 Ω to 10 MΩ, 1 W)
2 Transistor power supplies (0 to 30 V and 0 to 250 mA)
1 Square-wave generator (20 Hz to 200 kHz with $Z_0 = 600$ Ω)
1 Oscilloscope, dc time-base type; frequency response dc to 450 kHz; vertical sensitivity, 100 mV/cm

NOTE: **Gating circuits require two or more square-wave pulse inputs which may occur at the same time or at different times. Each of these input pulses is generated by a separate source but each of the sources is connected to the same time base. For this elementary technical exercise, however, a single square-wave generator is applied to one of the gating inputs. To simulate the simultaneous application of a number of gating inputs, connect all input terminals to the single square-wave generator. When a pulse is applied to one input only, the other input terminals should be connected to ground, in order to simulate zero input pulses.**

PROCEDURE:

PART A   Two-input diode-transistor logic OR gate

1. By use of the schematic shown in Fig. 19-1X, design a two-input diode-transistor Logic OR Gate from the following specifications:

$$\text{Output-pulse amplitude} = +15 \text{ V}$$
$$\text{Input-gate voltages} = +8 \text{ V, square wave, f} = 5 \text{ kHz}$$
$$I_E = 10 \text{ mA}$$

Given:

$$V_{cc} = +15 \text{ V}$$
$$V_{bb} = -15 \text{ V}$$
$$D_1 = D_2 = \text{silicon junction diode}$$
$$T_1 = \text{NPN silicon transistor}$$

Assume:

$$V_{D_{on}} = 0.7 \text{ V}$$
$$V_{BE_{off}} = -1 \text{ V}$$
$$\text{Standard junction voltages}$$

2. Design the two-input diode-transistor logic OR gate for the conditions specified. Include all calculations. Explain why any assumptions which have been made are valid.

298

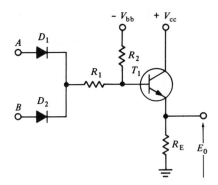

Figure 19-1X 2-Input DTL OR Gate

3. Draw a schematic of the circuit designed in Step 1, and label it with the practical standard color code values.
4. From the schematic drawn in Step 2, analyze the circuit designed in Step 1; prove that it will operate satisfactorily, for the required specifications, when standard value resistors are used.
5. Connect the circuit designed. Simulate an absence of input pulses by grounding input A and input B. Check the operation of the circuit. Record all necessary data.
6. Remove the ground from input A. Apply a positive 8 V, 5000 Hz square wave of voltage to input A. Input B should remain grounded. Check the operation of the circuit. Record all necessary data.
7. Reverse inputs A and B (ground input A and apply a positive, 8 V square wave of voltage to input B). Check the operation of the circuit. Record all necessary data.
8. Simulate the simultaneous application of pulses to input A and input B by connecting the positive, 8 V square-wave voltage souce to input A and input B. Check the operation of the circuit. Record all necessary data.
9. Compare the measured values with the computed values and explain any discrepancies.

PART B   Two-input resistor-transistor logic AND gate
1. Design a two-input resistor-transistor logic AND gate, as shown in Fig. 19-2X, from the following specifications:

$$\text{Output-pulse amplitude} = +12 \text{ V}$$
$$\text{Input-gate voltages} = +5 \text{ V, square wave}$$
$$f = 10 \text{ kHz}$$
$$I_{C_1} = I_{C_2} = 12 \text{ mA}$$
$$R_{L_1} = R_{L_2}$$
$$R_1 = R_2$$
$$R_3 = R_4$$

Given:
$$V_{cc} = +12 \text{ V}$$
$$V_{bb} = -12 \text{ V}$$
$$T_1 = T_2 = T_3 = \text{NPN silicon transistor}$$

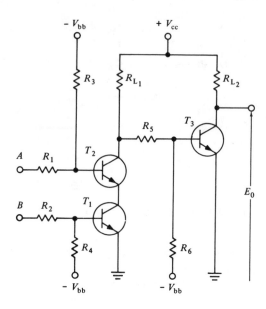

Figure 19-2X 2-Input RTL AND Gate

Assume:

$$V_{BE\text{off}} = -1 \text{ V}$$
Standard junction voltages

2. Design the two-input resistor-transistor logic AND gate for the conditions specified. Include all calculations. Explain why any assumptions which have been made are valid.

3. Draw a schematic of the circuit designed in Step 1, and label it with practical standard color code values.

4. From the schematic drawn in Step 2, analyze the circuit designed in Step 1; prove that it will operate satisfactorily, for the required specifications, when standard-value resistors are used.

5. Connect the circuit designed. Simulate an absence of input pulses by grounding input A and input B. Check the operation of the circuit. Record all necessary data.

6. Remove the ground from input A. Apply a positive, 5 V, 10 kHz square wave of voltage to input A. Input B should remain grounded. Check the operation of the circuit. Record all necessary data.

7. Reverse inputs A and B (ground input A and apply a positive, 5 V square wave of voltage to input B). Check the operation of the circuit. Record all necessary data.

8. Simulate the simultaneous application of input A and input B by connecting the positive 5 V square-wave voltage source to input A and input B. Check the operation of the circuit. Record all necessary data.

9. Compare the measured values with the computed values and explain any discrepancies.

QUESTIONS AND EXERCISES

1. Explain the relationship between an OR gate and a NOR gate.
2. Explain the operation of the diode-transistor logic OR gate shown in Fig. 19-1X. Draw a pulse-timing diagram to be used as an aid in the explanation.
3. Explain the operation of the integrated transistor-transistor logic NAND gate shown in Fig. 19-14. Draw a pulse-timing diagram to be used as an aid in the explanation.
4. Is any one system of logic circuitry superior to all the other systems? Explain.
5. What is meant by "fan-in" and "fan-out"? Explain.
6. Draw a schematic of a resistor-transistor logic NOR gate in which 1 is defined as a negative pulse with reference to ground and 0 is defined as the absence of a pulse. Explain the operation of this circuit.
7. Explain the operation of the current-mode logic NOR gate shown in Fig. 19-11(b). Draw a pulse-timing diagram to be used as an aid in the explanation.
8. Name the type of logic circuit shown in Fig. 19-3X. Draw and label a pulse-timing diagram of this circuit, and explain the operation of the circuit.

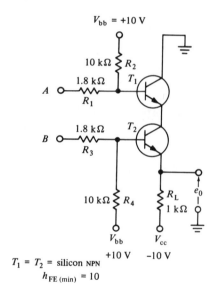

Figure 19-3X

9. Determine the minimum amplitude of input-gating pulses required for proper operation of the circuit shown in Fig. 19-3X.
10. Refer to Fig. 19-15. Analyze this circuit when

$$V_{cc} = +10 \text{ V}, \qquad V_{bb} = -5 \text{ V}, \qquad R_1 = 22 \text{ k}\Omega$$
$$R_2 = 560 \text{ k}\Omega, \qquad R_E = 1.8 \text{ k}\Omega, \qquad h_{FE\text{min}} = 30$$
$$E_A = E_B = 5 \text{ V positive pulses}$$

Prove that this circuit will or will not operate properly as a two-input DTL OR gate.

11. In the circuit of Prob. 10, which component, components, or voltages must be changed and to what values must they be changed in order to ensure proper circuit operation?

12. Refer to Fig. 19-18. Determine the minimum amplitude of the input pulses E_A and E_B necessary for proper operation of the two-input RTL AND gate when

$$T_1 = T_2 = T_3, \qquad R_{L_1} = R_{L_2} = 1.8 \text{ k}\Omega, \qquad R_1 = R_2 = 15 \text{ k}\Omega$$
$$R_3 = R_4 = 330 \text{ k}\Omega, \qquad R_5 = 15 \text{ k}\Omega, \qquad R_6 = 180 \text{ k}\Omega$$
$$V_{cc} = +20 \text{ V}, \qquad V_{bb} = -20 \text{ V}, \qquad h_{FE_{min}} = 10$$

13. What effect would an open-voltage source (V_{bb}) have on the operation of this circuit? Refer to Fig. 19-18.

chapter 20

SAMPLING GATES

Recall the two types of gating circuits: the logic gate (chapter 19) and the sampling gate. The sampling gate is a circuit in which an exact reproduction of the input-signal voltage is permitted to appear as an output signal, for a specified interval—sample of time.

20.1 DEFINITION

Sampling gate, linear gate, transmission gate, and time-selection circuit are names which may be used interchangeably for this type of circuit: each refers to a circuit characteristic. The term *sampling gate* is used in this manual. The sampling gate is further classified as unidirectional and bidirectional: The unidirectional sampling gate is a circuit which passes signal voltages of a single polarity; the bidirectional sampling gate is a circuit which passes signal voltages of both polarities.

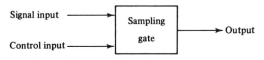

Figure 20-1 Block Diagram-Sampling Gate

Basically, the sampling gate consists of a signal input, a control input, and an output, as shown in Fig. 20-1. The control voltage is the "gate" which may prevent the signal voltage from passing through the "gate" or which may permit the signal voltage to continue its flow through the "gate" to emerge as the output voltage.

A diode or a transistor may be used for the switch in the sampling gate.

When the switch is closed (diode or transistor conducting), the input-signal voltage becomes the output voltage as it passes through the gate. When the control voltage is of the polarity to forward bias the diode and/or to forward bias the emitter-base junction of the transistor, the active element (switch) conducts. When the switch is ON (active element conducting), the input-signal voltage is permitted to pass through the "open gate" to appear as the output voltage. The control input is usually a square wave of voltage, generated from an astable or from a monostable multivibrator. The signal voltage is frequently a square wave, the frequency of which is either less than or greater than that of the control voltage.

20.2 APPLICATIONS

The applications of the sampling-gate circuit are many and varied. Some of the most common are the multiplex circuit, the chopper circuit, the telemetering circuit, and storage scopes.

The sampling gate is used in a multiplex circuit similar to that found in commercial FM stereo. In this application, the sampling gate alternately samples the two channels. The sampling rate is very high; therefore, the resultant signal from the stereo output appears to be continuous.

The chopper circuit is used to overcome the drift of the Q point, in a dc coupled amplifier. When extremely long time-constant circuits are used in conjunction with a dc amplifier, the transistor cannot distinguish between the long time-constant signal voltage and the Q-point drift. This defect may be rectified by breaking the long time-constant signal voltage into segments of ac voltage so that each resembles an ac voltage and, therefore, may be distinguished from the Q-point drift. The Q-point drift may be caused by the temperature-dependent leakage current and by the associated biasing circuit. The sampling gate is used to "chop" this extremely long time-constant ac signal into segments of short duration; hence, the name "chopper" circuit.

In a telemetry circuit, sampling gates sample and transmit, in sequence, a number of different transducer outputs. In this manner, a single transmitter is used to transmit, in sequence, a number of different sample signals.

In a storage scope, the interval to be recorded or "stored" is selected from the input signal by the sampling gate.

20.3 OPERATION—UNIDIRECTIONAL DIODE SAMPLING GATE

To trace the operation of a unidirectional diode sampling gate, refer to Fig. 20-2. Observe that diode D is reverse biased by the negative control voltage v_c. While the diode is reverse biased, it acts as an open switch; hence, there is no output voltage from the circuit. During the interval t_1 to t_4, the negative control-voltage pulse is removed; hence, the reverse bias from diode D

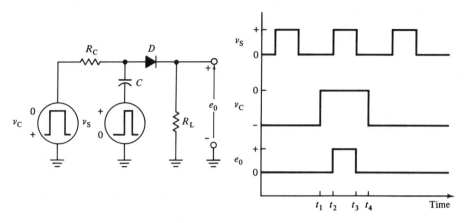

Figure 20-2 Unidirectional Diode Sampling Gate

is removed. If there is a positive-going signal voltage present during the interval t_1 to t_4, it forward biases diode D; hence, diode D conducts. Thus, if an ideal diode is assumed, the output voltage equals the input-signal voltage v_s.

When the signal voltage is omitted from the diode sampling gate, the circuit is a clipper; when the control voltage is omitted, the circuit is a clamper. The value of load resistance for either circuit is determined in the same manner, by solving for the geometric mean of the forward resistance of the diode and the reverse resistance of the diode. See chapter 4, Eq. 18.

The sampling-gate circuit has an advantage over the logic AND gate, which it resembles; it draws current from the output circuit, only when sampling occurs. The interaction between the signal voltage and the control voltage is a disadvantage of this circuit.

20.4 OPERATION—UNIDIRECTIONAL TRANSISTOR SAMPLING GATE

To follow this discussion of the operation of a unidirectional transistor sampling gate, refer to Fig. 20-3. This circuit is basically a bias-stabilized, grounded-emitter amplifier. The control-voltage pulse, which holds the sampling gate at cutoff, is in series with the emitter resistor and ground. This positive-control pulse provides the reverse bias, necessary to hold transistor T_1 at cutoff, independent of the signal voltage v_s. When the control-voltage pulse is removed, the transistor conducts at saturation. A negative-going signal voltage provides reverse bias, which causes the transistor to conduct in the active region. In the interval during which the control voltage is zero, the gate acts as a common-emitter, class B amplifier. Hence, the signal voltage appears, amplified and inverted, as the output voltage.

The unidirectional transistor sampling gate, shown in Fig. 20-3, may be

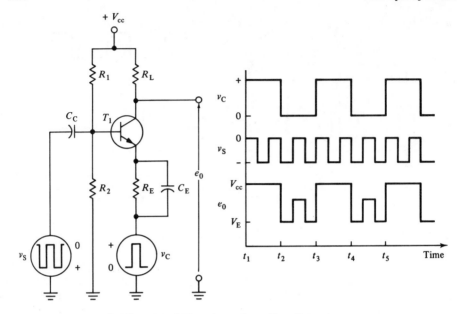

Figure 20-3   Unidirectional Transistor Sampling Gate

converted to a bidirectional gate by changing the values of resistors R_1 and R_2 so that the Q point is in the middle of the load line when the control pulse is removed. See Fig. 20-4.

There are two advantages of the transistor sampling gate over the diode sampling gate. One is the elimination of the interaction between the signal

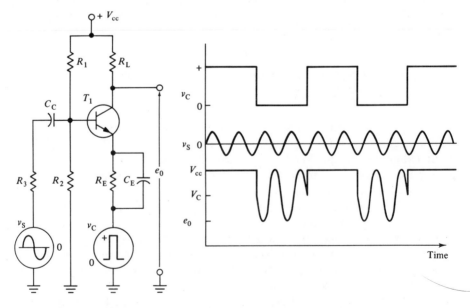

Figure 20-4   Bidirectional Transistor Sampling Gate

voltage and the control voltage. The isolation of the signal voltage from the control voltage is provided by the transistor amplifier. The fact that this circuit provides amplification for the signal voltage is another advantage.

20.5 SAMPLE DESIGN PROBLEM

Design a bidirectional transistor sampling gate, as shown in Fig. 20-5, from the following specifications:

Signal voltage is sampled every 200 μsec, for a duration of 200 μsec, at a frequency of 10 kHz.

Given:

$$V_{cc} = 15 \text{ V}$$
$$v_s = 5 \text{ V (rms) sine wave}; f = 10 \text{ kHz}$$
$$T_1 = \text{NPN silicon transistor}$$
$$h_{FE(min)} \approx h_{fe} = 50$$
$$h_{ie} = 2 \text{ k}\Omega$$

Assume (with $v_c = 0$ V):

$$I_C = 5 \text{ mA}$$
$$V_C = 10 \text{ V}$$
$$V_E = 2 \text{ V}$$
$$S = 5$$

Determine

$$R_L, R_E, R_1, R_2, R_3, C_E, C_C, v_{c(min)}, \text{ and } f_c$$

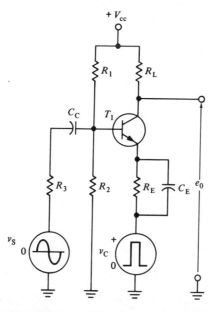

Figure 20-5 Bidirectional Transistor Sampling Gate

Solution:

1. Design a common-emitter amplifier circuit in which neither signal-voltage input nor control-voltage input is assumed. Determine the value of R_L by use of the dc load line.
2. Determine the value of base current necessary to produce the required 5 mA of collector current. Refer to Fig. 20-6.
3. Determine the value of emitter current by use of the value of h_{FE} and that of the given base current.
4. Determine the value of R_E by use of Ohm's law.
5. Simplify the schematic of the biasing network by the application of Thevenin's theorem, as shown in Fig. 20-7.
6. Determine the value of R_B by solving the standard dc biasing equation:

$$S = \frac{1 + \dfrac{R_E}{R_B}}{1 - \alpha + \dfrac{R_E}{R_B}}$$

7. Write Kirchhoff's voltage loop equation from the schematic shown in Fig. 20-7, and solve for E_{oc}. Neglect the emitter-base junction voltage.
8. By use of the equations for E_{oc} and for R_B, shown in Fig. 20-7, solve for R_1 in terms of R_B, V_{cc}, and E_{oc}. Make the necessary numerical substitutions, and determine the value for R_1.
9. By use of the value of R_1, obtained in Step 8, determine the numerical value of R_2 from the equation for R_B, shown in Fig. 20-7.
10. Select standard color code values for R_1 and R_2. The preceding steps fulfill the dc requirements for the circuit.

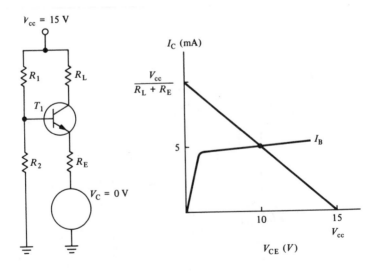

Figure 20-6

11. Determine the value of C_E. To ensure that the emitter is at ac ground, the reactance of the capacitor must be $0.1R_E$ at the lowest frequency to be amplified.
12. Determine the value of C_C. The reactance of the capacitor must be 0.1 of the input impedance of the amplifier, to ensure that the coupling capacitor does not attenuate the signal voltage at the lowest frequency to be amplified.
13. The control-voltage pulse must have sufficient amplitude to cut off the transistor; therefore, the minimum value of control voltage—worst case—must be used.
14. Determine the value of R_3 necessary to ensure that the input-signal voltage can prevent the transistor from being driven into saturation or cutoff.

$$R_L = \frac{V_{cc} - V_C}{I_C} = \frac{15 - 10}{5 \text{ mA}} = \frac{5}{5 \text{ mA}} = 1 \text{ k}\Omega$$

$$I_B = \frac{I_C}{h_{FE}} = \frac{5 \text{ mA}}{50} = 0.1 \text{ mA}$$

$$I_E = (h_{FE} + 1)I_B = (50 + 1)0.1 \text{ mA}$$
$$I_E = 5.12 \text{ mA}$$

$$R_E = \frac{V_E}{I_E} = \frac{2}{5.12 \text{ mA}} = 390 \ \Omega$$

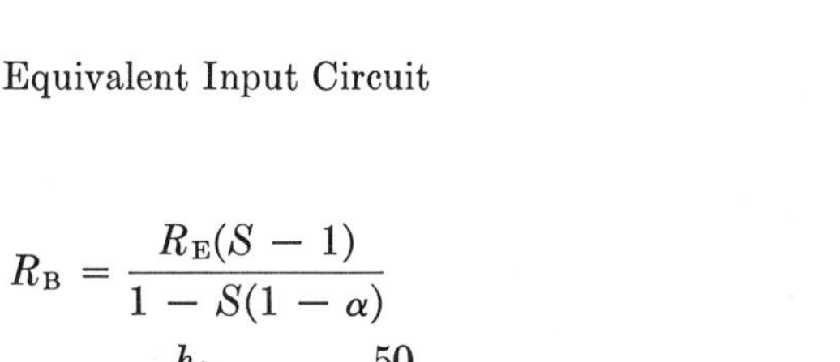

$$R_B = \frac{R_1 R_2}{R_1 + R_2} \quad T_1$$

$$E_{oc} = \frac{R_2 V_{cc}}{R_1 + R_2}$$

$$R_E (h_{FE} + 1)$$

Figure 20-7 Equivalent Input Circuit

$$R_B = \frac{R_E(S - 1)}{1 - S(1 - \alpha)}$$

$$\alpha = \frac{h_{fe}}{h_{fe} + 1} = \frac{50}{50 + 1} = 0.98$$

$$R_B = \frac{390(5 - 1)}{1 - 5(1 - 0.98)}$$

$$R_B = 1.73 \text{ k}\Omega$$
$$E_{oc} = V_E + I_B R_B + V_{BE}$$

Assume:

$$V_{BE} \approx 0$$
$$E_{oc} = 2 + (0.1 \text{ mA})(1.73 \text{ k}\Omega)$$
$$E_{oc} = 2.17 \text{ V}$$
$$R_B = \frac{R_1 R_2}{R_1 + R_2}$$
$$R_2 = \frac{R_1 R_B}{R_1 - R_B}$$
$$E_{oc} = \frac{R_2 V_{cc}}{R_1 + R_2}$$

Substituting the value of R_2,

$$E_{oc} = \frac{\dfrac{R_1 R_B}{R_1 - R_B} V_{cc}}{R_1 + \dfrac{R_1 R_B}{R_1 - R_B}}$$

Solving for R_1,

$$R_1 = \frac{R_B V_{cc}}{E_{oc}} = \frac{(1.73 \text{ k}\Omega)(15)}{2.17}$$
$$R_1 = 12 \text{ k}\Omega$$
$$R_2 = \frac{R_1 R_B}{R_1 - R_B} = \frac{(12 \text{ k}\Omega) - (1.73 \text{ k}\Omega)}{12 \text{ k}\Omega - 1.73 \text{ k}\Omega}$$
$$R_2 = 2.02 \text{ k}\Omega$$

Use
$$R_1 = 12 \text{ k}\Omega$$
$$R_2 = 2 \text{ k}\Omega$$

This completes the dc requirements for the circuit.

$$C_E = \frac{1}{2\pi f_c X_c}$$
$$X_c = \frac{R_E}{10} = \frac{390}{10} = 39 \ \Omega$$
$$f_c = \frac{1}{T_c} = \frac{1}{t_1 + t_2} = \frac{1}{200 \ \mu\text{sec} + 200 \ \mu\text{sec}} = \frac{1}{400 \ \mu\text{sec}}$$
$$f_c = 2.5 \text{ kHz}$$
$$C_E = \frac{1}{6.28 \times 2.5 \times 10^3 \times 39}$$
$$C_E = 2.31 \ \mu\text{F}$$

Use
$$C = 3 \ \mu\text{F}$$
$$v_{c_{min}} = V_E + V_{BE_{off}}$$
$$v_{c_{min}} = 2 + 0.5$$
$$v_{c_{min}} = 2.5 \text{ V}$$

$$I_{C_{sat}} = \frac{V_{cc}}{R_L + R_E} = \frac{15}{1 \text{ k} + 0.39 \text{ k}} = 10.8 \text{ mA}$$

$$I_{C_Q} = 5 \text{ mA}$$

$$i_{c_{+max}} = I_{C_{sat}} - I_{C_Q} = 10.8 \text{ mA} - 5 \text{ mA} = 5.8 \text{ mA}$$

$$I_{C_{-max}} = I_{C_Q} - 0 = 5 \text{ mA} - 0 = 5 \text{ mA}$$

$$i_c = 5.8 \text{ mA} - 0 - 5 \text{ mA}$$

Hence, 5 mA is the controlling value.

$$i_{s_{max}} \approx \frac{i_{c_{max}}}{h_{fe}} = \frac{5 \text{ mA}}{50} = 0.1 \text{ mA}$$

$$v_{in_{max}} = i_{s_{max}} Z_{in}$$

$$Z_{in} = \frac{R_B R_{in}}{R_B + R_{in}}$$

$$R_{in} \approx h_{ie} = 2 \text{ k}\Omega$$

$$Z_{in} \approx \frac{R_B R_{in}}{R_B + R_{in}} \approx \frac{(1.73 \text{ k})(2 \text{ k})}{(1.73 \text{ k}) + 2 \text{ k}}$$

$$Z_{in} \approx 0.93 \text{ k}\Omega$$

$$v_{in_{max}} = i_{s_{max}} Z_{in} = (0.1 \text{ mA})(0.93 \text{ k})$$

$$v_{in_{max}} = 93 \text{ mV}$$

$$v_{in_{rms}} = 0.707 v_{in_{max}} = 0.707 \times 93 \text{ mV}$$

$$v_{in_{rms}} = 65.8 \text{ mV}$$

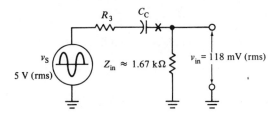

Figure 20-8   ac Input Circuit

$$i_s = \frac{v_{in}}{Z_{in}} = \frac{65.8 \text{ mV}}{0.93 \text{ k}\Omega} = 70.7 \text{ } \mu\text{A}$$

$$Z_T = \frac{v_s}{i_s} = \frac{5}{70.7 \text{ } \mu\text{A}} = 70.7 \text{ k}\Omega$$

$$R_3 = Z_T - Z_{in} = 70.7 \text{ k}\Omega - 0.93 \text{ k}\Omega$$

$$R_3 = 69.7 \text{ k}\Omega$$

Use $$R_3 = 68 \text{ k}\Omega$$

$$C_c = \frac{1}{2\pi f_s X_c}$$

$$X_c = \frac{Z_T}{10} = \frac{70.7 \text{ k}\Omega}{10} = 7.07 \text{ k}\Omega$$

$$C_c = \frac{1}{6.28 \times 10^4 \times 7.07 \times 10^3}$$

$$C_c = 0.0022 \ \mu\text{F}$$

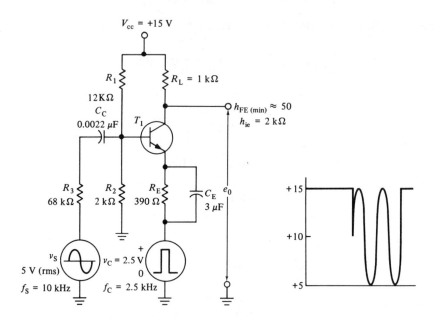

Figure 20-9 Bidirectional Transistor Sampling Gate

20.6 SAMPLE CIRCUIT ANALYSIS PROBLEM

Prove that the circuit, shown in Fig. 20.9, operates properly as a bidirectional transistor sampling gate. Determine the amplitude of the output-signal voltage.

Analysis of dc circuit:

1. Assume $v_c = 0$ V. By the application of Thevenin's theorem, simplify the dc biasing network so that it resembles the schematic shown in Fig. 20-10.
2. Write Kirchhoff's voltage loop equation for the dc input circuit shown in Fig. 20-10. Solve the input loop equation for I_B in terms of R_B, R_E, E_{oc}, and h_{FE}.
3. Determine the value of V_E by use of Ohm's law.
4. Determine the value of V_C by use of Ohm's law.

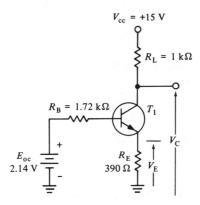

Figure 20-10

Analysis of ac circuit:

5. Determine the minimum value of control voltage necessary to ensure cutoff of transistor T_1. Assume $V_{BE_{off}} = 0.5$ V.
6. Determine the input impedance of the amplifier. Assume $R_{in} \approx h_{ie} = 2$ kΩ.
7. Draw the equivalent ac input circuit from the schematic of the amplifier shown in Fig. 20-11.
8. Determine v_{in} by use of the voltage divider equation.
9. Determine the amplitude of the output-signal voltage. Assume:

$$e_o = A_v v_{in}$$

and
$$A_V = A_i \frac{R_L}{Z_{in}}$$

and
$$A_i \approx h_{fe}$$

10. Determine whether the value of C_E is of a capacitance sufficient to prevent attenuation of the control voltage. Use the capacitive-reactance equation.
11. Determine whether the value of C_c is of a capacitance sufficient to prevent attenuation of the signal voltage. Use the capacitive-reactance equation.

$$R_B = \frac{R_1 R_2}{R_1 + R_2} = \frac{(12 \text{ k})(2 \text{ k})}{(12 \text{ k}) + (2 \text{ k})} = 1.72 \text{ kΩ}$$

$$E_{oc} = \frac{R_2 V_{cc}}{R_1 + R_2} = \frac{(2 \text{ k})(15)}{12 \text{ k} + 2 \text{ k}} = 2.14 \text{ V}$$

$$E_{oc} = E_{R_B} + V_{BE} + E_{R_E}$$

$$E_{oc} = I_B R_B + V_{BE} + I_E R_E$$

But $I_E = I_B + I_C$

and $I_C = h_{FE} I_B$

$$I_E = I_B + h_{FE} I_B$$

$$I_E = (h_{FE} + 1) I_B$$

Assume:

$$V_{BE} \approx 0$$

$$E_{oc} = I_B R_B + (h_{FE} + 1)I_B R_E$$

$$I_B = \frac{E_{oc}}{R_B + (h_{FE} + 1)R_E} = \frac{2.14}{1.72 \text{ k} + (50 + 1)(0.39 \text{ k})}$$

$$I_B = 0.099 \text{ mA}$$

$$V_E = I_E R_E = (h_{FE} + 1)I_B R_E = (50 + 1)(0.099 \text{ mA})(0.39 \text{ k})$$

$$V_E = 1.97 \text{ V}$$

$$V_C = V_{cc} - E_{RL} = V_{cc} - I_C R_L = V_{cc} - h_{FE}I_B R_L$$

$$V_C = 15 - (50)(0.099 \text{ mA})(1 \text{ k}) = 15 - 4.95$$

$$V_C = 10.05 \text{ V}$$

$$v_{c_{min}} > V_E + V_{BE_{off}}$$

$$2.5 > 1.97 + 0.5$$

$$2.5 > 2.47$$

$$R_{in} \approx h_{ie} = 2 \text{ k}\Omega$$

$$Z_{in} = \frac{R_B R_{in}}{R_B + R_{in}} = \frac{(1.72 \text{ k})(2 \text{ k})}{1.72 \text{ k} + 2 \text{ k}} = 0.925 \text{ k}\Omega$$

$$v_{in} = \frac{Z_{in}v_s}{Z_{in} + R_3} = \frac{(5)(0.925 \text{ k})}{0.925 \text{ k} + 68 \text{ k}} = 66.2 \text{ mV}$$

$$A_i \approx h_{fe} \approx 50$$

$$A_v = A_i \frac{R_L}{Z_{in}} = \frac{50 \times 1 \text{ k}}{0.93 \text{ k}} = 53.8$$

$$e_o = A_v v_{in} = (53.8)(66.2 \text{ mV})$$

$$e_o = 3.56 \text{ V} \qquad \text{(rms)}$$

$$\Delta e_{o_{p-p}} = 10 \text{ V}$$

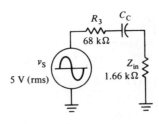

Figure 20-11   ac Input Circuit

SAMPLING GATES

OBJECT:
1. To familiarize the student with the unidirectional diode sampling gate
2. To design and analyze a bidirectional transistor sampling gate
3. To design and analyze a unidirectional transistor sampling gate

MATERIALS:
3   Silicon, switching diodes (example—1N914)
1   Silicon, switching transistor (example—2N3646 NPN silicon)
4   Resistor substitution boxes (10 Ω to 10 MΩ, 1 W)
2   Capacitor substitution boxes (0.0001 to 0.22 μF, 450 V)
1   Transistor power supply (0 to 30 V and 0 to 250 mA)
2   Sine square-wave generators (20 Hz to 200 kHz with $Z_0 = 600\ \Omega$)
1   Oscilloscope, dc time-base type; frequency response dc to 450 kHz; vertical sensitivity, 100 mV/cm

PROCEDURE:

PART A   Diode sampling gate
1. Connect the circuit shown in Fig. 20-2, in which $R_C = 4.7$ kΩ, $R_L = 100$ kΩ, $C = 0.01\ \mu$F, $v_s = 15$ V peak square wave at a frequency of 500 Hz, and $v_c = 0$ to 30 V transistor power supply (to act as the control voltage).
2. Vary the control voltage; observe and record the resultant output-signal voltage.
3. Explain the operation of the circuit in detail.
4. Analyze the circuit conditions observed in Step 2, and verify experimental results. (State all assumptions made.)

PART B   Transistor sampling gate
1. Design a bidirectional transistor sampling gate, as shown in Fig. 20-1X, from the following specifications:

$$I_C = 2\text{ mA}, \qquad V_{cc} = 20\text{ V}, \qquad V_C = 10\text{ V}, \qquad V_E = 2\text{ V}, \qquad S = 5$$

Frequency of control voltage = 1 kHz
Frequency of signal voltage = 6 kHz

2. Connect the circuit designed in Step 1. Observe, measure, and record all necessary data.
3. Explain any differences between the design values and the experimental values.
4. Redesign the circuit, designed in Step 1, for unidirectional (saturated) operation. Assume $I_{C_{sat}} = 4$ mA. All other specifications are to remain the same. Show all work, and explain how all values have been obtained.
5. Connect the unidirectionally designed circuit, and measure and record any and all necessary experimental values.
6. Compare the experimental results obtained in Step 5 with results obtained in Step 4. Note any differences and explain.

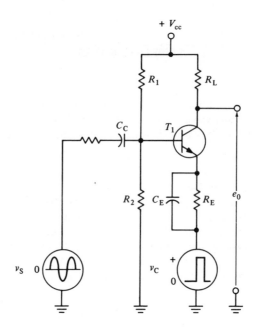

Figure 20-1X

QUESTIONS AND EXERCISES

1. Explain the operation of the unidirectional diode sampling gate shown in Fig. 20-2.
2. Explain the operation of the unidirectional transistor sampling gate shown in Fig. 20-3.
3. What is the difference between a unidirectional sampling gate and a bidirectional sampling gate? Explain.
4. Explain how the transistor sampling gate overcomes the interaction between the signal voltage and the control voltage.
5. The transistor sampling gate circuit, shown in Fig. 20-1X, is an emitter-gated circuit. Draw the schematic of a base-gated transistor sampling gate, and include the signal-voltage waveform, the control-voltage waveform, and the output-voltage waveform.
6. Draw and label a block diagram of the circuits necessary to construct a transistorized electronic switch for an oscilloscope. Assume that two trace lines—each varied by a different voltage input to this system of circuits—would appear on a scope face.
7. Determine whether the circuit, shown in Fig. 20-2X, operates satisfactorily as a unidirectional transistor sampling gate when

$$v_s = -10 \text{ V square-wave pulse}$$
$$f_s = 1 \text{ kHz}$$
$$v_c = 4 \text{ V square wave}$$
$$v_c = 1 \text{ kHz}$$

Show all work. Assume C_c and C_E to be of proper value.

Questions and Exercises

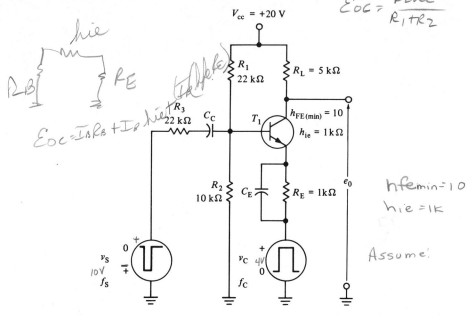

$$\mathcal{E}_{OC} = \frac{R_2 V_{cc}}{R_1 + R_2} \quad 317$$

$\mathcal{E}_{OC} = I_B R_B + I_B \cdot \text{hie}$

$\text{hfemin} = 10$
$\text{hie} = 1k$

Assume:

Figure 20-2X Unidirectional Transistor Sampling Gate

8. Determine the value of capacitor C_c appropriate for the circuit shown in Fig. 20-2X and for the circuit conditions specified in Problem 7. Show all work.
9. Determine the value of capacitor C_E appropriate for the circuit shown in Fig. 20-2X and for the circuit conditions specified in Problem 7. Show all work.
10. Determine the amplitude of the resultant output voltage of the circuit shown in Fig. 20-2X, for the circuit conditions specified in Problem 7. Show all work.
11. Draw and label a schematic of a unidirectional transistor sampling gate in which an NPN transistor in a common-base configuration is used.
12. Explain the operation of the circuit drawn in Problem 11. Draw and label an input–output pulse-timing diagram to aid in your explanation.
13. Refer to Fig. 20-5. Analyze the circuit when

$V_{cc} = +10$ V,	$R_L = 500 \ \Omega$,	$R_E = 100\Omega$,	$R_1 = 3.9 \ k\Omega$
$R_2 = 560 \ \Omega$,	$R_3 = 10 \ k\Omega$,	$C_E = 2 \ \mu F$,	$C_c = 0.01 \ \mu F$
$h_{FE_{min}} \approx h_{fe} = 20$	$h_{ie} = 500 \ \Omega$,	$S = 5$	

Determine the minimum amplitude of the control voltage. Assume that the control voltage is a positive square wave, the frequency of which is 5 kHz. Determine the maximum amplitude of the sine wave signal voltage when the frequency of the signal input is 20 kHz.
14. Redesign the circuit of Problem 13 for proper operation as a unidirectional sampling gate.

REFERENCES

Babb, D. S., *Pulse Circuits: Switching and Shaping*. Prentice-Hall, Inc., Englewood Cliffs, N.J., 1964.

Blitzer, R., *Basic Pulse Circuits*. McGraw-Hill Book Company, New York, 1964.

Burroughs Corporation, *Digital Computer Principles*. McGraw-Hill Book Company, New York, 1962.

Chirlian, P. M., *Analysis and Design of Electronic Circuits*. McGraw-Hill Book Company, New York, 1965.

Cutler, P., *Electronic Circuit Analysis (Passive Networks)*. McGraw-Hill Book Company, New York, 1960.

Doyle, J. M., *Pulse Fundamentals*. Prentice-Hall, Inc., Englewood Cliffs, N.J., 1963.

GE Transistor Manual, Semiconductor Products Dept., General Electric Company, Syracuse, N.Y., 1964.

Gillie, A. C., *Electrical Principles of Electronics*. McGraw-Hill Book Company, New York, 1961.

Houpis, C. H. and Lubelfeld, J., *Outline of Pulse Circuits*. Regents Publishing Co., Inc., New York, 1966.

Hurley, R. B., *Junction Transistor Electronics*. John Wiley & Sons, Inc., New York, 1958.

Ketchum, D. J. and Alvarez, E. C., *Pulse and Switching Circuits*. McGraw-Hill Book Company, New York, 1965.

Millman, J. and Taub, H., *Pulse, Digital, and Switching Waveforms*. McGraw-Hill Book Company, New York, 1965.

Oppenheimer, A. L., *Semiconductor Logic and Switching Circuits*. Charles E. Merrill Books, Inc., Columbus, Ohio, 1966.

Romanowitz, H. A., *Electrical Fundamentals and Circuit Analysis*. John Wiley & Sons, Inc., New York, 1966.

Stanton, W. A., *Pulse Technology*. John Wiley & Sons, Inc., New York, 1964.

Suriana, T. and Herrick, C., *Semiconductor Electronics*. Holt, Rinehart and Winston, Inc., New York, 1964.

Uzunoglu, V., *Semiconductor Network Analysis and Design*. McGraw-Hill Book Company, New York, 1964.

Wright, R. R. and Skutt, H. R., *Electronics Circuits and Devices*. The Ronald Press Company, New York, 1965.

appendix A

TABLE OF LOGARITHMS

N	0	1	2	3	4	5	6	7	8	9
10	0000	0043	0086	0128	0170	0212	0253	0297	0334	0374
11	0414	0453	0492	0531	0569	0607	0645	0682	0719	0755
12	0792	0828	0864	0899	0934	0969	1004	1038	1072	1106
13	1139	1173	1206	1239	1271	1303	1335	1367	1399	1430
14	1461	1492	1523	1553	1584	1614	1644	1673	1703	1732
15	1761	1790	1818	1847	1875	1903	1931	1959	1987	2014
16	2041	2068	2095	2122	2148	2175	2201	2227	2253	2279
17	2304	2330	2355	2380	2405	2430	2455	2480	2504	2529
18	2553	2577	2601	2625	2648	2672	2695	2718	2742	2765
19	2788	2810	2833	2856	2878	2900	2923	2945	2967	2989
20	3010	3032	3054	3075	3096	3118	3139	3160	3181	3201
21	3222	3243	3263	3284	3304	3324	3345	3365	3385	3404
22	3424	3444	3464	3483	3502	3522	3541	3560	3579	3598
23	3617	3636	3655	3674	3692	3711	3729	3747	3766	3784
24	3802	3820	3838	3856	3874	3892	3909	3927	3945	3962
25	3979	3997	4014	4031	4048	4065	4082	4099	4116	4133
26	4150	4166	4183	4200	4216	4232	4249	4265	4281	4298
27	4314	4330	4346	4362	4278	4393	4409	4425	4440	4456
28	4472	4487	4502	4518	4533	4548	4564	4579	4594	4609
29	4624	4639	4654	4669	4683	4698	4713	4728	4742	4757
30	4771	4786	4800	4814	4829	4843	4857	4871	4886	4900
31	4914	4928	4942	4955	4969	4983	4997	5011	5024	5038
32	5051	5065	5079	5092	5105	5119	5132	5145	5159	5172
33	5185	5198	5211	5224	5237	5250	5263	5276	5289	5302
34	5315	5328	5340	5353	5366	5378	5391	5403	5416	5428
35	5441	5453	5465	5478	5490	5502	5514	5527	5539	5551
36	5563	5575	5587	5599	5611	5623	5635	5647	5658	5670
37	5682	5694	5705	5717	5729	5740	5752	5763	5775	5786
38	5798	5809	5821	5832	5843	5855	5866	5877	5888	5899
39	5911	5922	5933	5944	5955	5966	5977	5988	5999	6010
40	6021	6031	6042	6053	6064	6075	6085	6096	6107	6117
41	6128	6138	6149	6160	6170	6180	6191	6201	6212	6222
42	6232	6243	6253	6263	6274	6284	6294	6304	6314	6325
43	6335	6345	6355	6365	6375	6385	6395	6405	6415	6425
44	6435	6444	6454	6464	6474	6484	6403	6503	6513	6522
45	6532	6532	6551	6561	6571	6580	6590	6599	6609	6618
46	6628	6637	6646	6656	6665	6675	6684	6693	6702	6712
47	6721	7630	6739	6749	6758	6767	6776	6785	6794	6803
48	6821	6821	6830	6839	6848	6857	6866	6875	6884	6893
49	6902	6911	6920	6928	6937	6946	6955	6964	6972	6981
50	6990	6998	7007	7016	7024	7033	7042	7050	7059	7067
51	7076	7084	7093	7101	7110	7118	7126	7135	7143	7152
52	7160	7168	7177	7185	7193	7202	7210	7218	7226	7235
53	7243	7251	7259	7267	7275	7284	7292	7300	7308	7316
54	7324	7332	7340	7348	7356	7364	7372	7380	7388	7396

TABLE OF LOGARITHMS (*Continued*)

N	0	1	2	3	4	5	6	7	8	9
55	7404	7412	7419	7427	7435	7443	7451	7459	7466	7474
56	7482	7490	7497	7505	7513	7520	7528	7536	7543	7551
57	7559	7566	7574	7582	7589	7597	7604	7612	7619	7627
58	7634	7642	7649	7657	7664	7672	7679	7686	7694	7701
59	7709	7716	7723	7731	7738	7745	7752	7760	7767	7774
60	7782	7789	7796	7803	7810	7818	7825	7832	7839	7846
61	7853	7860	7868	7875	7882	7889	7896	7903	7910	7917
62	7924	7931	7938	7945	7952	7959	7966	7973	6980	7987
63	7993	8000	8007	8014	8021	8028	8035	8041	8048	8055
64	8062	8069	8075	8082	8089	8096	8102	8109	8116	8122
65	8129	8136	8142	8149	8156	8162	8169	8176	8182	8189
66	8195	8202	8209	8216	8222	8228	8235	8241	8248	8254
67	8261	8267	8274	8280	8287	8293	9299	8306	8312	8319
68	8325	8331	8338	8344	8351	8357	8363	8370	8376	8382
69	8388	8395	8401	8407	8414	8420	8426	8432	8439	8445
70	8451	8457	8463	8470	8476	8482	8488	8494	8500	8506
71	8513	8519	8525	8531	8537	8548	8549	8555	8561	8567
72	8573	8579	8585	8591	8597	8603	8609	8615	8621	8627
73	8633	8639	8645	8651	8657	8663	8669	8675	8681	8686
74	8692	8698	8704	8610	8716	8622	8627	8633	8639	8645
75	8751	8656	8762	8768	8774	8779	8785	8691	8697	8802
76	8808	8814	8820	8825	8831	8836	8842	8848	8854	8859
77	8865	8871	8876	8882	8887	8893	8899	8904	8910	8915
78	8921	8927	8932	8938	8943	8949	8954	8960	8965	8971
79	8976	8982	8987	8993	8998	9004	9009	9015	9020	9025
80	9031	9036	9042	9047	9053	9058	9063	9069	9074	9079
81	9085	9090	9096	9101	9106	9112	9117	9122	9128	9133
82	9183	9143	9149	9154	9159	9165	9170	9175	9180	9186
83	9191	9196	9201	9206	9212	9217	9222	9227	9232	9238
84	9243	9248	9253	9258	9263	9269	9274	9279	9284	9289
85	9294	9299	9304	9309	9315	9320	9325	9330	9335	9340
86	9345	9350	9355	9360	9365	9370	9375	9380	9385	9390
87	9395	9400	9405	9410	9415	9420	9425	9430	9435	9440
88	9445	9450	9455	9460	9465	9469	9474	9479	9484	9489
89	9494	9499	9504	9509	9513	9518	9523	9528	9533	9538
90	9542	9547	9552	9557	9562	9566	9571	9576	9581	9586
91	9590	9595	9600	9605	9609	9614	9619	9624	9628	9633
92	9638	9643	9647	9652	9657	9661	9666	9671	9675	9680
93	9685	9689	9694	9699	9703	9708	9713	9717	1922	9627
94	9731	9736	9741	1945	9750	9754	9759	9763	9768	9773
95	9777	9782	9686	9791	9695	9800	9805	9809	9814	9818
96	9823	9827	9832	9836	9841	9845	9850	9854	9859	9863
97	9868	9872	9877	9881	9886	9890	9894	9899	9903	9908
98	9912	9917	9921	9926	9930	9934	9939	9943	9948	9952
99	9956	9961	9965	9969	9974	9978	9983	9987	9991	9996

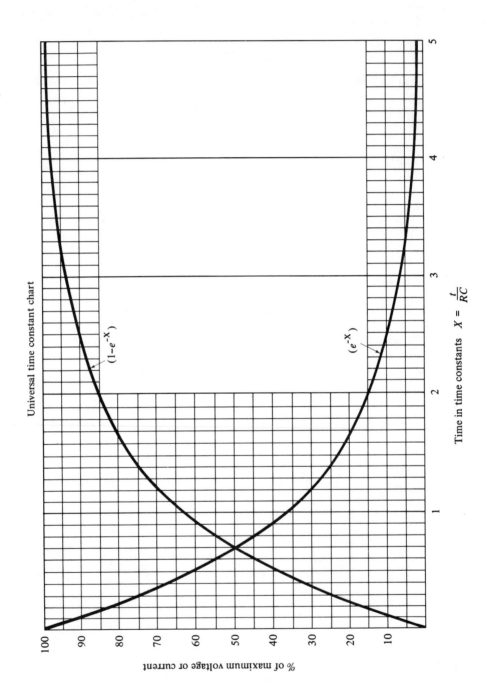

Universal time constant chart

$(1-e^{-x})$

(e^{-x})

% of maximum voltage or current

100 90 80 70 60 50 40 30 20 10

Time in time constants $X = \dfrac{t}{RC}$

MANUFACTURER'S SPECIFICATION SHEETS FOR THE FOLLOWING SEMICONDUCTOR DEVICES

Transistors

2N3638 and 2N3638A	PNP, Silicon, Switching transistor
2N3646	NPN, Silicon, Switching transistor
2N3903–2N3904	NPN, Silicon, Switching transistor
2N3905–2N3906	PNP, Silicon, Switching transistor
TIS-49	NPN, Silicon, Switching transistor
TIS-50	PNP, Silicon, Switching transistor

Unijunction

TIS-43	PN, Silicon, Unijunction
2N4851-2N4853	PN, Silicon, Unijunction
2N2646–2N2647	PN, Silicon, Unijunction

Zener Diode

    1N4728 to 1N4764 Silicon, Zener Diode

Diode

    1N914 to 1N917 Silicon, Diode

Tunnel Diode

    1N3712 to 1N3721 Germanium, Tunnel Diode

2N3638 · 2N3638A
PNP HIGH CURRENT SWITCHES
DIFFUSED SILICON PLANAR* EPITAXIAL TRANSISTORS

- **FAST SWITCHING** - - t_{on} = 75 ns (max.) @ 300 mA
  - - t_{off} = 170 ns (max.) @ 300 mA
- **HIGH BETA** - - h_{FE} 100 (min.) @ I_C = 50 mA
- **HIGH CURRENT** - - Up to 500 mA
- **LOW** V_{CE}(sat) - - 1.0 Volt (max.) @ 300 mA
- **LOW COST IN ALL QUANTITIES**

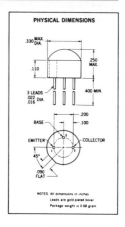

PHYSICAL DIMENSIONS

ABSOLUTE MAXIMUM RATINGS (Note 1)

Maximum Temperatures

Storage Temperature	−55°C to +125°C
Operating Junction Temperature	+125°C Maximum
Lead Temperature (Soldering, 10 sec time limit)	+260°C Maximum

Maximum Power Dissipation

Total Dissipation at 25°C Case Temperature	(Notes 2 and 3)	0.7 Watt
at 25°C Free Air Temperature	(Notes 2 and 3)	0.3 Watt

Maximum Voltages and Current

V_{CBO}	Collector to Base Voltage		− 25 Volts
V_{CES}	Collector to Emitter Voltage		− 25 Volts
V_{CEO}	Collector to Emitter Voltage	(Note 4)	− 25 Volts
V_{EBO}	Emitter to Base Voltage		−4.0 Volts
I_C	Collector Current	(Note 2)	500 mA

ELECTRICAL CHARACTERISTICS (25°C Free Air Temperature unless otherwise noted)

SYMBOL	CHARACTERISTIC	2N3638 MIN.	TYP.	MAX.	2N3638A MIN.	TYP.	MAX.	UNITS	TEST CONDITIONS	
h_{FE}	DC Pulse Current Gain (Note 5)				80	140			I_C = 1.0 mA	V_{CE} = − 10V
h_{FE}	DC Pulse Current Gain (Note 5)	20	70		100	160			I_C = 10 mA	V_{CE} = − 10V
h_{FE}	**DC Pulse Current Gain** (Note 5)	30	67		**100**	**130**			I_C **= 50 mA**	V_{CE} **= − 1.0 V**
h_{FE}	DC Pulse Current Gain (Note 5)	20	40		20	50			I_C = 300 mA	V_{CE} = − 2.0 V
V_{CE}(sat)	Pulsed Collector Saturation Voltage (Note 5)		−0.08	−0.25		−0.08	−0.25	Volt	I_C = 50 mA	I_B = 2.5 mA
V_{CE}(sat)	**Pulsed Collector Saturation Voltage** (Note 5)		−0.38	−1.0		**−0.38**	**−1.0**	**Volt**	I_C **= 300 mA**	I_B **= 30 mA**
V_{CEO}(sust)	Collector to Emitter Sustaining Voltage (Notes 4 & 5)	−25			−25			Volts	I_C = 10 mA (pulsed)	I_B = 0
BV_{CBO}	Collector to Base Breakdown Voltage	−25			−25			Volts	I_C = 100 μA	I_E = 0
BV_{CES}	Collector to Emitter Breakdown Voltage	−25			−25			Volts	I_C = 100 μA	V_{EB} = 0
t_{on}	**Turn On Time** (Note 6)		28	75		**28**	**75**	ns	I_C ≈ 300 mA	I_{B1} ≈ 30 mA
t_{off}	**Turn Off Time** (Note 6)		110	170		**110**	**170**	ns	I_C ≈ 300 mA	I_{B1} ≈ 30 mA I_{B2} ≈ − 30 mA
h_{fe}	**High Frequency Current Gain** (f = 100 MHz)	1.0	1.9		**1.5**	**1.9**			I_C **= 50 mA**	V_{CE} **= − 3.0 V**
C_{obo}	Common-Base, Open-Circuit Output Capacitance		6.0	20		**6.0**	**10**	pF	I_E = 0	V_{CB} = − 10 V
C_{ibo}	Common-Base, Open-Circuit Input Capacitance		18	65		18	25	pF	I_C = 0	V_{EB} = − 0.5 V

*Planar is a patented Fairchild process.

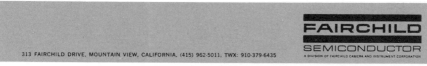

313 FAIRCHILD DRIVE, MOUNTAIN VIEW, CALIFORNIA, (415) 962-5011, TWX: 910-379-6435

FAIRCHILD
SEMICONDUCTOR
A DIVISION OF FAIRCHILD CAMERA AND INSTRUMENT CORPORATION

FAIRCHILD TRANSISTORS 2N3638 · 2N3638A

ELECTRICAL CHARACTERISTICS (25°C Free Air Temperature unless otherwise noted)

SYMBOL	CHARACTERISTIC	MIN.	TYP.	MAX.	UNITS	TEST CONDITIONS	
$V_{BE}(sat)$	Base-Emitter Saturation Voltage (pulsed, Note 5)		−0.9	−1.1	Volts	$I_C = 50$ mA	$I_B = 2.5$ mA
$V_{BE}(sat)$	Base-Emitter Saturation Voltage (pulsed, Note 5)	−0.8	−1.25	−2.0	Volts	$I_C = 300$ mA	$I_B = 30$ mA
BV_{EBO}	Emitter to Base Breakdown Voltage	−4.0			Volts	$I_E = 100\ \mu A$	$I_C = 0$
I_{CES}	Collector Reverse Current		0.1	35	nA	$V_{CE} = -15$ V	$V_{EB} = 0$
$I_{CES}(65°C)$	Collector Reverse Current		0.002	2.0	μA	$V_{CE} = -15$ V	$V_{EB} = 0$

TYPICAL ELECTRICAL CHARACTERISTICS

2N3638

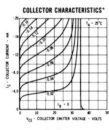

COLLECTOR CHARACTERISTICS*

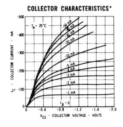

COLLECTOR CHARACTERISTICS*

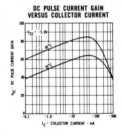

DC PULSE CURRENT GAIN
VERSUS COLLECTOR CURRENT

2N3638A

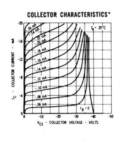

COLLECTOR CHARACTERISTICS*

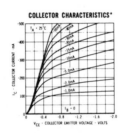

COLLECTOR CHARACTERISTICS*

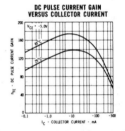

DC PULSE CURRENT GAIN
VERSUS COLLECTOR CURRENT

2N3638 · 2N3638A

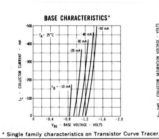

BASE CHARACTERISTICS*

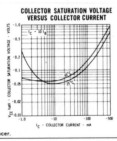

COLLECTOR SATURATION VOLTAGE
VERSUS COLLECTOR CURRENT

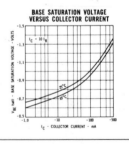

BASE SATURATION VOLTAGE
VERSUS COLLECTOR CURRENT

* Single family characteristics on Transistor Curve Tracer.

TYPICAL ELECTRICAL CHARACTERISTICS

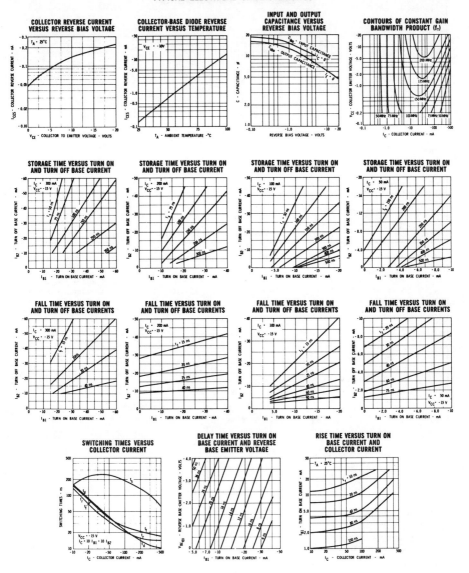

SMALL SIGNAL CHARACTERISTICS

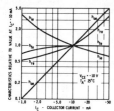 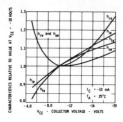 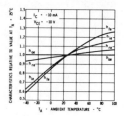

NOTES:

(1) These ratings are limiting values above which the serviceability of any individual semiconductor device may be impaired.

(2) These are steady state limits. The factory should be consulted on applications involving pulsed or low duty cycle operations.

(3) These ratings give a maximum junction temperature of 125°C and junction to case thermal resistance of 14 °C/Watt (derating factor of 7.0mW/°C); junction to ambient thermal resistance of 333°C/Watt (derating factor of 3.0 mW/°C).

(4) Rating refers to a high-current point where collector to emitter voltage is lowest. For more information send for Fairchild Publication APP-4/2.

(5) Pulse conditions: length = 300 μs; duty cycle = 1%.

(6) See switching circuit for exact values of I_C, I_{B1}, and I_{B2}.

h PARAMETERS (f = 1.0 kHz)

SYMBOL	CHARACTERISTIC	MIN.	2N3638 TYP.	2N3638 MAX.	MIN.	2N3638A TYP.	2N3638A MAX.	UNITS	TEST CONDITIONS	
h_{ie}	Input Resistance		200	2000		480	2000	ohms	$I_C = 10$ mA	$V_{CE} = -10$ V
h_{oe}	Output Conductance		80	1200		80	1200	μmhos	$I_C = 10$ mA	$V_{CE} = -10$ V
h_{re}	Voltage Feedback Ratio		162	2600		162	1500	$x10^{-6}$	$I_C = 10$ mA	$V_{CE} = -10$ V
h_{fe}	Small Signal Current Gain	25	74		100	180			$I_C = 10$ mA	$V_{CE} = -10$ V

T_{ON} and T_{OFF} TEST CIRCUIT

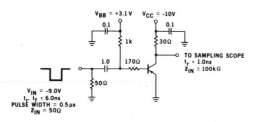

2N3646

NPN HIGH SPEED SATURATED SWITCH
DIFFUSED SILICON PLANAR* EPITAXIAL TRANSISTOR

- HIGH SPEED . . . . . . $\tau_S = 18$ ns (MAX) AT 10 mA
  $t_{on} = 18$ ns (MAX) AT 300 mA
  $t_{off} = 28$ ns (MAX) AT 300 mA
- MEDIUM VOLTAGE . . . $V_{CEO(sus)} = 15$ V (MIN)
- HIGH FREQUENCY . . . $f_T = 350$ MHz AT 30 mA

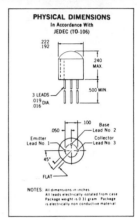

PHYSICAL DIMENSIONS
In Accordance With
JEDEC (TO-106)

NOTES: All dimensions in inches
All leads electrically isolated from case
Package weight is 0.31 gram Package
is electrically non conductive material

ABSOLUTE MAXIMUM RATINGS (Note 1)

Maximum Temperatures

Storage Temperature	−55°C to +125°C
Operating Junction Temperature	+125°C
Lead Temperature (Soldering, 10 sec. time limit)	+260°C

Maximum Power Dissipation

Total Dissipation at 25°C Case Temperature (Notes 2 and 3)	0.5 Watt
at 25°C Ambient Temperature (Notes 2 and 3)	0.2 Watt

Maximum Voltages

V_{CBO}	Collector to Base Voltage	40 Volts
V_{CES}	Collector to Emitter Voltage	40 Volts
V_{CEO}	**Collector to Emitter Voltage** (Note 4)	**15 Volts**
V_{EBO}	Emitter to Base Voltage	5.0 Volts

ELECTRICAL CHARACTERISTICS (25°C Free Air Temperature unless otherwise noted)

SYMBOL	CHARACTERISTIC	MIN.	TYP.	MAX.	UNITS	TEST CONDITIONS	
$V_{CE(sat)}$	Collector Saturation Voltage (Note 5)		0.16	0.2	Volts	$I_C = 30$ mA	$I_B = 3.0$ mA
$V_{CE(sat)}$	Collector Saturation Voltage (Note 5)		0.18	0.28	Volts	$I_C = 100$ mA	$I_B = 10$ mA
$V_{CE(sat)}$	Collector Saturation Voltage (Note 5) ($T_A = +65°C$)		0.18	0.3	Volts	$I_C = 30$ mA	$I_B = 3.0$ mA
$V_{CE(sat)}$	Collector Saturation Voltage (Note 5)		0.39	0.5	Volts	$I_C = 300$ mA	$I_B = 30$ mA
h_{fe}	High Frequency Current Gain (f = 100 MHz)	3.5	5.5			$I_C = 30$ mA	$V_{CE} = 10$ V
τ_s	Charge Storage Time Constant (Note 6)		8.0	18	ns	$I_C = I_{B1} = 10$ mA	$I_{B2} \approx -10$ mA
t_{on}	Turn On Time (Note 6)		9.0	18	ns	$I_C \approx 300$ mA	$I_{B1} \approx 30$ mA
t_{off}	Turn Off Time (Note 6)		15	28	ns	$I_C \approx 300$ mA, $I_{B1} \approx 30$ mA, $I_{B2} \approx -30$ mA	
C_{obo}	Common Base Open Circuit Output Capacitance		3.3	5.0	pF	$I_E = 0$	$V_{CB} = 5.0$ V
C_{ibo}	Common Base Open Circuit Input Capacitance		6.6	8.0	pF	$I_C = 0$	$V_{EB} = 0.5$ V
h_{FE}	DC Pulse Current Gain (Note 5)	30	60	120		$I_C = 30$ mA	$V_{CE} = 0.4$ V
h_{FE}	DC Pulse Current Gain (Note 5)	25	55			$I_C = 100$ mA	$V_{CE} = 0.5$ V
h_{FE}	DC Pulse Current Gain (Note 5)	15				$I_C = 300$ mA	$V_{CE} = 1.0$ V

Additional Electrical Characteristics on page 2
Notes on page 2

*Planar is a patented Fairchild process.

FAIRCHILD
SEMICONDUCTOR
A DIVISION OF FAIRCHILD CAMERA AND INSTRUMENT CORPORATION

313 FAIRCHILD DRIVE, MOUNTAIN VIEW, CALIFORNIA, (415) 962-5011, TWX: 910-379-6435

FAIRCHILD TRANSISTOR 2N3646

ELECTRICAL CHARACTERISTICS (25°C Free Air Temperature unless otherwise noted)

SYMBOL	CHARACTERISTIC	MIN.	TYP.	MAX.	UNITS	TEST CONDITIONS	
BV_{CBO}	Collector to Base Breakdown Voltage	40			Volts	$I_C = 100\ \mu A$	$I_E = 0$
BV_{CES}	Collector to Emitter Breakdown Voltage	40			Volts	$I_C = 100\ \mu A$	$V_{EB} = 0$
$V_{CEO(sus)}$	**Collector to Emitter Sustaining Voltage** (Notes 4 and 5)	15			**Volts**	$I_C = 10\ mA$	$I_B = 0$
BV_{EBO}	Emitter to Base Breakdown Voltage	5.0			Volts	$I_E = 100\ \mu A$	$I_C = 0$
$V_{BE(sat)}$	Base Saturation Voltage (Note 5)	0.75	0.82	0.95	Volts	$I_C = 30\ mA$	$I_B = 3.0\ mA$
$V_{BE(sat)}$	Base Saturation Voltage (Note 5)		0.97	1.2	Volts	$I_C = 100\ mA$	$I_B = 10\ mA$
$V_{BE(sat)}$	Base Saturation Voltage (Note 5)		1.3	1.7	Volts	$I_C = 300\ mA$	$I_B = 30\ mA$
I_{CES}	Collector Reverse Current		0.04	0.5	μA	$V_{CE} = 20\ V$	$V_{EB} = 0$
$I_{CES}(65°C)$	Collector Reverse Current		0.5	3.0	μA	$V_{CE} = 20\ V$	$V_{EB} = 0$

NOTES:
(1) These ratings are limiting values above which the serviceability of any individual semiconductor device may be impaired.
(2) These are steady state limits. The factory should be consulted on applications involving pulsed or low duty cycle operations.
(3) These ratings give a maximum junction temperature of 125°C and junction to case thermal resistance of 200°C/Watt (derating factor of 5.0 mW/°C). Junction to ambient thermal resistance of 500°C/Watt (derating factor of 2.0 mW/°C).
(4) Rating refers to a high current point where collector to emitter voltage is lowest.
(5) Pulse Conditions: length = 300 μs; duty cycle = 1%.
(6) See switching circuits for exact values of I_C, I_{B1}, and I_{B2}.

TYPICAL ELECTRICAL CHARACTERISTICS

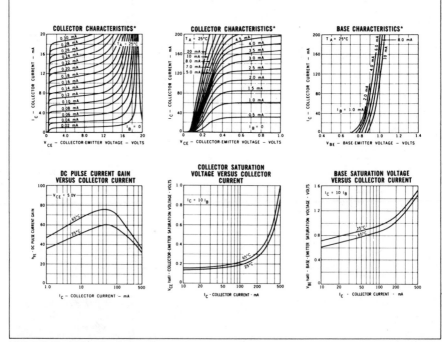

* Single family characteristics on Transistor Curve Tracer

334

TYPICAL ELECTRICAL CHARACTERISTICS

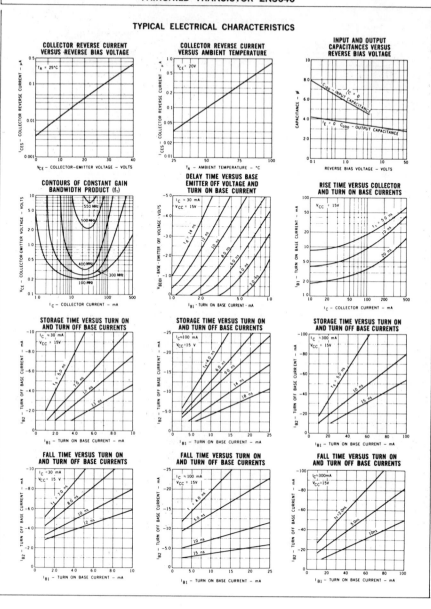

TYPICAL ELECTRICAL CHARACTERISTICS

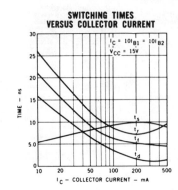

SWITCHING TIMES VERSUS COLLECTOR CURRENT

$I_C = 10 I_{B1} = 10 I_{B2}$
$V_{CC} = 15V$

TIME – ns

I_C – COLLECTOR CURRENT – mA

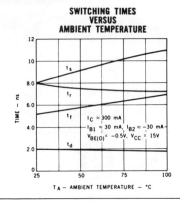

SWITCHING TIMES VERSUS AMBIENT TEMPERATURE

$I_C \approx 300$ mA
$I_{B1} \approx 30$ mA, $I_{B2} \approx -30$ mA
$V_{BE(0)} = -0.5V$, $V_{CC} = 15V$

TIME – ns

T_A – AMBIENT TEMPERATURE – °C

t_{ON} and t_{OFF} TEST CIRCUIT

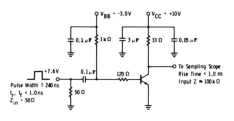

$V_{BB} = -3.0V$ $V_{CC} = +10V$

$0.1\,\mu F$ $1k\Omega$ $3\,\mu F$ $33\,\Omega$ $0.05\,\mu F$

+7.6V
Pulse Width ≥ 240ns
t_r, t_f < 1.0ns
$Z_{in} = 50\Omega$

$0.1\,\mu F$ $120\,\Omega$ $50\,\Omega$

To Sampling Scope
Rise Time < 1.0 ns
Input Z ≈ 100 kΩ

CHARGE STORAGE TIME MEASUREMENT CIRCUIT

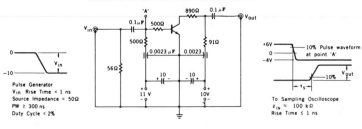

V_{in}

0
–10

V_{in}

Pulse Generator
V_{in} Rise Time < 1 ns
Source Impedance = 50Ω
PW ≥ 300 ns
Duty Cycle < 2%

$0.1\,\mu F$ 500Ω 890Ω $0.1\,\mu F$ V_{out}
500Ω 91Ω
$0.0023\,\mu F$ 0.0023
56Ω
10 10
11 V $10V$

+6V
0
–4V

10% Pulse waveform at point 'A'

V_{out}
10%
τ_s

To Sampling Oscilloscope
$z_{in} \approx 100$ kΩ
Rise Time ≤ 1 ns

MOTOROLA
Semiconductors
BOX 955 • PHOENIX 1, ARIZONA

2N3903
2N3904

NPN SILICON ANNULAR* TRANSISTORS

. . . designed for general purpose switching and amplifier applications and for complementary circuitry with types 2N3905 and 2N3906.

* One-Piece, Injection-Molded Unibloc† Package for High Reliability
* High Voltage Ratings — BV_{CEO} = 40 Volts minimum
* Current Gain Specified from 100 μA to 100 mA
* Complete Switching and Amplifier Specifications
* Low Capacitance — C_{obo} = 4.0 pf maximum

NPN SILICON
SWITCHING & AMPLIFIER
TRANSISTORS
AUGUST 1965 — DS 5127

"D" shape package lies flat or easy printed circuit mounting.

Rugged, one-piece, high temperature, pressure-molded, humidity resistant, plastic package.

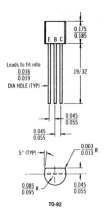

EBC configuration easily adaptable to standard TO–18 pin circle.

19/32 inch, gold-plated nickel, oval leads permit reliable solder connections.

Si HIGH FREQUENCY TRANSISTORS
2N3903, 2N3904
DS 5127

MAXIMUM RATINGS

Characteristic	Symbol	Rating	Unit
Collector-Base Voltage	V_{CB}	60	Vdc
Collector-Emitter Voltage	V_{CEO}	40	Vdc
Emitter-Base Voltage	V_{EB}	6	Vdc
Collector Current	I_C	200	mAdc
Total Device Dissipation @ T_A = 60°C	P_D	210	mW
Total Device Dissipation @ T_A = 25°C Derate above 25°C	P_D	310 2.81	mW mW/°C
Thermal Resistance, Junction to Ambient	θ_{JA}	0.357	°C/mW
Junction Operating Temperature	T_J	135	°C
Storage Temperature Range	T_{stg}	-55 to +135	°C

E B C

$\dfrac{0.175}{0.185}$

Leads to fit into $\dfrac{0.016}{0.019}$ DIA HOLE (TYP)

19/32

$\dfrac{0.045}{0.055}$

$\dfrac{0.045}{0.055}$

5° (TYP)

$\dfrac{0.003}{0.013}$ R.

$\dfrac{0.085}{0.095}$ R.

$\dfrac{0.045}{0.055}$

TO-92

*Annular semiconductors patented by Motorola Inc.
†Trademark of Motorola Inc.

MOTOROLA *Semiconductor Products Inc.* Ⓜ A SUBSIDIARY OF MOTOROLA INC

ELECTRICAL CHARACTERISTICS ($T_A = 25°C$ unless otherwise noted)

Characteristic		Fig. No.	Symbol	Min	Max	Unit
OFF CHARACTERISTICS						
Collector-Base Breakdown Voltage ($I_C = 10 \mu Adc$, $I_E = 0$)			BV_{CBO}	60	—	Vdc
Collector-Emitter Breakdown Voltage* ($I_C = 1$ mAdc)			$BV_{CEO}*$	40	—	Vdc
Emitter-Base Breakdown Voltage ($I_E = 10 \mu Adc$, $I_C = 0$)			BV_{EBO}	6	—	Vdc
Collector Cutoff Current ($V_{CE} = 40$ Vdc, $V_{OB} = 3$ Vdc)			I_{CEX}	—	50	nAdc
Base Cutoff Current ($V_{CE} = 40$ Vdc, $V_{OB} = 3$ Vdc)			I_{BL}	—	50	nAdc
ON CHARACTERISTICS						
DC Current Gain * ($I_C = 0.1$ mAdc, $V_{CE} = 1$ Vdc)	2N3903 2N3904	15	$h_{FE}*$	20 40	— —	—
($I_C = 1.0$ mAdc, $V_{CE} = 1$ Vdc)	2N3903 2N3904			35 70	— —	
($I_C = 10$ mAdc, $V_{CE} = 1$ Vdc)	2N3903 2N3904			50 100	150 300	
($I_C = 50$ mAdc, $V_{CE} = 1$ Vdc)	2N3903 2N3904			30 60	— —	
($I_C = 100$ mAdc, $V_{CE} = 1$ Vdc)	2N3903 2N3904			15 30	— —	
Collector-Emitter Saturation Voltage* ($I_C = 10$ mAdc, $I_B = 1$ mAdc)		16, 17	$V_{CE(sat)}*$	—	0.2	Vdc
($I_C = 50$ mAdc, $I_B = 5$ mAdc)				—	0.3	
Base-Emitter Saturation Voltage* ($I_C = 10$ mAdc, $I_B = 1$ mAdc)		17	$V_{BE(sat)}*$	0.65	0.85	Vdc
($I_C = 50$ mAdc, $I_B = 5$ mAdc)				—	0.95	
SMALL SIGNAL CHARACTERISTICS						
High Frequency Current Gain ($I_C = 10$ mA, $V_{CE} = 20$ V, $f = 100$ mc)	2N3903 2N3904		$|h_{fe}|$	2.5 3.0	— —	—
Current-Gain—Bandwidth Product ($I_C = 10$ mA, $V_{CE} = 20$ V, $f = 100$ mc)	2N3903 2N3904		f_T	250 300	— —	mc
Output Capacitance ($V_{CB} = 5$ Vdc, $I_E = 0$, $f = 100$ kc)		3	C_{obo}	—	4	pf
Input Capacitance ($V_{OB} = 0.5$ Vdc, $I_C = 0$, $f = 100$ kc)		3	C_{ibo}	—	8	pf
Small Signal Current Gain ($I_C = 1.0$ mA, $V_{CE} = 10$ V, $f = 1$ kc)	2N3903 2N3904	11	h_{fe}	50 100	200 400	—
Voltage Feedback Ratio ($I_C = 1.0$ mA, $V_{CE} = 10$ V, $f = 1$ kc)	2N3903 2N3904	14	h_{re}	0.1 0.5	5.0 8.0	$X10^{-4}$
Input Impedance ($I_C = 1.0$ mA, $V_{CE} = 10$ V, $f = 1$ kc)	2N3903 2N3904	13	h_{ie}	0.5 1.0	8 10	Kohms
Output Admittance ($I_C = 1.0$ mA, $V_{CE} = 10$ V, $f = 1$ kc)	Both Types	12	h_{oe}	1.0	40	μmhos
Noise Figure ($I_C = 100 \mu A$, $V_{CE} = 5$ V, $R_g = 1$ Kohms, Noise Bandwidth = 10 cps to 15.7 kc)	2N3903 2N3904	9, 10	NF	— —	6 5	db
SWITCHING CHARACTERISTICS						
Delay Time $V_{CC} = 3$ Vdc, $V_{OB} = 0.5$ Vdc, $I_C = 10$ mAdc, $I_{B1} = 1$ mA		1, 5	t_d	—	35	nsec
Rise Time		1, 5, 6	t_r	—	35	nsec
Storage Time $V_{CC} = 3$ Vdc, $I_C = 10$ mAdc, $I_{B1} = I_{B2} = 1$ mAdc	2N3903 2N3904	2, 7	t_s	— —	175 200	nsec
Fall Time		2, 8	t_f	—	50	nsec

*Pulse Test: Pulse Width = 300 μsec, Duty Cycle = 2% V_{OB} = Base Emitter Reverse Bias

FIGURE 1 — DELAY AND RISE TIME EQUIVALENT TEST CIRCUIT

FIGURE 2 — STORAGE AND FALL TIME EQUIVALENT TEST CIRCUIT

*Total shunt capacitance of test jig and connectors

338

TRANSIENT CHARACTERISTICS
— $T_J = 25°C$ —$T_J = 125°C$

FIGURE 3 — CAPACITANCE

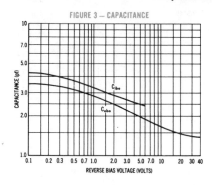

FIGURE 4 — CHARGE DATA

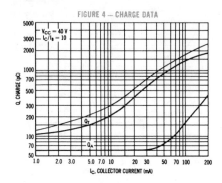

FIGURE 5 — TURN-ON TIME

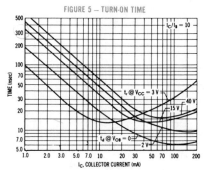

FIGURE 6 — RISE TIME

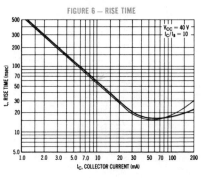

FIGURE 7 — STORAGE TIME

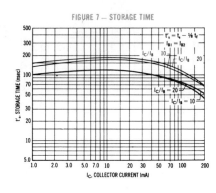

FIGURE 8 — FALL TIME

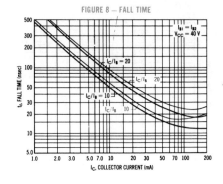

AUDIO SMALL SIGNAL CHARACTERISTICS

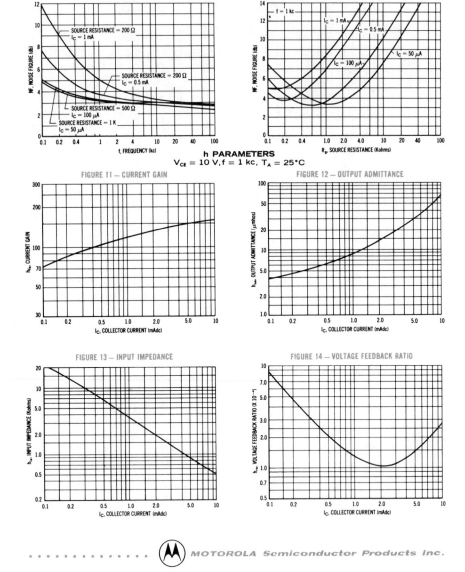

NOISE FIGURE VARIATIONS
$V_{CE} = 5$ Vdc, $T_A = 25°C$

FIGURE 9

FIGURE 10

h PARAMETERS
$V_{CE} = 10$ V, $f = 1$ kc, $T_A = 25°C$

FIGURE 11 — CURRENT GAIN

FIGURE 12 — OUTPUT ADMITTANCE

FIGURE 13 — INPUT IMPEDANCE

FIGURE 14 — VOLTAGE FEEDBACK RATIO

MOTOROLA *Semiconductor Products Inc.*

STATIC CHARACTERISTICS

FIGURE 15 -- NORMALIZED CURRENT GAIN

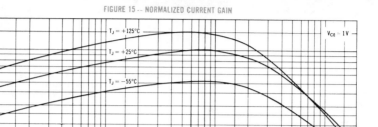

FIGURE 16 — COLLECTOR SATURATION REGION

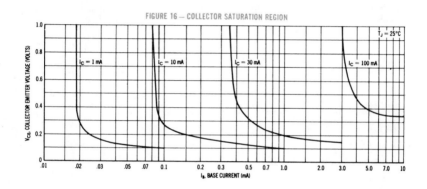

FIGURE 17 — "ON" VOLTAGES

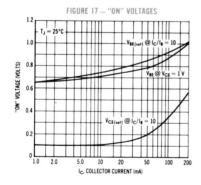

FIGURE 18 — TEMPERATURE COEFFICIENTS

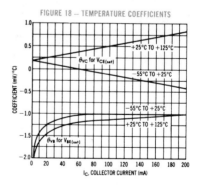

341

MOTOROLA
Semiconductors
BOX 955 • PHOENIX 1, ARIZONA

2N3905
2N3906

PNP SILICON ANNULAR* TRANSISTORS

. . . designed for general purpose switching and amplifier applications and for complementary circuitry with types 2N3903 and 2N3904.

- One-Piece, Injection-Molded Unibloc† Package for High Reliability
- High Voltage Ratings — BV_{CEO} = 40 Volts minimum
- Current Gain Specified from 100 μA to 100 mA
- Complete Switching and Amplifier Specifications
- Low Capacitance — C_{obo} = 4.5 pf maximum

PNP SILICON SWITCHING & AMPLIFIER TRANSISTORS

AUGUST 1965 — DS 5128

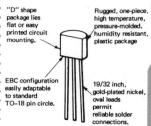

"D" shape package lies flat or easy printed circuit mounting.

Rugged, one-piece, high temperature, pressure-molded, humidity resistant, plastic package

EBC configuration easily adaptable to standard TO–18 pin circle.

19/32 inch, gold-plated nickel, oval leads permit reliable solder connections.

MAXIMUM RATINGS

Characteristic	Symbol	Rating	Unit
Collector-Base Voltage	V_{CB}	40	Vdc
Collector-Emitter Voltage	V_{CEO}	40	Vdc
Emitter-Base Voltage	V_{EB}	5	Vdc
Collector Current	I_C	200	mAdc
Total Device Dissipation @ T_A = 60°C	P_D	210	mW
Total Device Dissipation @ T_A = 25°C Derate above 25°C	P_D	310 2.81	mW mW/°C
Thermal Resistance, Junction to Ambient	θ_{JA}	0.357	°C/mW
Junction Operating Temperature	T_J	135	°C
Storage Temperature Range	T_{stg}	-55 to +135	°C

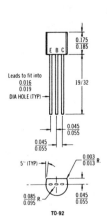

E B C

0.175
0.185

Leads to fit into
0.016
0.019
DIA HOLE (TYP)

19/32

0.045
0.055

0.045
0.055

5° (TYP)

0.003
0.013 R.

0.085 R.
0.095

0.045
0.055

TO-92

*Annular semiconductors patented by Motorola Inc.
†Trademark of Motorola Inc.

MOTOROLA Semiconductor Products Inc. A SUBSIDIARY OF MOTOROLA INC

ELECTRICAL CHARACTERISTICS ($T_A = 25°C$ unless otherwise noted)

Characteristic	Fig. No.	Symbol	Min	Max	Unit
OFF CHARACTERISTICS					
Collector-Base Breakdown Voltage ($I_C = 10 \mu Adc$, $I_E = 0$)		BV_{CBO}	40	—	Vdc
Collector-Emitter Breakdown Voltage* ($I_C = 1$ mAdc)		$BV_{CEO}*$	40	—	Vdc
Emitter-Base Breakdown Voltage ($I_E = 10 \mu Adc$, $I_C = 0$)		BV_{EBO}	5	—	Vdc
Collector Cutoff Current ($V_{CE} = 40$ Vdc, $V_{OB} = 3$ Vdc)		I_{CEX}	—	50	nAdc
Base Cutoff Current ($V_{CE} = 40$ Vdc, $V_{OB} = 3$ Vdc)		I_{BL}	—	50	nAdc
ON CHARACTERISTICS					
DC Current Gain* ($I_C = 0.1$ mAdc, $V_{CE} = 1$ Vdc) 2N3905 / 2N3906	15	$h_{FE}*$	30 / 60	— / —	—
($I_C = 1.0$ mAdc, $V_{CE} = 1$ Vdc) 2N3905 / 2N3906			40 / 80	— / —	
($I_C = 10$ mAdc, $V_{CE} = 1$ Vdc) 2N3905 / 2N3906			50 / 100	150 / 300	
($I_C = 50$ mAdc, $V_{CE} = 1$ Vdc) 2N3905 / 2N3906			30 / 60	— / —	
($I_C = 100$ mAdc, $V_{CE} = 1$ Vdc) 2N3905 / 2N3906			15 / 30	— / —	
Collector-Emitter Saturation Voltage* ($I_C = 10$ mAdc, $I_B = 1$ mAdc)	16, 17	$V_{CE(sat)}*$	—	0.25	Vdc
($I_C = 50$ mAdc, $I_B = 5$ mAdc)			—	0.4	
Base-Emitter Saturation Voltage* ($I_C = 10$ mAdc, $I_B = 1$ mAdc)	17	$V_{BE(sat)}*$	0.65	0.85	Vdc
($I_C = 50$ mAdc, $I_B = 5$ mAdc)			—	0.95	
SMALL SIGNAL CHARACTERISTICS					
High-Frequency Current Gain ($I_C = 10$ mAdc, $V_{CE} = 20$ Vdc, $f = 100$ mc) 2N3905 / 2N3906		$|h_{fe}|$	2.0 / 2.5	— / —	—
Current-Gain—Bandwidth Product ($I_C = 10$ mAdc, $V_{CE} = 20$ Vdc, $f = 100$ mc) 2N3905 / 2N3906		f_T	200 / 250	— / —	mc
Output Capacitance ($V_{CB} = 5$ Vdc, $I_E = 0$, $f = 100$ kc)	3	C_{obo}	—	4.5	pf
Input Capacitance ($V_{OB} = 0.5$ Vdc, $I_C = 0$, $f = 100$ kc)	3	C_{ibo}	—	10	pf
Small Signal Current Gain ($I_C = 1.0$ mA, $V_{CE} = 10$ V, $f = 1$ kc) 2N3905 / 2N3906	11	h_{fe}	50 / 100	200 / 400	—
Voltage Feedback Ratio ($I_C = 1.0$ mA, $V_{CE} = 10$ V, $f = 1$ kc) 2N3905 / 2N3906	14	h_{re}	0.1 / 1.0	5 / 10	$\times 10^{-4}$
Input Impedance ($I_C = 1.0$ mA, $V_{CE} = 10$ V, $f = 1$ kc) 2N3905 / 2N3906	13	h_{ie}	0.5 / 2.0	8 / 12	Kohms
Output Admittance ($I_C = 1.0$ mA, $V_{CE} = 10$ V, $f = 1$ kc) 2N3905 / 2N3906	12	h_{oe}	1.0 / 3.0	40 / 60	μ mhos
Noise Figure ($I_C = 100 \mu A$, $V_{CE} = 5$ V, $R_g = 1$ Kohms, Noise Bandwidth = 10 cps to 15.7 kc) 2N3905 / 2N3906	9, 10	NF	— / —	5.0 / 4.0	db
SWITCHING CHARACTERISTICS					
Delay Time $V_{CC} = 3$ Vdc, $V_{OB} = 0.5$ Vdc, $I_C = 10$ mAdc, $I_{B1} = 1$ mA	1, 5	t_d	—	35	nsec
Rise Time	1, 5, 6	t_r	—	35	nsec
Storage Time $V_{CC} = 3$ Vdc, $I_C = 10$ mAdc, $I_{B1} = I_{B2} = 1$ mAdc 2N3905 / 2N3906	2, 7	t_s	— / —	200 / 225	nsec
Fall Time 2N3905 / 2N3906	2, 8	t_f	— / —	60 / 75	nsec

*Pulse Test: Pulse Width = 300 μsec, Duty Cycle = 2% V_{OB} = Base Emitter Reverse Bias

FIGURE 1 — DELAY AND RISE TIME EQUIVALENT TEST CIRCUIT

*Total shunt capacitance of test jig and connectors

FIGURE 2 — STORAGE AND FALL TIME EQUIVALENT TEST CIRCUIT

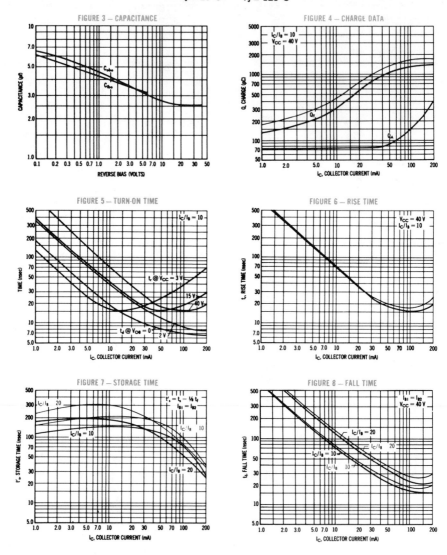

FIGURE 3 — CAPACITANCE

FIGURE 4 — CHARGE DATA

FIGURE 5 — TURN-ON TIME

FIGURE 6 — RISE TIME

FIGURE 7 — STORAGE TIME

FIGURE 8 — FALL TIME

AUDIO SMALL SIGNAL CHARACTERISTICS

NOISE FIGURE VARIATIONS
$V_{CE} = 5$ Vdc, $T_A = 25°C$

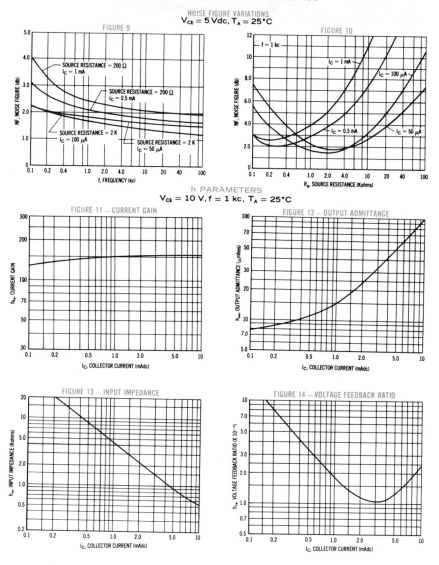

FIGURE 9

SOURCE RESISTANCE = 200 Ω
$I_C = 1$ mA

SOURCE RESISTANCE = 200 Ω
$I_C = 0.5$ mA

SOURCE RESISTANCE = 2 K
$I_C = 100$ μA

SOURCE RESISTANCE = 2 K
$I_C = 50$ μA

NF, NOISE FIGURE (db)

f, FREQUENCY (kc)

FIGURE 10

f = 1 kc

$I_C = 1$ mA

$I_C = 100$ μA

$I_C = 0.5$ mA

$I_C = 50$ μA

NF, NOISE FIGURE (db)

R_g, SOURCE RESISTANCE (Kohms)

h PARAMETERS
$V_{CE} = 10$ V, f = 1 kc, $T_A = 25°C$

FIGURE 11 — CURRENT GAIN

h_{fe}, CURRENT GAIN

I_C, COLLECTOR CURRENT (mAdc)

FIGURE 12 — OUTPUT ADMITTANCE

h_{oe}, OUTPUT ADMITTANCE (μmhos)

I_C, COLLECTOR CURRENT (mAdc)

FIGURE 13 — INPUT IMPEDANCE

h_{ie}, INPUT IMPEDANCE (Kohms)

I_C, COLLECTOR CURRENT (mAdc)

FIGURE 14 — VOLTAGE FEEDBACK RATIO

h_{re}, VOLTAGE FEEDBACK RATIO (X 10^{-4})

I_C, COLLECTOR CURRENT (mAdc)

MOTOROLA *Semiconductor Products Inc.*

STATIC CHARACTERISTICS

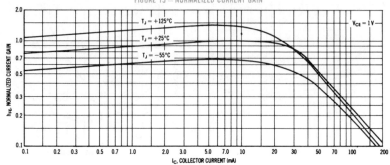

FIGURE 15 — NORMALIZED CURRENT GAIN

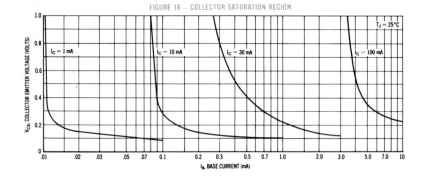

FIGURE 16 — COLLECTOR SATURATION REGION

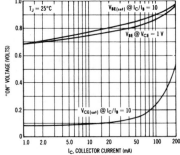

FIGURE 17 — "ON" VOLTAGES

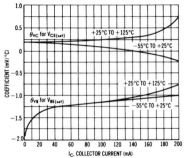

FIGURE 18 — TEMPERATURE COEFFICIENTS

TYPE TIS49
BULLETIN NO. DL-S 668591, MARCH 1966

SILECT† TRANSISTOR FOR VERY-HIGH-SPEED SWITCHING APPLICATIONS REQUIRING AN ECONOMY TRANSISTOR ELECTRICALLY SIMILAR TO THE 2N2369A

mechanical data

These transistors are encapsulated in a plastic compound specifically designed for this purpose, using a highly mechanized process‡ developed by Texas Instruments. The case will withstand soldering temperatures without deformation. These devices exhibit stable characteristics under high-humidity conditions and are capable of meeting MIL-STD-202C method 106B. The transistors are insensitive to light.

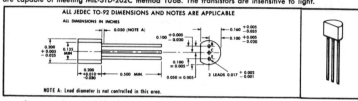

ALL JEDEC TO-92 DIMENSIONS AND NOTES ARE APPLICABLE
ALL DIMENSIONS IN INCHES

NOTE A: Lead diameter is not controlled in this area.

absolute maximum ratings at 25°C free-air temperature (unless otherwise noted)

Collector-Base Voltage	40 V
Collector-Emitter Voltage (See Note 1)	40 V
Collector-Emitter Voltage (See Note 2)	15 V
Emitter-Base Voltage	4.5 V
Continuous Collector Current	200 mA
Peak Collector Current (See Note 3)	500 mA
Continuous Device Dissipation at (or below) 25°C Free-Air Temperature (See Note 4)	250 mW
Storage Temperature Range	−55°C to 150°C
Lead Temperature ⅛ Inch from Case for 10 Seconds	260°C

electrical characteristics at 25°C free-air temperature (unless otherwise noted)

	PARAMETER	TEST CONDITIONS			MIN	MAX	UNIT
$V_{(BR)CBO}$	Collector-Base Breakdown Voltage	$I_C = 10\ \mu A,$	$I_E = 0$		40		V
$V_{(BR)CEO}$	Collector-Emitter Breakdown Voltage	$I_C = 10\ mA,$	$I_B = 0,$	See Note 5	15		V
$V_{(BR)CES}$	Collector-Emitter Breakdown Voltage	$I_C = 10\ \mu A,$	$V_{BE} = 0$		40		V
$V_{(BR)EBO}$	Emitter-Base Breakdown Voltage	$I_E = 10\ \mu A,$	$I_C = 0$		4.5		V
I_{CBO}	Collector Cutoff Current	$V_{CB} = 20\ V,$	$I_E = 0,$	$T_A = 70°C$		3	μA
I_{CES}	Collector Cutoff Current	$V_{CE} = 20\ V,$	$V_{BE} = 0$			0.4	μA
I_B	Base Current	$V_{CE} = 20\ V,$	$V_{BE} = 0$			−0.4	μA
h_{FE}	Static Forward Current Transfer Ratio	$V_{CE} = 0.35\ V,$	$I_C = 10\ mA,$	See Note 5	40		
		$V_{CE} = 1\ V,$	$I_C = 10\ mA,$	See Note 5		120	
		$V_{CE} = 0.4\ V,$	$I_C = 30\ mA,$	See Note 5	30		
		$V_{CE} = 1\ V,$	$I_C = 100\ mA,$	See Note 5	20		
V_{BE}	Base-Emitter Voltage	$I_B = 1\ mA,$	$I_C = 10\ mA$		0.72	0.87	V
		$I_B = 3\ mA,$	$I_C = 30\ mA$			1.15	V
		$I_B = 10\ mA,$	$I_C = 100\ mA$			1.6	V
		$I_B = 1\ mA,$	$I_C = 10\ mA,$	$T_A = 70°C$	0.6		V
$V_{CE(sat)}$	Collector-Emitter Saturation Voltage	$I_B = 1\ mA,$	$I_C = 10\ mA$			0.2	V
		$I_B = 3\ mA,$	$I_C = 30\ mA$			0.25	V
		$I_B = 10\ mA,$	$I_C = 100\ mA$			0.5	V
		$I_B = 1\ mA,$	$I_C = 10\ mA,$	$T_A = 70°C$		0.3	V
$|h_{fe}|$	Small-Signal Common-Emitter Forward Current Transfer Ratio	$V_{CE} = 10\ V,$	$I_C = 20\ mA,$	$f = 100\ Mc/s$	5		
C_{obo}	Common-Base Open-Circuit Output Capacitance	$V_{CB} = 5\ V,$	$I_E = 0,$	$f = 140\ kc/s$		4	pF

NOTES: 1. This value applies when the base-emitter diode is short-circuited.
2. This value applies between 10 μA and 10 mA collector current when the base-emitter diode is open-circuited.
3. This value applies for $t_p \leq 10\ \mu s$.
4. Derate linearly to 125°C free-air temperature at the rate of 2.5 mW/deg.
5. These parameters must be measured using pulse techniques. $t_p = 300\ \mu s$, duty cycle $\leq 2\%$.

†Trademark of Texas Instruments
‡Patent Pending

TEXAS INSTRUMENTS
INCORPORATED
SEMICONDUCTOR-COMPONENTS DIVISION
POST OFFICE BOX 5012 • DALLAS 22, TEXAS

TYPE TIS49 (2369A)
N-P-N EPITAXIAL PLANAR SILICON TRANSISTOR

switching characteristics at 25°C free-air temperature

	PARAMETER	TEST CONDITIONS†			MAX	UNIT
t_{on}	Turn-On Time	$I_C = 10$ mA,	$I_{B(1)} = 3$ mA,	$I_{B(2)} = -1.5$ mA,	12	ns
t_{off}	Turn-Off Time	$V_{BE(off)} = -1.5$ V,	$R_L = 250$ Ω,	See Figure 1	23	ns
t_s	Storage Time	$I_C = I_{B(1)} = 10$ mA,	$I_{B(2)} = -10$ mA,	See Figure 2	18	ns

†Voltage and current values shown are nominal; exact values vary slightly with transistor parameters.

PARAMETER MEASUREMENT INFORMATION

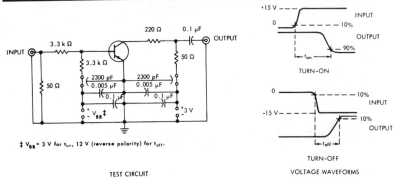

‡ $V_{BB} = 3$ V for t_{on}, 12 V (reverse polarity) for t_{off}.

TEST CIRCUIT

TURN-ON

TURN-OFF

VOLTAGE WAVEFORMS

FIGURE 1 — TURN-ON AND TURN-OFF TIMES

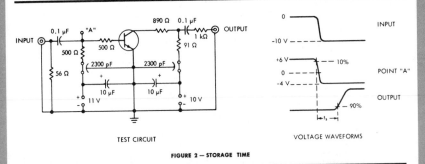

TEST CIRCUIT

VOLTAGE WAVEFORMS

FIGURE 2 — STORAGE TIME

NOTES: a. The input waveforms are supplied by a generator with the following characteristics: $Z_{out} = 50$ Ω, $t_r \leq 1$ ns, $t_p \geq 300$ ns, duty cycle $\leq 2\%$.
b. Output waveforms are monitored on an oscilloscope with the following characteristics: $t_r \leq 1$ ns, $Z_{in} = 50$ Ω.

PRINTED IN U.S.A.
TI cannot assume any responsibility for any circuits shown
or represent that they are free from patent infringement.

TEXAS INSTRUMENTS
INCORPORATED
SEMICONDUCTOR-COMPONENTS DIVISION
POST OFFICE BOX 5012 • DALLAS 22, TEXAS

TEXAS INSTRUMENTS RESERVES THE RIGHT TO MAKE CHANGES AT ANY TIME
IN ORDER TO IMPROVE DESIGN AND TO SUPPLY THE BEST PRODUCT POSSIBLE.

TYPE TIS50
BULLETIN NO. DL-S 648601, MARCH 1966

SILECT† TRANSISTOR FOR HIGH-SPEED SWITCHING APPLICATIONS
REQUIRING AN ECONOMY TRANSISTOR ELECTRICALLY SIMILAR TO THE 2N2894

mechanical data

These transistors are encapsulated in a plastic compound specifically designed for this purpose, using a highly mechanized process‡ developed by Texas Instruments. The case will withstand soldering temperatures without deformation. These devices exhibit stable characteristics under high-humidity conditions and are capable of meeting MIL-STD-202C method 106B. The transistors are insensitive to light.

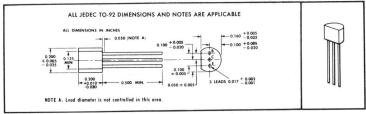

ALL JEDEC TO-92 DIMENSIONS AND NOTES ARE APPLICABLE

NOTE A: Lead diameter is not controlled in this area.

absolute maximum ratings at 25°C free-air temperature (unless otherwise noted)

Collector-Base Voltage	−12 V
Collector-Emitter Voltage (See Note 1)	−12 V
Emitter-Base Voltage	−4 V
Continuous Collector Current	−200 mA
Continuous Device Dissipation at (or below) 25°C Free-Air Temperature (See Note 2)	250 mW
Storage Temperature Range	−55°C to 150°C
Lead Temperature 1/16 Inch from Case for 10 Seconds	260°C

electrical characteristics at 25°C free-air temperature (unless otherwise noted)

	PARAMETER	TEST CONDITIONS		MIN	MAX	UNIT
$V_{(BR)CBO}$	Collector-Base Breakdown Voltage	$I_C = -10\ \mu A$, $I_E = 0$		−12		V
$V_{(BR)CEO}$	Collector-Emitter Breakdown Voltage	$I_C = -10$ mA, $I_B = 0$,	See Note 3	−12		V
$V_{(BR)CES}$	Collector-Emitter Breakdown Voltage	$I_C = -10\ \mu A$, $V_{BE} = 0$		−12		V
$V_{(BR)EBO}$	Emitter-Base Breakdown Voltage	$I_E = -100\ \mu A$, $I_C = 0$		−4		V
I_{CBO}	Collector Cutoff Current	$V_{CB} = -6$ V, $I_E = 0$,	$T_A = 70°C$		−1	μA
I_{CES}	Collector Cutoff Current	$V_{CE} = -6$ V, $V_{BE} = 0$			−80	nA
I_B	Base Current	$V_{CE} = -6$ V, $V_{BE} = 0$			80	nA
h_{FE}	Static Forward Current Transfer Ratio	$V_{CE} = -0.3$ V, $I_C = -10$ mA, See Note 3		30		
		$V_{CE} = -0.5$ V, $I_C = -30$ mA, See Note 3		40	150	
		$V_{CE} = -1$ V, $I_C = -100$ mA, See Note 3		20		
V_{BE}	Base-Emitter Voltage	$I_B = -1$ mA, $I_C = -10$ mA, See Note 3		−0.76	−0.98	V
		$I_B = -3$ mA, $I_C = -30$ mA, See Note 3		−0.82	−1.2	V
		$I_B = -10$ mA, $I_C = -100$ mA, See Note 3			−1.7	V
$V_{CE(sat)}$	Collector-Emitter Saturation Voltage	$I_B = -1$ mA, $I_C = -10$ mA, See Note 3			−0.15	V
		$I_B = -3$ mA, $I_C = -30$ mA, See Note 3			−0.2	V
		$I_B = -10$ mA, $I_C = -100$ mA, See Note 3			−0.5	V

NOTES: 1. This value applies between 10 μA and 10 mA collector current when the base-emitter diode is open-circuited.

2. Derate linearly to 125°C free-air temperature at the rate of 2.5 mW/deg.

3. These parameters must be measured using pulse techniques. $t_p = 300\ \mu s$, duty cycle ≤ 2%.

†Trademark of Texas Instruments
‡Patent Pending

TEXAS INSTRUMENTS
INCORPORATED
SEMICONDUCTOR-COMPONENTS DIVISION
POST OFFICE BOX 5012 • DALLAS 22, TEXAS

TYPE TIS50 (2894)
P-N-P EPITAXIAL PLANAR SILICON TRANSISTOR

electrical characteristics at 25°C free-air temperature

	PARAMETER	TEST CONDITIONS	MIN	MAX	UNIT
$|h_{fe}|$	Small-Signal Common-Emitter Forward Current Transfer Ratio	$V_{CE} = -5$ V, $I_C = -30$ mA, f = 100 Mc/s	4		
C_{obo}	Common-Base Open-Circuit Output Capacitance	$V_{CB} = -5$ V, $I_E = 0$, f = 140 kc/s		6	pF
C_{ibo}	Common-Base Open-Circuit Input Capacitance	$V_{EB} = -0.5$ V, $I_C = 0$, f = 140 kc/s		6	pF

switching characteristics at 25°C free-air temperature

	PARAMETER	TEST CONDITIONS†	MAX	UNIT
t_{on}	Turn-On Time	$I_C = -30$ mA, $I_{B(1)} = -1.5$ mA, $V_{BE(off)} = 3$ V, $R_L = 62$ Ω, See Figure 1	60	ns
t_{off}	Turn-Off Time	$I_C = -30$ mA, $I_{B(1)} = -1.5$ mA, $I_{B(2)} = 1.5$ mA, $R_L = 62$ Ω, See Figure 1	90	ns

†Voltage and current values shown are nominal; exact values vary slightly with transistor parameters.

PARAMETER MEASUREMENT INFORMATION

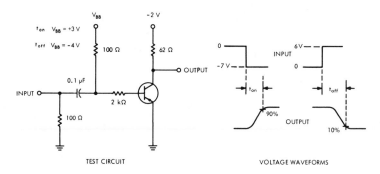

TEST CIRCUIT   VOLTAGE WAVEFORMS

FIGURE 1 — TURN-ON AND TURN-OFF TIMES

NOTES: a. The input waveforms are supplied by a generator with the following characteristics: $Z_{out} = 50$ Ω, $t_r \leq 1$ ns, $t_p > 200$ ns.

   b. Waveforms are monitored on an oscilloscope with the following characteristics: $t_r \leq 1$ ns, $R_{in} \geq 100$ kΩ, $C_{in} \leq 10$ pF.

TEXAS INSTRUMENTS
INCORPORATED
SEMICONDUCTOR-COMPONENTS DIVISION
POST OFFICE BOX 5012 • DALLAS 22, TEXAS

TYPE TIS43
BULLETIN NO. DL-S 646618, MARCH 1966

PLANAR UNIJUNCTION SILECT† TRANSISTOR
FOR APPLICATION IN SCR DRIVERS, MOTOR-SPEED CONTROLS, TIMERS, WAVEFORM GENERATORS, MULTIVIBRATORS, RING COUNTERS, ELECTRONIC ORGANS, AND MILITARY FUSES

- **Extremely Low Leakage Allows More Accurate Timing Circuit Design**
- **High Performance Capability at Low Driving Currents**
- **Provides Wider Range of Design Applications than Conventional Unijunction Transistors**
- **Improved Reliability and Performance Ensured by Planar Technology**

mechanical data

These transistors are encapsulated in a plastic compound specifically designed for this purpose, using a highly mechanized process‡ developed by Texas Instruments. The case will withstand soldering temperatures without deformation. These devices exhibit stable characteristics under high-humidity conditions and are capable of meeting MIL-STD-202C method 106B. The transistors are insensitive to light.

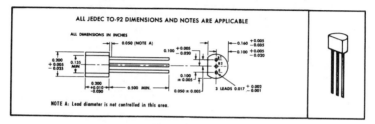

ALL JEDEC TO-92 DIMENSIONS AND NOTES ARE APPLICABLE

ALL DIMENSIONS IN INCHES

NOTE A: Lead diameter is not controlled in this area.

absolute maximum ratings at 25°C free-air temperature (unless otherwise noted)

Emitter–Base-Two Reverse Voltage	30 V
Interbase Voltage	See Note 1
Continuous Emitter Current	50 mA
Peak Emitter Current (See Note 2)	1 A
Continuous Device Dissipation at (or below) 25°C Free-Air Temperature (See Note 3)	300 mW
Storage Temperature Range	−55°C to 150°C
Lead Temperature ⅟₁₆ Inch from Case for 10 Seconds	260°C

NOTES: 1. Interbase voltage is limited solely by power dissipation, $V_{B2-B1} = \sqrt{r_{BB} \cdot P_T}$. The r_{BB} range specified gives maximum values ranging from 35 V to 52 V.

2. This value applies for a capacitor discharge through the emitter–base-one diode. Current must fall to 0.37 A within 3 ms and pulse-repetition rate must not exceed 10 pps.

3. Derate linearly to 125°C free-air temperature at the rate of 3 mW/deg.

†Trademark of Texas Instruments
‡Patent Pending.

TEXAS INSTRUMENTS
INCORPORATED
SEMICONDUCTOR-COMPONENTS DIVISION
POST OFFICE BOX 5012 • DALLAS 22, TEXAS

electrical characteristics at 25°C free-air temperature (unless otherwise noted)

	PARAMETER	TEST CONDITIONS	MIN	MAX	UNIT
r_{BB}	Static Interbase Resistance	$V_{B1-B2} = 3$ V, $I_E = 0$	4	9.1	k Ω
α_{rBB}	Interbase Resistance Temperature Coefficient	$V_{B2-B1} = 3$ V, $I_E = 0$, $T_A = -55°C$ to $100°C$, See Note 5	0.1	0.9	%/deg
η	Intrinsic Standoff Ratio	$V_{B2-B1} = 10$ V, See Figure 1	0.55	0.82	
$I_{B2(mod)}$	Modulated Interbase Current	$V_{B2-B1} = 10$ V, $I_E = 50$ mA	10		mA
I_{EB2O}	Emitter Reverse Current	$V_{B2-E} = 30$ V, $I_{B1} = 0$		-10	nA
I_P	Peak-Point Emitter Current	$V_{B2-B1} = 25$ V		5	μA
$V_{EB1(sat)}$	Emitter — Base-One Saturation Voltage	$V_{B2-B1} = 10$ V, $I_E = 50$ mA, See Note 4		4	V
I_V	Valley-Point Emitter Current	$V_{B2-B1} = 20$ V,	2		mA
V_{OB1}	Base-One Peak Pulse Voltage	See Figure 2	3		V

NOTES: 4. This parameter is measured using pulse techniques. $t_p = 300$ μs, duty cycle ≤ 2%.

5. Temperature coefficient, α_{rBB}, is determined by the following formula:

$$\alpha_{rBB} = \left[\frac{(r_{BB} @ 100°C) - (r_{BB} @ -55°C)}{(r_{BB} @ 25°C)} \right] \frac{100\%}{155 \text{ deg}}$$

To obtain r_{BB} for a given temperature $T_{A(2)}$, use the following formula:

$$r_{BB(2)} = [r_{BB} @ 25°C] [1 + (\alpha_{rBB}/100) (T_{A(2)} - 25°C)]$$

PARAMETER MEASUREMENT INFORMATION

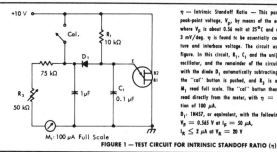

η — Intrinsic Standoff Ratio — This parameter is defined in terms of the peak-point voltage, V_p, by means of the equation: $V_p = \eta V_{B2-B1} + V_F$, where V_F is about 0.56 volt at 25°C and decreases with temperature at about 3 mV/deg. η is found to be essentially constant over wide ranges of temperature and interbase voltage. The circuit used to measure η is shown in the figure. In this circuit, R_1, C_1 and the unijunction transistor form a relaxation oscillator, and the remainder of the circuit serves as a peak-voltage detector with the diode D_1 automatically subtracting the voltage V_F. To use the circuit, the "cal" button is pushed, and R_3 is adjusted to make the current meter M_1 read full scale. The "cal" button then is released and the value of η is read directly from the meter, with $\eta = 1$ corresponding to full-scale deflection of 100 μA.

D_1: 1N457, or equivalent, with the following characteristics:

$V_F = 0.565$ V at $I_F = 50$ μA,

$I_R \leq 2$ μA at $V_R = 20$ V

FIGURE 1 — TEST CIRCUIT FOR INTRINSIC STANDOFF RATIO (η)

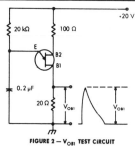

FIGURE 2 — V_{OB1} TEST CIRCUIT

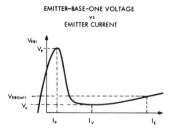

EMITTER–BASE-ONE VOLTAGE
vs
EMITTER CURRENT

FIGURE 3 — GENERAL STATIC EMITTER CHARACTERISTIC CURVE

MOTOROLA
Semiconductors
BOX 955 • PHOENIX, ARIZONA 85001

SILICON ANNULAR† UNIJUNCTION TRANSISTORS

. . . designed for pulse and timing circuits, sensing circuits, and thyristor trigger circuits.

- Low Peak-Point Current — $I_P = 0.4\ \mu A$ Max
- Low Emitter Reverse Current — $I_{EO} = 50$ nA Max
- Fast Switching — 1.0 MHz Min·

PN UNIJUNCTION TRANSISTORS

MAY 1967 — DS 2502

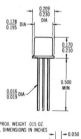

MAXIMUM RATINGS ($T_A = 25°C$ unless otherwise noted)

Rating	Symbol	Value	Unit
RMS Power Dissipation*	P_D*	300	mW
RMS Emitter Current	I_e	50	mA
Peak-Pulse Emitter Current **	i_e**	1.5	Amp
Emitter Reverse Voltage	V_{B2E}	30	Volts
Interbase Voltage †	V_{B2B1}†	35	Volts
Operating Junction Temperature Range	T_J	-65 to +125	°C
Storage Temperature Range	T_{stg}	-65 to +200	°C

* Derate 3.0 mW/°C increase in ambient temperature.
** Duty cycle ≤ 1%, PRR = 10 PPS (see figure 6)
† Based upon power dissipation at $T_A = 25°C$

FIGURE 1 — UNIJUNCTION TRANSISTOR
SYMBOL AND NOMENCLATURE

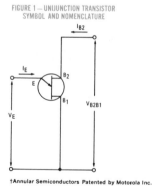

FIGURE 2 — STATIC EMITTER
CHARACTERISTICS CURVES

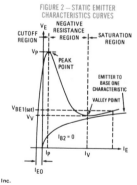

†Annular Semiconductors Patented by Motorola Inc.

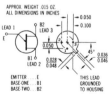

APPROX. WEIGHT .015 OZ.
ALL DIMENSIONS IN INCHES

EMITTER . . . E
BASE-ONE . . . B1
BASE-TWO . . . B2

THIS LEAD
GROUNDED
TO HOUSING

Lead 3 connected to case

CASE 22A

(TO-18 Outline
Except for Lead Position)

MOTOROLA *Semiconductor Products Inc.* A SUBSIDIARY OF MOTOROLA INC.

ELECTRICAL CHARACTERISTICS (T_A = 25°C unless otherwise noted)

Rating		Figure No.	Symbol	Min	Typ	Max	Unit
Intrinsic Standoff Ratio* (V_{B2B1} = 10 V)	2N4851	4, 8	η*	0.56	—	0.75	—
	2N4852, 2N4853			0.70	—	0.85	
Interbase Resistance (V_{B2B1} = 3.0 V, I_E = 0)		11, 12	R_{BB}	4.7	—	9.1	k ohms
Interbase Resistance Temperature Coefficient (V_{B2B1} = 3.0 V, I_E = 0, T_A = -65 to +125°C)		12	αR_{BB}	0.2	—	0.8	%/°C
Emitter Saturation Voltage** (V_{B2B1} = 10 V, I_E = 50 mA)			$V_{EB1(sat)}$**	—	2.5	—	Volts
Modulated Interbase Current (V_{B2B1} = 10 V, I_E = 50 mA)			$I_{B2(mod)}$	—	15	—	mA
Emitter Reverse Current (V_{B2E} = 30 V, I_{B1} = 0)	2N4851, 2N4852	7	I_{EB2O}	—	—	0.1	μA
	2N4853			—	—	0.05	
Peak-Point Emitter Current (V_{B2B1} = 25 V)	2N4851, 2N4852	9, 10	I_P	—	—	2.0	μA
	2N4853			—	—	0.4	
Valley-Point Current** (V_{B2B1} = 20 V, R_{B2} = 100 ohms)	2N4851	13, 14	I_V**	2.0	—	—	mA
	2N4852			4.0	—	—	
	2N4853			6.0	—	—	
Base-One Peak Pulse Voltage	2N4851	3, 17	V_{OB1}	3.0	—	—	Volts
	2N4852			5.0	—	—	
	2N4853			6.0	—	—	
Maximum Frequency of Oscillation		5	$f_{(max)}$	1.0	1.25	—	MHz

*η, Intrinsic standoff ratio, is defined in terms of the peak-point voltage, V_P, by means of the equation: $V_P = \eta V_{B2B1} + V_F$, where V_F is about 0.49 volt at 25°C @ I_F = 10 μA and decreases with temperature at about 2.5 mV, °C. The test circuit is shown in Figure 4. Components R_1, C_1, and the UJT form a relaxation oscillator; the remaining circuitry serves as a peak-voltage detector. The forward drop of Diode D_1 compensates for V_F. To use, the "cal" button is pushed, and R_3 is adjusted to make the current meter, M_1, read full scale. When the "cal" button is released, the value of η is read directly from the meter, if full scale on the meter reads 1.0.

**Use pulse techniques: PW ≈ 300 μs, duty cycle ≤ 2.0% to avoid internal heating, which may result in erroneous readings.

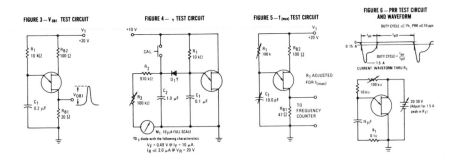

FIGURE 3 — V_{OB1} TEST CIRCUIT   FIGURE 4 — η TEST CIRCUIT   FIGURE 5 — $f_{(max)}$ TEST CIRCUIT   FIGURE 6 — PRR TEST CIRCUIT AND WAVEFORM

TYPICAL CHARACTERISTICS

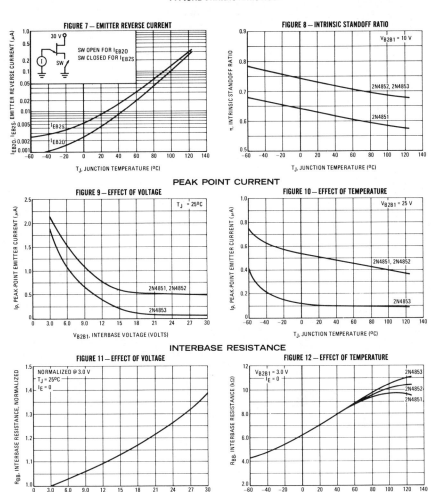

FIGURE 7 — EMITTER REVERSE CURRENT

FIGURE 8 — INTRINSIC STANDOFF RATIO

PEAK POINT CURRENT

FIGURE 9 — EFFECT OF VOLTAGE

FIGURE 10 — EFFECT OF TEMPERATURE

INTERBASE RESISTANCE

FIGURE 11 — EFFECT OF VOLTAGE

FIGURE 12 — EFFECT OF TEMPERATURE

MOTOROLA *Semiconductor Products Inc.*

TYPICAL CHARACTERISTICS

VALLEY CURRENT

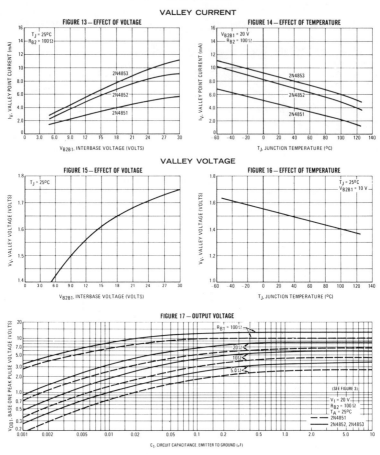

FIGURE 13 — EFFECT OF VOLTAGE

FIGURE 14 — EFFECT OF TEMPERATURE

VALLEY VOLTAGE

FIGURE 15 — EFFECT OF VOLTAGE

FIGURE 16 — EFFECT OF TEMPERATURE

FIGURE 17 — OUTPUT VOLTAGE

MOTOROLA Semiconductor Products Inc.

BOX 955 • PHOENIX ARIZONA 85001 • A SUBSIDIARY OF MOTOROLA INC

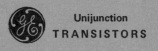

Unijunction
TRANSISTORS

SILICON TYPES
2N2646
2N2647

The General Electric 2N2646 and 2N2647 Silicon Unijunction Transistors have an entirely new structure resulting in lower saturation voltage, peak-point current and valley current as well as much as higher base-one peak pulse voltage. In addition these devices are much faster switches.

The 2N2646 is intended for general purpose industrial applica-

tions where circuit economy is of primary importance, and is ideal for use in firing circuits for Silicon Controlled Rectifiers and other applications where a guaranteed minimum pulse amplitude is required, The 2N2647 is intended for applications where a low emitter leakage current and a low peak point emitter current (trigger current) are required (i.e. long timing applications), and also for triggering high power SCR's.

NOTE 1: Max. diameter leads at a gaging plane .054 + .001 below base seat to be within .007 of their true location relative to max. 230 diameter measured with a suitable gage. When gage is not used, measurement will be made at base seat.

NOTE 2: Lead diameter is controlled in the zone between .050 and .250 from the base seat. Between .250 and the end of lead a max. of .021 is held.

NOTE 3: Calculated by measuring flange diameter, including tab and excluding tab and subtracting the smaller diameter from the larger diameter.

absolute maximum ratings: (25° C)

Power Dissipation (Note 1)	300 mw
RMS Emitter Current	50 ma
Peak Emitter Current (Note 2)	2 amperes
Emitter Reverse Voltage	30 volts
Interbase Voltage	35 volts
Operating Temperature Range	−65°C to +125°C
Storage Temperature Range	−65°C to +150°C

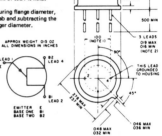

electrical characteristics: (25° C)

PARAMETER	2N2646			2N2647			
	Min.	Typ.	Max.	Min.	Typ.	Max.	
Intrinsic Standoff Ratio ($V_{BB} = 10V$) η	0.56	0.65	0.75	0.68	0.75	0.82	
Interbase Resistance ($V_{BB} = 3V$, $I_E = 0$) R_{BBO}	4.7	7	9.1	4.7	7	9.1	KΩ
Emitter Saturation Voltage ($V_{BB} = 10V$, $I_E = 50$ ma) $V_{E(SAT)}$		2			2		volts
Modulated Interbase Current ($V_{BB} = 10V$, $I_E = 50$ ma) $I_{B2(MOD)}$		12			12		ma
Emitter Reverse Current ($V_{B2E} = 30V$, $I_{B1} = 0$) I_{EO}		0.05	12		0.01	0.2	μa
Peak Point Emitter Current ($V_{BB} = 25V$) I_P		0.4	5		0.4	2	μa
Valley Point Current ($V_{BB} = 20V$, $R_{B2} = 100\Omega$) I_V	4	6		8	11	18	ma
Base-One Peak Pulse Voltage (Note 3) V_{OB1}	3.0	6.5		6.0	7.5		volts
SCR Firing Conditions (See Figure 26, back page)							

NOTES:
1. Derate 3.0 MW/°C increase in ambient temperature. The total power dissipation (available power to Emitter and Base-Two) must be limited by the external circuitry.
2. Capacitor discharge—10μfd or less, 30 volts or less.
3. The Base-One Peak Pulse Voltage is measured in the circuit below. This specification on the 2N2646 and 2N2647 is used to ensure a minimum pulse amplitude for applications in SCR firing circuits and other types of pulse circuits.

4. The intrinsic standoff ratio, η, is essentially constant with temperature and interbase voltage. η is defined by the equation:

$$V_P = \eta \ V_{BB} + V_D$$

Where V_P = Peak Point Emitter Voltage
V_{BB} = Interbase Voltage

$$V_D = \text{Junction Diode Drop (Approx. .5V)}$$

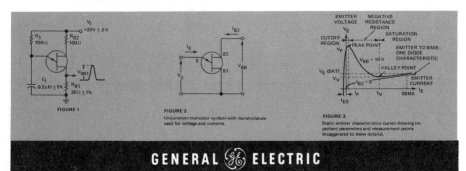

FIGURE 1

FIGURE 2
Unijunction transistor symbol with nomenclature used for voltage and currents.

FIGURE 3
Static emitter characteristics curves showing important parameters and measurement points (exaggerated to show details).

GENERAL (GE) **ELECTRIC**

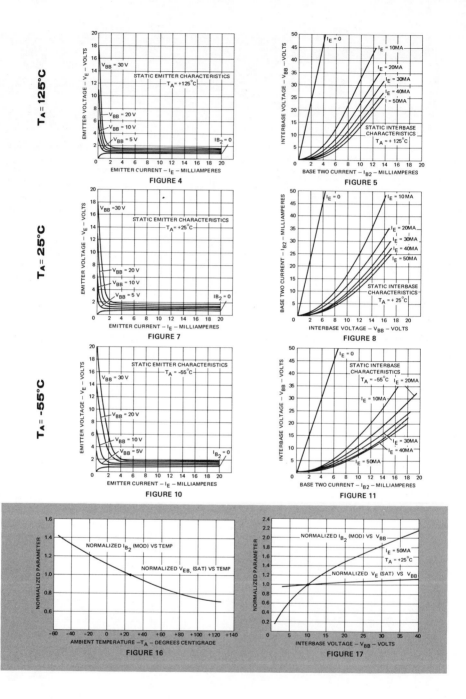

TA = 125°C

EMITTER VOLTAGE – VE – VOLTS

VBB = 30 V

STATIC EMITTER CHARACTERISTICS
TA = +125°C

VBB = 20 V

VBB = 10 V

VBB = 5 V

IB2 = 0

EMITTER CURRENT – IE – MILLIAMPERES

FIGURE 4

INTERBASE VOLTAGE – VBB – VOLTS

IE = 0

IE = 10MA
IE = 20MA
IE = 30MA
IE = 40MA
I = 50MA

STATIC INTERBASE
CHARACTERISTICS
TA = + 125°C

BASE TWO CURRENT – IB2 – MILLIAMPERES

FIGURE 5

TA = 25°C

EMITTER VOLTAGE – VE – VOLTS

VBB = 30 V

STATIC EMITTER CHARACTERISTICS
TA = +25°C

VBB = 20 V

VBB = 10 V

VBB = 5 V

IB2 = 0

EMITTER CURRENT – IE – MILLIAMPERES

FIGURE 7

BASE TWO CURRENT – IB2 – MILLIAMPERES

IE = 0

IE = 10MA

IE = 20MA
IE = 30MA
IE = 40MA
IE = 50MA

STATIC INTERBASE
CHARACTERISTICS
TA = + 25°C

INTERBASE VOLTAGE – VBB – VOLTS

FIGURE 8

TA = -55°C

EMITTER VOLTAGE – VE – VOLTS

STATIC EMITTER CHARACTERISTICS
TA = -55°C

VBB = 30 V

VBB = 20 V

VBB = 10 V

VBB = 5V

IB2 = 0

EMITTER CURRENT – IE – MILLIAMPERES

FIGURE 10

INTERBASE VOLTAGE – VBB – VOLTS

IE = 0

STATIC INTERBASE
CHARACTERISTICS
TA = -55°C   IE = 20MA

IE = 10MA

IE = 30MA
IE = 40MA

IE = 50MA

BASE TWO CURRENT – IB2 – MILLIAMPERES

FIGURE 11

NORMALIZED PARAMETER

NORMALIZED IB2 (MOD) VS TEMP

NORMALIZED VEB, (SAT) VS TEMP

AMBIENT TEMPERATURE –TA – DEGREES CENTIGRADE

FIGURE 16

NORMALIZED PARAMETER

NORMALIZED IB2 (MOD) VS VBB

IE = 50MA
TA = +25°C

NORMALIZED VE (SAT) VS VBB

INTERBASE VOLTAGE – VBB – VOLTS

FIGURE 17

358

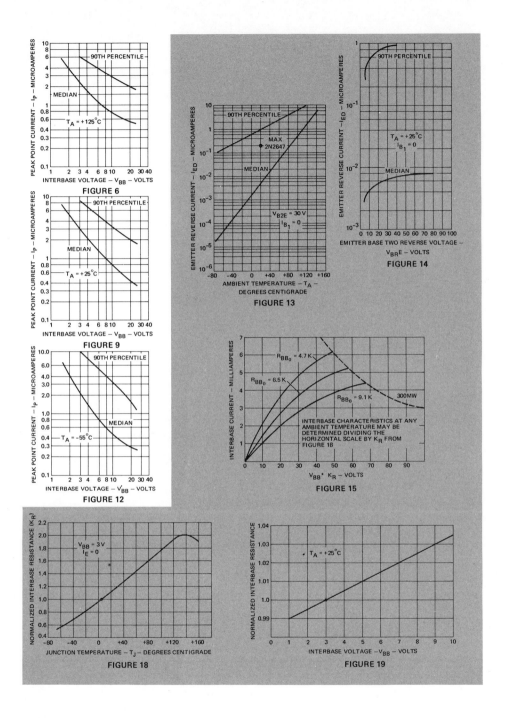

FIGURE 6

FIGURE 9

FIGURE 12

FIGURE 13

FIGURE 14

FIGURE 15

FIGURE 18

FIGURE 19

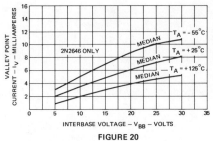

FIGURE 20

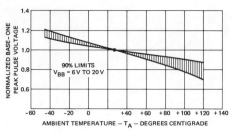

FIGURE 21

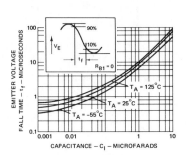

FIGURE 22

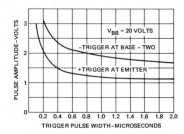

MINIMUM TRIGGER AMPLITUDE AS A FUNCTION
OF TRIGGER PULSE WIDTH FOR TURN ON
OF UNIJUNCTION TRANSISTOR

FIGURE 23

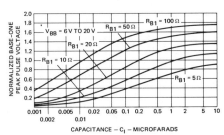

FIGURE 24

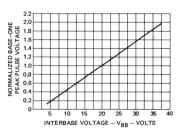

FIGURE 25

CURVE	SCR TYPE	R_{B1}	V_1(MAX)
A	C35, C36	27Ω ± 10%	35 V
B	C35, C36	47Ω ± 10%	20 V
B	C10, C11	27Ω ± 10%	32 V
C	C10, C11	47Ω ± 10%	18 V
C	C35, C36	SPRAGUE 3J 204	35 V
D	C10, C11	PULSE TRANS.	35 V

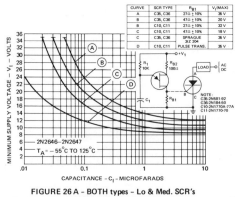

NOTE:
C35-2N581-92
C36-2N184-60
C10-2N1770A-77A
C11-2N1770-70

FIGURE 26 A - BOTH types - Lo & Med. SCR's

SCR TYPE	CURVE	Rt	V_1(MAX)
C80	A	27Ω ± 10%	20 V
	D	PULSE TRANS. PE223I	35 V
C60(2N2023-30)	B	27Ω ± 10%	35 V
C55, C56	C	47Ω ± 10%	20 V
C52(2N1792-98)	C	PULSE TRANS. PE223I	35 V
C50(2N1909-16)			
C45 and C46			
ZJ257			
ZJ285			

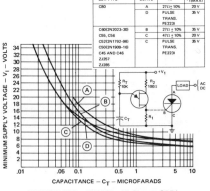

FIGURE 26 B - 2N2647 - Hi Current SCR's

SEMICONDUCTOR PRODUCTS DEPARTMENT
ELECTRONICS PARK   SYRACUSE 1, N. Y.

GENERAL ⓰ ELECTRIC

(In Canada, Canadian General Electric Company, Ltd., Toronto, Ont.
Outside the U.S.A. and Canada, by Electronic Component Sales,
I.G.E. Export Division, 159 Madison Avenue, New York, N. Y. 10016)

MOTOROLA
Semiconductors
BOX 955 • PHOENIX, ARIZONA 85001

1 WATT SURMETIC° SILICON ZENER DIODES

(1M3.3ZS10 thru 1M100ZS10)

. . . designed for constant voltage reference from 3.3 thru 100 volts, with 10% and 5% tolerances. These diodes are packaged in a void-free silicone polymer case which is no larger than the conventional 400 mW glass package. The diodes are subjected to 100% oscilloscope zener knee testing for reliable operation. They feature:

**1 WATT
ZENER DIODES**

**SILICON
OXIDE PASSIVATED**

3.3-100 VOLTS

JUNE 1965 — DS 7030 R1

• Guaranteed Reverse Leakage Current
• Temperature Range from −65 to +200°C
• Nail-Head Construction

*Trademark of Motorola Inc.

ABSOLUTE MAXIMUM RATINGS

Junction and Storage Temperature: −65°C to +200°C

Lead Temperature not less than 1/16″ from the case for 10 seconds: 230°C

D-C Power Dissipation: 1 Watt
(Derate 6.67 mW/°C above 50°C)

MECHANICAL CHARACTERISTICS

CASE: Void free, transfer molded, thermosetting silicone polymer.

FINISH: All external surfaces are corrosion resistant. Leads are readily solderable.

POLARITY: Cathode, indicated by color band. When operated in zener mode, cathode will be positive with respect to anode.

MOUNTING POSITION: Any.

WEIGHT: 0.42 gram (approximately).

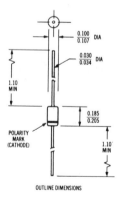

"SURMETIC" Package

FIGURE 1 — POWER RATING versus AMBIENT TEMPERATURE

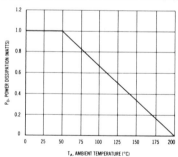

OUTLINE DIMENSIONS

0.100 / 0.107 DIA
0.030 / 0.034 DIA
1.10 MIN
0.185 / 0.205
POLARITY MARK (CATHODE)
1.10 MIN

SILICON ZENER DIODES
1N4728 THRU 1N4764
DS 7030 R1

ELECTRICAL CHARACTERISTICS (At 25°C ambient temperature unless otherwise specified) $V_F = 1.5$ V max @ 200 mA on all types

JEDEC Type No. (Note 1)	Motorola Type No.	Nominal Zener Voltage V_Z @ I_{ZT} Volts (Note 2)	Test Current I_{ZT} mA	Max Zener Impedance (Note 3)			I_R μA Max	V_R @ Volts	Surge Current @ $T_A = 25°C$ (Note 5)	Max DC Zener Current I_{ZM} mA (Note 4)	Typical Zener Voltage Temp. Coeff. %/°C
				Z_{ZT} @ I_{ZT} Ohms	Z_{ZK} @ I_{ZK} Ohms	I_{ZK} mA					
1N4728	1M3.3ZS10	3.3	76	10	400	1.0	100	1	1380	276	-.075
1N4729	1M3.6ZS10	3.6	69	10	400	1.0	100	1	1260	252	-.065
1N4730	1M3.9ZS10	3.9	64	9	400	1.0	50	1	1190	234	-.055
1N4731	1M4.3ZS10	4.3	58	9	400	1.0	10	1	1070	217	-.040
1N4732	1M4.7ZS10	4.7	53	8	500	1.0	10	1	970	193	-.020
1N4733	1M5.1ZS10	5.1	49	7	550	1.0	10	1	890	178	.005
1N4734	1M5.6ZS10	5.6	45	5	600	1.0	10	2	810	162	.020
1N4735	1M6.2ZS10	6.2	41	2	700	1.0	10	3	730	146	.035
1N4736	1M6.8ZS10	6.8	37	3.5	700	1.0	10	4	660	133	.040
1N4737	1M7.5ZS10	7.5	34	4.0	700	0.5	10	5	605	121	.045
1N4738	1M8.2ZS10	8.2	31	4.5	700	0.5	10	6	550	110	.048
1N4739	1M9.1ZS10	9.1	28	5.0	700	0.5	10	7	500	100	.051
1N4740	1M10ZS10	10	25	7	700	0.25	10	7.6	454	91	.055
1N4741	1M11ZS10	11	23	8	700	0.25	5	8.4	414	83	.060
1N4742	1M12ZS10	12	21	9	700	0.25	5	9.1	380	76	.065
1N4743	1M13ZS10	13	19	10	700	0.25	5	9.9	344	69	.065
1N4744	1M15ZS10	15	17	14	700	0.25	5	11.4	304	61	.070
1N4745	1M16ZS10	16	15.5	16	700	0.25	5	12.2	285	57	.070
1N4746	1M18ZS10	18	14	20	750	0.25	5	13.7	250	50	.075
1N4747	1M20ZS10	20	12.5	22	750	0.25	5	15.2	225	45	.075
1N4748	1M22ZS10	22	11.5	23	750	0.25	5	16.7	205	41	.080
1N4749	1M24ZS10	24	10.5	25	750	0.25	5	18.2	190	38	.080
1N4750	1M27ZS10	27	9.5	35	750	0.25	5	20.6	170	34	.085
1N4751	1M30ZS10	30	8.5	40	1,000	0.25	5	22.8	150	30	.085
1N4752	1M33ZS10	33	7.5	45	1,000	0.25	5	25.1	135	27	.085
1N4753	1M36ZS10	36	7.0	50	1,000	0.25	5	27.4	125	25	.085
1N4754	1M39ZS10	39	6.5	60	1,000	0.25	5	29.7	115	23	.090
1N4755	1M43ZS10	43	6.0	70	1,500	0.25	5	32.7	110	22	.090
1N4756	1M47ZS10	47	5.5	80	1,500	0.25	5	35.8	95	19	.090
1N4757	1M51ZS10	51	5.0	95	1,500	0.25	5	38.8	90	18	.090
1N4758	1M56ZS10	56	4.5	110	2,000	0.25	5	42.6	80	16	.090
1N4759	1M62ZS10	62	4.0	125	2,000	0.25	5	47.1	70	14	.090
1N4760	1M68ZS10	68	3.7	150	2,000	0.25	5	51.7	65	13	.090
1N4761	1M75ZS10	75	3.3	175	2,000	0.25	5	56.0	60	12	.090
1N4762	1M82ZS10	82	3.0	200	3,000	0.25	5	62.2	55	11	.090
1N4763	1M91ZS10	91	2.8	250	3,000	0.25	5	69.2	50	10	.090
1N4764	1M100ZS10	100	2.5	350	3,000	0.25	5	76.0	45	9	.090

Standard Tolerances are ±5%, ±2%, and ±1%.

SPECIAL SELECTIONS AVAILABLE INCLUDE:

1—Nominal zener voltages between those shown.

2—Matched sets:
 a. Two or more units for series connection with specified tolerance on total voltage
 b. Two or more units matched to one another with any specified tolerance

NOTE 1 — TOLERANCE AND VOLTAGE DESIGNATION
Tolerance designation

The JEDEC type numbers shown have a standard tolerance of ±10% on the nominal zener voltage. Suffix "A" for ±5% units.

Non-standard voltage designation

To designate units with zener voltages other than those assigned JEDEC numbers, the Motorola type number should be used.

EXAMPLE:

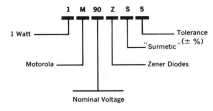

1 Watt

Motorola

Nominal Voltage

Tolerance (± %)

"Surmetic"

Zener Diodes

Matched sets for closer tolerances or higher voltages

Series matched sets make zener voltages in excess of 100 volts or tolerances of less than 5% possible as well as providing lower temperature coefficients, lower dynamic impedance and greater power handling ability.

For other special circuit requirements, contact your Motorola District Sales Manager.

NOTE 2 — ZENER VOLTAGE (V_Z) MEASUREMENT

Nominal zener voltage is measured with the device junction in thermal equilibrium with 25°C ambient temperature. Standard test currents (I_{ZT}) have been selected to most nearly approximate those commonly encountered in equipment design so that at nominal voltages, the dissipation is a constant 0.25 watts (¼ maximum power rating). This results in a nominal junction temperature rise of 25°C. Zener voltages may, therefore, be checked rapidly by heating units in an oven to 50°C with zero dissipation and then testing at a low duty cycle. Second order sources of error will usually be well under 1%.

NOTE 3 — ZENER IMPEDANCE (Z_Z) DERIVATION

The zener impedance is derived from the 60 cycle ac voltage, which results when an ac current having an rms value equal to 10% of the dc zener current (I_{ZT} or I_{ZK}) is superimposed on I_{ZT} or I_{ZK}. Zener impedance is measured at 2 points on all Motorola zener diodes to insure a sharp knee on the breakdown curve and to eliminate unstable units. When making zener impedance measurements at a zener current of $250\mu a$, it may be necessary to insert a 60 cps band pass filter between diode and voltmeter in order to avoid errors resulting from low level noise signals.

A 100% cathode ray tube curve trace test is used to insure that each zener diode breakdown region begins at a current lower than I_{ZK} and continues at nearly constant voltage to a current level in excess of I_{ZM}.

NOTE 4 — MAXIMUM ZENER CURRENT RATINGS (I_{ZM})

Maximum zener current ratings are based on maximum voltage of a 10% tolerance unit. For closer tolerance units (5%) or units where the actual zener voltage (V_Z) is known at the operating point, the maximum zener current may be increased and is limited by the derating curve.

NOTE 5 — SURGE CURRENT (i_s)

The rating given is maximum peak, non-recurrent, reverse surge current. The current ratings are for maximum surge of ½ square wave or equivalent sine wave pulse of 1/120 second duration superimposed on the test current, I_{ZT}.

FIGURE 2 — TYPICAL ZENER DIODE CHARACTERISTICS and SYMBOL IDENTIFICATION

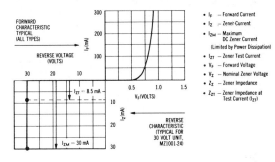

FORWARD CHARACTERISTIC TYPICAL (ALL TYPES)

REVERSE VOLTAGE (VOLTS)

$I_{ZT} = 8.5$ mA

$I_{ZM} = 30$ mA

V_F (VOLTS)

REVERSE CHARACTERISTIC (TYPICAL FOR 30 VOLT UNIT, MZ1001-24)

- I_F — Forward Current
- I_Z — Zener Current
- I_{ZM} — Maximum DC Zener Current (Limited by Power Dissipation)
- I_{ZT} — Zener Test Current
- V_F — Forward Voltage
- V_Z — Nominal Zener Voltage
- Z_Z — Zener Impedance
- Z_{ZT} — Zener Impedance at Test Current (I_{ZT})

 MOTOROLA *Semiconductor Products Inc.*

TYPES 1N914, USN 1N914, 1N914A, 1N914B, 1N915, 1N916, 1N916A, 1N916B, 1N917 BULLETIN NO. DL-S 63424, JANUARY 1963

- Extremely Stable and Reliable High Speed Switching Diodes
- Meet All Requirements of MIL-S-19500C

mechanical data

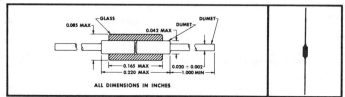

GLASS — 0.085 MAX — 0.042 MAX — DUMET — DUMET
0.165 MAX — 0.020 ± 0.002
0.220 MAX — 1.000 MIN

ALL DIMENSIONS IN INCHES

absolute maximum ratings at 25°C ambient temperature (unless otherwise noted)

		:1N914	1N914A	1N914B	1N915	1N916	1N916A	1N916B	1N917	Units
V_R	Reverse Voltage at — 65 to + 150°C	75	75	75	50	75	75	75	30	v
I_o	Average Rectified Fwd. Current	75	75	75	75	75	75	75	50	ma
I_o	Average Rectified Fwd. Current at + 150°C	10	10	10	10	10	10	10	10	ma
i_f	Recurrent Peak Fwd. Current	225	225	225	225	225	225	225	150	ma
$i_{f(surge)}$	Surge Current, 1 sec	500	500	500	500	500	500	500	300	ma
P	Power Dissipation	250	250	250	250	250	250	250	250	mw
T_A	Operating Temperature Range	— 65 to + 175								°C
T_{stg}	Storage Temperature Range	200								°C

maximum electrical characteristics at 25°C ambient (unless otherwise noted)

BV_R	Min Breakdown Voltage at 100 μa	100	100	100	65	100	100	100	40	v
I_R	Reverse Current at V_R	5	5	5	5	5	5	5		μa
I_R	Reverse Current at — 20v	0.025	0.025	0.025		0.025	0.025	0.025		μa
I_R	Reverse Current at — 20v at 100°C	3	3	3	5	3	3	3	25	μa
I_R	Reverse Current at — 20v at + 150°C	50	50	50			50	50	50		μa
I_R	Reverse Current at — 10v				0.025				0.05	μa
I_R	Reverse Current at — 10v at 125°C									μa
I_F	Min Fwd Current at V_F = 1 vdc	10	20	100	50	10	20	30	10	ma
V_F	at 250μa								0.64	v
V_F	at 1.5ma								0.74	v
V_F	at 3.5ma								0.83	v
V_F	at 5ma			0.72	0.73			0.73		v
V_F	Min. at 5ma				0.60					v
C	Capacitance at V_R = 0vdc	4	4	4	4	2	2	2	2.5	pf

operating characteristics at 25°C ambient temperature (unless otherwise noted)

t_{rr}	Max Reverse Recovery Time	**4 °8	**4 °8	**4 °8	°10	**4 °8	**4 °8	**4 °8	°3	nsec nsec
V_f	Fwd Recovery Voltage (50ma Peak Sq. wave, 0.1 μsec pulse width, 10 nsec rise time, 5Kc-100Kc rep. rate)	2.5	2.5	2.5	2.5	2.5	2.5	2.5	2.5	v

° Lumatron (10ma I_F, 10ma I_R, recover to 1ma)
** EG&G (10ma I_F, 6v V_R, recover to 1ma)
: Military Specifications MIL-S-19500/116 USN

*Trademark of Texas Instruments Incorporated

TEXAS INSTRUMENTS
INCORPORATED
SEMICONDUCTOR-COMPONENTS DIVISION
POST OFFICE BOX 5012 • DALLAS 22, TEXAS

PRINTED IN U.S.A.

TEXAS INSTRUMENTS RESERVES THE RIGHT TO MAKE CHANGES AT ANY TIME IN ORDER TO IMPROVE DESIGN AND TO SUPPLY THE BEST PRODUCT POSSIBLE.

Germanium
Computer — Industrial
Tunnel Diodes

1N3712-20
1N3713-21

The General Electric 1N3712 through 1N3720 and 1N3713 through 1N3721 are Germanium Tunnel Diodes offering peak currents of 1.0, 2.2, 4.7, 10, and 22 ma. These devices, which make use of the quantum mechanical tunneling phenomenon to obtain a negative conductance characteristic, are designed for low level switching and small signal applications at very high frequencies. All 1N3713-1N3721 version parameters are closely controlled for use in critical applications such as level detection, frequency converters, etc. These devices are housed in General Electric's new hermetically sealed subminiature axial package.

FEATURES:

▶ V_FS Specified for more accurate designing of load lines
▶ Low capacitance
▶ Fast speed

absolute maximum ratings

AXIAL DIODE OUTLINE

	1N3712 1N3713	1N3714 1N3715	1N3716 1N3717	1N3718 1N3719	1N3720 1N3721	
Forward Current*	5	10	25	50	100	ma
Reverse Current*	10	20	50	50	100	ma
Storage Temperature	←		−55 to +100		→	°C
Lead Temperature 1/16″ ± 1/32″ from case for 10 seconds	←		260		→	°C

*Derate maximum currents 1 % per °C ambient temperature above 25°C.

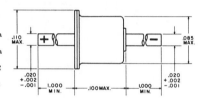

ALL DIMENSIONS IN INCHES.
DIMENSIONS ARE REFERENCE UNLESS TOLERANCED.

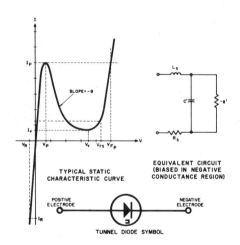

TYPICAL STATIC
CHARACTERISTIC CURVE

EQUIVALENT CIRCUIT
(BIASED IN NEGATIVE
CONDUCTANCE REGION)

POSITIVE
ELECTRODE

NEGATIVE
ELECTRODE

TUNNEL DIODE SYMBOL

GENERAL *GE* ELECTRIC

electrical characteristics:

STATIC CHARACTERISTICS		1N3712			1N3713			1N3714			1N3715		
		Min.	Typ.	Max.	Min.	Typ.	Max.	Min.	Typ.	Max.	Min.	Typ.	Max.
Peak Point Current	I_P	0.9	1.0	1.1	0.975	1.000	1.025	2.0	2.2	2.4	2.15	2.20	2.25
Valley Point Current	I_V		0.12	0.18	.075	.095	.140		0.29	0.48	.165	.210	.310
Peak Point Voltage	V_P		65		58	65	72		65		58	65	72
Valley Point Voltage	V_V		350		315	355	395		350		315	355	395
Reverse Voltage ($I_R = I_P$ typ.)	V_R			40		20	40			40		20	40
Forward Voltage ($I_F = I_P$ typ.)	V_{FP}		500		475	510	535		500		475	510	535
► ($I_F = .25\ I_P$ typ.)	V_{FS}*				410	450					410	450	
DYNAMIC CHARACTERISTICS													
Total Series Inductance	L_S		0.5			0.5			0.5			0.5	
Total Series Resistance	R_S		1.5	4.0		1.7	4.0		1.0	3.0		1.1	3.0
► Valley Point Terminal Capacitance	C		5	10		3.5	5.0		10	25		7.0	10.0
Max. Negative Terminal Conductance	-G		8		7.5	8.5	9.5		18		16	19	22
Resistive Cutoff Frequency	f_{ro}		2.3			3.2			2.2			3.0	
Self-Resonant Frequency	f_{xo}		3.2			3.8			2.2			2.7	
Frequency of Oscillation	F_{ose}**		3.2			3.8			2.2			2.7	
► Rise Time	t_r***					1.7						1.6	

*V_{FS} is defined as the value of forward voltage at a forward current of one quarter the typical peak current.
**The frequency of oscillation (under short circuit conditions) for steady state large signal sinusoidal oscillation is given by equation (3) which is the maximum frequency attainable without capacitance compensation.

***Switching speed with constant current drive. $t_r \approx \dfrac{V_{FP} - V_P}{I_P - I_V}\ C$

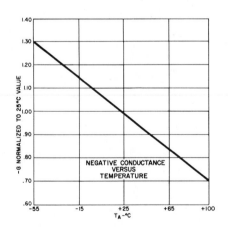

NEGATIVE CONDUCTANCE
VERSUS
TEMPERATURE

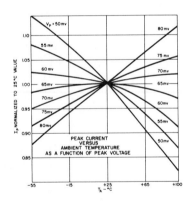

PEAK CURRENT
VERSUS
AMBIENT TEMPERATURE
AS A FUNCTION OF PEAK VOLTAGE

| 1N3716 | | | 1N3717 | | | 1N3718 | | | 1N3719 | | | 1N3720 | | | 1N3721 | | | |
Min.	Typ.	Max.	Min.	Typ.	Max.	Min.	Typ.	Max.	Min.	Typ.	Max.	Min.	Typ.	Max.	Min.	Typ.	Max.	
4.2	4.7	5.2	4.58	4.70	4.82	9.0	10.0	11.0	9.75	10.00	10.25	20	22	24	21.5	22	22.5	ma
	0.60	1.04	.350	.45	.60		1.3	2.2	.75	.95	1.40		2.9	4.8	1.65	2.10	3.10	ma
	65		58	65	72		65		58	65	72		65		58	65	72	mv
	350		315	355	395		350		315	355	395		350		315	355	395	mv
		40		20	40			40		20	40			40		20	40	mv
	500		475	510	535		500		475	510	535		500		475	510	575	mv
			410	450					410	450					410	450		
	0.5			0.5			0.5			0.5			0.5			0.5		nh
	.50	2.0		.52	2.0		.30	1.5		.36	1.5		.20	1.0		.22	1.0	ohms
	25	50		13	25		50	90		27	50		90	150		55	100	pf
	40		36	41	46		80		75	85	95		180		160	190	220	10^{-3} mho
	1.8			3.4			1.6			2.8			1.6			2.6		KMC
	1.4			1.9			.97			1.3			.67			.78		KMC
	1.4			2.0			1.0			1.4			.74			.95		KMC
				1.4						1.3						1.2		nsec

$$f_{ro} = \frac{|g'|}{2\pi\,C'} \sqrt{\frac{1}{R_S|g'|} - 1} \quad (1) \qquad f_{xo} = \frac{1}{2\pi} \sqrt{\frac{1}{L_S C'} - \left(\frac{|g'|}{C'}\right)^2} \quad (2) \qquad f_{osc} = \frac{1}{2\pi} \sqrt{\frac{1}{L_S C} - \left(\frac{R_T}{L}\right)^2} \quad (3)$$

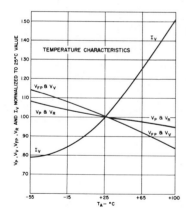

TEMPERATURE CHARACTERISTICS

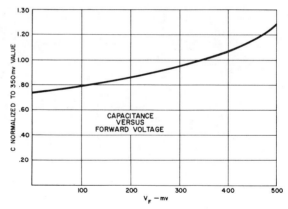

CAPACITANCE VERSUS FORWARD VOLTAGE

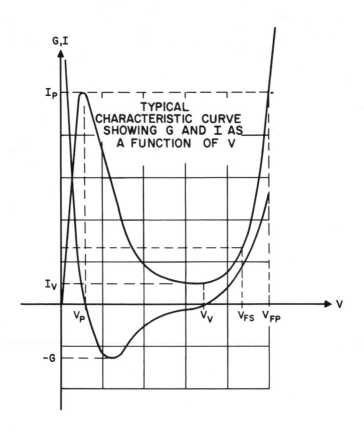

TYPICAL
CHARACTERISTIC CURVE
SHOWING G AND I AS
A FUNCTION OF V

appendix C

PARTS AND EQUIPMENT LIST

SEMICONDUCTORS REQUIRED FOR EXPERIMENTS

Quantity	Type
3	NPN, Silicon, switching, with manufacturer's specification sheet (Specification sheets, for the following transistor types are included in Appendix B)
	2N3646
	2N3903
	TIS-49
2	PNP, Silicon, switching, with manufacturer's specification sheet (Specification sheets, for the following transistor types are included in Appendix B)
	2N3638
	2N3905
	TIS-50
1	Unijunction, PN, silicon, with manufacturer's specification sheet (Specification sheets, for the following unijunction types are included in Appendix B)
	TIS-43
	2N4851
	2N2646

Quantity	Type
2	Zener Diodes, silicon, with manufacturer's specification sheets (Specification sheets, for the following Zener Diode types are included in Appendix B)

1N4728 to 1N4764 1 W
　　(Suggested values: 3.3, 3.9, 4.7, 5.1 V)

4　　Diodes, silicon, junction, with manufacturer's specification sheets
　　(Specification sheets, for the following diode types are included
　　in Appendix B)
　　1N914 to 1N917

4　　Diodes, germanium, point contact type
　　1N34 or equivalent

1　　Tunnel Diode, germanium, with manufacturer's specification sheets
　　(Specification sheets, for the following tunnel diode types are
　　included in Appendix B)
　　1N3712 to 1N3721

PARTS LIST

Parts needed for the experiments in this text, if resistor substitution boxes and capacitor substitution boxes are to be used, are listed under Method A. Parts needed for the experiments, if separate components are to be used, are listed under Method B.

METHOD A

Quantity	Description
9	Resistor substitution boxes (10 Ω to 10 MΩ, 1 W)
5	Capacitor substitution boxes (0.0001 μF to 0.22 μF, 400 V)
1	1 μF 50 V capacitor, electrolytic
2	3 μF 50 V capacitor, electrolytic
2	10 μF 50 V capacitor, electrolytic
2	25 μF 50 V capacitor, electrolytic
1	1 kΩ potentiometer, linear, 1 W
1	5 kΩ potentiometer, linear, 1 W
1	10 kΩ potentiometer, linear, 1 W
1	50 kΩ potentiometer, linear, 1 W
1	100 kΩ potentiometer, linear, 1 W
1	1 mH 100 mA choke
1	vector board and frame, approximately 5 in. $\times$ 8 in.
3	transistor sockets
20	flea clips
20	clip leads, approximate length 9 in.

METHOD B

Quantity	Description	Quantity	Description
Resistors ($\frac{1}{2}$ W)		Capacitors (50 V)	
1	4.7 Ω	5	100 pF
1	10 Ω	4	220 pF
1	15 Ω	2	680 pF
1	47 Ω	2	0.001 μF
1	100 Ω	2	0.0022 μF
1	150 Ω	2	0.0033 μF
1	270 Ω	2	0.005 μF
1	300 Ω	2	0.01 μF
2	470 Ω	2	0.02 μF
4	1 kΩ	2	0.1 μF
2	2 kΩ	2	0.22 μF
2	3.3 kΩ	1	0.5 μF
2	4.7 kΩ	1	1 μF electrolytic
2	6.8 kΩ	1	3 μF electrolytic
4	10 kΩ	2	10 μF electrolytic
2	12 kΩ	2	25 μF electrolytic
2	15 kΩ		
2	22 kΩ		

Quantity	Description	Quantity	Description
Resistors ($\frac{1}{2}$ W)		Potentiometers (linear, 1 W)	
2	27 kΩ	1	1 kΩ
4	47 kΩ	1	5 kΩ
2	68 kΩ	1	10 kΩ
2	100 kΩ	1	50 kΩ
2	270 kΩ	1	100 kΩ
2	1 MΩ		
Choke			
1	1 mH, 100 mA		
1	vector board and frame, approximately 5 in. $\times$ 8 in.		
3	transistor sockets		
20	flea clips		
20	clip leads, approximate length, 9 in.		

EQUIPMENT LIST

Quantity

1	Audio signal generator, sine-square-wave frequency range, 20 Hz to 200 kHz, $Z_0 = 600\ \Omega$ output voltage, 0–10 V rms; 0–10 V p–p
1	Oscilloscope, dc time base type frequency response, dc to 450 kHz vertical sensitivity, 100 mV/cm (ac and dc)
1	Oscilloscope probe—10X attenuator, 10 MΩ
1	Vacuum-tube voltmeter, ac/dc type voltage range, 1.5 to 500 V full scale dc and ac rms resistance range, 1 Ω to 100 MΩ
2	Semiconductor power supplies output voltage, 0–30 V output current, 0–250 mA

Equation Index

Index